AF509464

CFD Modeling of Complex Chemical Processes

CFD Modeling of Complex Chemical Processes: Multiscale and Multiphysics Challenges

Editors

Li Xi
De-Wei Yin
Jae Sung Park

MDPI • Basel • Beijing • Wuhan • Barcelona • Belgrade • Manchester • Tokyo • Cluj • Tianjin

Editors

Li Xi
Department of Chemical
Engineering, School of
Computational Science and
Engineering, McMaster
University
Canada

De-Wei Yin
The Dow Chemical Company
USA

Jae Sung Park
Department of Mechanical and
Materials Engineering,
University of Nebraska-Lincoln
USA

Editorial Office
MDPI
St. Alban-Anlage 66
4052 Basel, Switzerland

This is a reprint of articles from the Special Issue published online in the open access journal *Actuators* (ISSN 2076-0825) (available at: https://www.mdpi.com/journal/processes/special_issues/CFD_model).

For citation purposes, cite each article independently as indicated on the article page online and as indicated below:

LastName, A.A.; LastName, B.B.; LastName, C.C. Article Title. *Journal Name* **Year**, *Volume Number*, Page Range.

ISBN 978-3-0365-1266-2 (Hbk)
ISBN 978-3-0365-1267-9 (PDF)

Contents

About the Editors

Li Xi obtained his Ph.D. from the University of Wisconsin-Madison. After completing his postdoctoral training at the Massachusetts Institute of Technology, he joined the Department of Chemical Engineering at McMaster University as a faculty member. He has broad research interests in chemical and materials engineering, with recent focuses on the modeling and analysis of chemical and pharmaceutical reactors and the development of sustainable polymer production processes. In addition, he also actively contributes to fundamental research in turbulent flows, polymer physics, and rheology.

De-Wei Yin is a research engineer at the Dow Chemical Company in Midland, Michigan, U.S.A. As a member of the Fluid Mechanics & Mixing Discipline in the Core R&D Engineering & Process Sciences Laboratory at Dow, he has conducted research on porous media flows, multiphase stirred-tank reactor design, flow measurement techniques, characterization and optimization of mixing processes, numerical algorithm development, and the application of dimensional analysis in design of experiments. He is an active member of the American Institute of Chemical Engineers and of the North American Mixing Forum.

Jae Sung Park is currently an assistant professor in the Department of Mechanical and Materials Engineering at the University of Nebraska-Lincoln. He was a postdoctoral associate in the Department of Chemical and Biological Engineering at the University of Wisconsin-Madison. He received his Ph.D. in Mechanical Engineering from the University of Illinois at Urbana-Champaign. His research interests encompass a wide range of transport phenomena, including nano/micro/biofluids, electrokinetics, turbulent flows, and flow control. He has developed advanced computational algorithms to investigate various applications in fluid mechanics.

Editorial

Special Issue "CFD Modeling of Complex Chemical Processes: Multiscale and Multiphysics Challenges"

Li Xi [1,2,*], De-Wei Yin [3] and Jae Sung Park [4]

1 Department of Chemical Engineering, McMaster University, Hamilton, ON L8S 4L7, Canada
2 School of Computational Science and Engineering, McMaster University, Hamilton, ON L8S 4K1, Canada
3 The Dow Chemical Company, Midland, MI 48667-1776, USA; dwyin@dow.com
4 Department of Mechanical and Materials Engineering, University of Nebraska-Lincoln, Lincoln, NE 68588-0526, USA; jaesung.park@unl.edu
* Correspondence: xili@mcmaster.ca

Citation: Xi, L.; Yin, D.-W.; Park, J.S. Special Issue "CFD Modeling of Complex Chemical Processes: Multiscale and Multiphysics Challenges". *Processes* **2021**, *9*, 775. https://doi.org/10.3390/pr9050775

Received: 15 April 2021
Accepted: 16 April 2021
Published: 28 April 2021

Publisher's Note: MDPI stays neutral with regard to jurisdictional claims in published maps and institutional affiliations.

After decades of development, computational fluid dynamics (CFD), which solves fluid mechanics and, more generally, transport phenomena problems using numerical analysis, has become a main-stream tool in many areas of engineering practice. In chemical processes, CFD is widely used not only in predicting fluid flow and mixing patterns in chemical reactors, but also in the simulation and modeling of various unit operations where complex flow and transport processes make process outcomes hard to predict and control. Although CFD models typically still require experimental validation, their application brings many advantages in process research and development. First, numerical solutions from CFD contain fully detailed information about the flow field and distribution profiles of other quantities of interest (temperature, pressure, chemical species concentration, etc.), which are often hard to measure directly from experiments, especially in situ. The availability of such detailed data on the fluid dynamics is often critical to process analysis and improvement. Second, a carefully validated CFD model can be used to predict process outcomes in new operating conditions and process design, with which untested process parameters can be prescreened. As such, the number of experiments and trials and errors can be greatly reduced. This not only reduces the cost in, e.g., process design, adaption, and scale-up, but also reduces the risk of potential safety, environmental, and health hazards associated with many chemical processes.

With the increasing power of high-performance computing (HPC) facilities and continued development of numerical techniques, the accuracy and sophistication of CFD models have improved. Obtaining reliable and practically useful predictions, however, is far from a simple click of a button, owing to the intrinsic complexities in many chemical processes. Most chemical processes involve complex coupling of fluid flow with other physicochemical processes, such as heat and mass transfer, chemical reactions, and particle/bubble/droplet dynamics. In addition, dynamics over vastly different length and time scales also interact in nontrivial ways. For one thing, many chemical processes involve flow turbulence, which is intrinsically multiscale. For another, multiphase flow processes where phase separation and fluid flow occur at different scales are also common. Overcoming multiscale and multiphysics challenges is thus often a central theme and requires innovative modeling approaches that best suit the particular systems.

This special issue "CFD Modeling of Complex Chemical Processes: Multiscale and Multiphysics Challenges" highlights some recent advances in the application of CFD in chemical processes. We take a very liberal approach and define chemical processes as those where principles of chemical engineering, such as transport phenomena, are important, even though the direct application may not fall within the traditional scope of the chemical industry. A total of 16 papers are collected, including one review, 14 original research articles, and one correction. The topics range from pure methodology development to the application of existing CFD tools in practical systems, and everything in between.

1

The review paper by Nakhaei et al. [1] offered a comprehensive and in-depth account of CFD application in gas–solid cyclone separators. It covered the fundamentals of gas–solid cyclone separator systems including their key performance metrics and how they are influenced by operating conditions. This was followed by an extensive review of available approaches for the numerical modeling of such systems and a summary of existing CFD studies at both ambient and elevated temperatures. Capabilities and limitations of existing approaches were also discussed.

Among the original research articles, Vachaparambil and Einarsrud [2] had a focus on modeling methodology. Based on simple benchmark problems, the study compared different surface tension models used in the volume of fluids method for gas–liquid two-phase flows. Special attention was given to the occurrence of spurious currents due to model inaccuracies.

Other studies all targeted more complex and practical engineering systems. Several studies applied CFD to different types of chemical and biochemical reactors. Ebrahimi et al. [3] studied the flow and mixing performance of double impellers in a stirred-tank reactor. Effects of impeller configuration and speed were reported. Han et al. [4] studied the combustion performance in a pulverized coal furnace with burners installed on the front wall by coupling the momentum balance (with turbulence model), species transport, and energy balance equations. The goal was to optimize the process for reduced nitrogen emission and air pollution. Reactors with multiphase flow were also represented. Khezri et al. [5] studied a bubbling fluidized-bed reactor for biomass gasification. The gas–solid flow was modeled with an Eulerian–Eulerian approach and various turbulence and drag models were compared. Effects of air distributors with different pore sizes were also examined. Tao et al. [6] again used an Eulerian–Eulerian two-fluid model to study a large-scale high-pressure gas–liquid bubble column reactor, with a focus on the effects of operating parameters on the gas holdup of bubbles of different sizes. An Eulerian–Eulerian two-fluid model was also used in Sarkizi Shams Hajian et al. [7] for the gas–liquid two-phase system in the production process of baker's yeast at the industrial scale. The study combined turbulence modeling with species transport and bio-kinetic models to analyze species distribution in the reactor.

CFD studies of other unit operations are also included. Wang, Z. et al. [8] studied a rotating packed bed process for gas desulfurization, where a liquid solution was dispersed in the gas phase for the selective absorption of H_2S. An Eulerian–Lagrangian approach was used where the dynamics of liquid droplets were described by coupling force balance with kinetic models for their coalescence and breakup, with which characteristics of droplets were studied. Qadir et al. [9] numerically simulated the gas separation process in membrane modules and studied the effects of membrane configuration and operating parameters. Landauer and Foerst [10] experimentally studied the triboelectric separation of starch and protein particles. After passing through a charging tube, particles carrying different triboelectric charges settled to different distances in a separation chamber under an electrical field. Effects of particle size and material on the settling distance were studied. The gas flow field in the chamber was modeled with CFD, and simulated particle trajectories in the flow and electrical fields were used to estimate the minimum charge for particle separation. Along the lines of particle processes, Yang et al. [11] used an Eulerian–Lagrangian discrete particle model to simulate the pneumatic conveying of coal particles, studied the effects of operating conditions on the pressure drop, and identified the range of model applicability by comparison with experiments.

Applications in transport processes not in the traditionally-defined scope of chemical engineering, which nonetheless were built on the same principles, are also represented in this special issue. Gao et al. [12] used CFD to model the hydrodynamics in the lateral inlet/outlet part of a pumped hydroelectric storage system. Surrogate models were then built based on CFD results and used for the geometric optimization of the device. Wang, W. et al. [13] measured the air flow, temperature, and relative humidity in the blind heading of an underground mine in situ to study the heat emission by equipment. CFD was conducted to obtain the detailed flow and temperature fields and to optimize the design of the ventilation system. Also in mining, Zhou et al. [14] numerically solved the air flow and species transport equations in a typical stone-coal mine laneway to study the effects of forced ventilation on the radon concentration. Numerical analysis of coupled fluid flow and transport processes was also the theme of Mousavi et al. [15], where the air flow and heat transfer in a novel sustainable farming compartment (SFC) was modeled. This SFC was designed with an evaporative cooling system where the vaporization of treated wastewater was used to absorb heat and the cooling effect was distributed across the domain by forced convection. An experimental prototype was also set up to demonstrate the concept, and CFD was used to optimize the system design and operating conditions.

Finally, we would like to express our sincere gratitude towards all authors for contributing to this special issue. We look forward to the continued development of CFD methodology and its application in chemical process engineering in years to come.

Conflicts of Interest: The authors declare no conflict of interest.

References

1. Nakhaei, M.; Lu, B.; Tian, Y.; Wang, W.; Dam-Johansen, K.; Wu, H. CFD Modeling of gas–solid cyclone separators at ambient and elevated temperatures. *Processes* **2020**, *8*, 228. [CrossRef]
2. Vachaparambil, K.J.; Einarsrud, K.E. Comparison of surface tension models for the volume of fluid method. *Processes* **2019**, *7*, 542. [CrossRef]
3. Ebrahimi, M.; Tamer, M.; Villegas, R.M.; Chiappetta, A.; Ein-Mozaffari, F. Application of CFD to Analyze the Hydrodynamic Behaviour of a Bioreactor with a Double Impeller. *Processes* **2019**, *7*, 694. [CrossRef]
4. Han, J.; Zhu, L.; Lu, Y.; Mu, Y.; Mustafa, A.; Ge, Y. Numerical Simulation of Combustion in 35 t/h Industrial Pulverized Coal Furnace with Burners Arranged on Front Wall. *Processes* **2020**, *8*, 1272. [CrossRef]
5. Khezri, R.; Wan Ab Karim Ghani, W.A.; Masoudi Soltani, S.; Awang Biak, D.R.; Yunus, R.; Silas, K.; Shahbaz, M.; Rezaei Motlagh, S. Computational Fluid Dynamics Simulation of Gas–Solid Hydrodynamics in a Bubbling Fluidized-Bed Reactor: Effects of Air Distributor, Viscous and Drag Models. *Processes* **2019**, *7*, 524. [CrossRef]
6. Tao, F.; Ning, S.; Zhang, B.; Jin, H.; He, G. Simulation Study on Gas Holdup of Large and Small Bubbles in a High Pressure Gas–Liquid Bubble Column. *Processes* **2019**, *7*, 594. [CrossRef]
7. Sarkizi Shams Hajian, C.; Haringa, C.; Noorman, H.; Takors, R. Predicting By-Product Gradients of Baker's Yeast Production at Industrial Scale: A Practical Simulation Approach. *Processes* **2020**, *8*, 1554. [CrossRef]
8. Wang, Z.; Wu, X.; Yang, T.; Wang, S.; Liu, Z.; Dan, X. Droplet Characteristics of Rotating Packed Bed in H2S Absorption: A Computational Fluid Dynamics Analysis. *Processes* **2019**, *7*, 724. [CrossRef]
9. Qadir, S.; Hussain, A.; Ahsan, M. A Computational Fluid Dynamics Approach for the Modeling of Gas Separation in Membrane Modules. *Processes* **2019**, *7*, 420. [CrossRef]
10. Landauer, J.; Foerst, P. Influence of particle charge and size distribution on triboelectric separation—New evidence revealed by in situ particle size measurements. *Processes* **2019**, *7*, 381. [CrossRef]
11. Yang, D.; Wang, Y.; Hu, Z. Research on the Pressure Dropin Horizontal Pneumatic Conveying for Large Coal Particles. *Processes* **2020**, *8*, 650. [CrossRef]
12. Gao, X.; Zhu, H.; Zhang, H.; Sun, B.; Qin, Z.; Tian, Y. CFD optimization process of a lateral inlet/outlet diffusion part of a pumped hydroelectric storage based on optimal surrogate models. *Processes* **2019**, *7*, 204. [CrossRef]
13. Wang, W.; Zhang, C.; Yang, W.; Xu, H.; Li, S.; Li, C.; Ma, H.; Qi, G. In situ measurements and CFD numerical simulations of thermal environment in blind headings of underground mines. *Processes* **2019**, *7*, 313. [CrossRef]
14. Zhou, B.; Chang, P.; Xu, G. Computational Fluid Dynamic Simulation of Inhaled Radon Dilution by Auxiliary Ventilation in a Stone-Coal Mine Laneway and Dosage Assessment of Miners. *Processes* **2019**, *7*, 515. [CrossRef]
15. Mousavi, M.; Mirfendereski, S.; Park, J.S.; Eun, J. Experimental and Numerical Analysis of a Sustainable Farming Compartment with Evaporative Cooling System. *Processes* **2019**, *7*, 823. [CrossRef]

Review

CFD Modeling of Gas–Solid Cyclone Separators at Ambient and Elevated Temperatures

Mohammadhadi Nakhaei [1,2,*], Bona Lu [3,4], Yujie Tian [3,4], Wei Wang [3,4], Kim Dam-Johansen [1] and Hao Wu [1]

[1] Department of Chemical and Biochemical Engineering, Technical University of Denmark, Lyngby Campus, 2800 Kgs. Lyngby, Denmark; KDJ@kt.dtu.dk (K.D.-J.); haw@kt.dtu.dk (H.W.)
[2] Sino-Danish Center for Education and Research, Beijing 100093, China
[3] State Key Laboratory of Multiphase Complex Systems, Institute of Process Engineering, Chinese Academy of Sciences, Beijing 100190, China; bnlu@ipe.ac.cn (B.L.); tianyj@ipe.ac.cn (Y.T.); wangwei@ipe.ac.cn (W.W.)
[4] Sino-Danish College, University of Chinese Academy of Sciences, Beijing 100049, China
* Correspondence: mnak@kt.dtu.dk

Received: 15 January 2020; Accepted: 7 February 2020; Published: 15 February 2020

Abstract: Gas–solid cyclone separators are widely utilized in many industrial applications and usually involve complex multi-physics of gas–solid flow and heat transfer. In recent years, there has been a progressive interest in the application of computational fluid dynamics (CFD) to understand the gas–solid flow behavior of cyclones and predict their performance. In this paper, a review of the existing CFD studies of cyclone separators, operating in a wide range of solids loadings and at ambient and elevated temperatures, is presented. In the first part, a brief background on the important performance parameters of cyclones, namely pressure drop and separation efficiency, as well as how they are affected by the solids loading and operating temperature, is described. This is followed by a summary of the existing CFD simulation studies of cyclones at ambient temperature, with an emphasis on the high mass loading of particles, and at elevated temperatures. The capabilities as well as the challenges and limitations of the existing CFD approaches in predicting the performance of cyclones operating in such conditions are evaluated. Finally, an outlook on the prospects of CFD simulation of cyclone separators is provided.

Keywords: gas–solid; cyclone separator; computational fluid dynamics; elevated temperature process

---

1. Introduction

Gas–solid cyclones are frequently used in industrial processes with the primary purpose of two–phase flow separation, i.e., separation of a high-density phase from a lower-density carrier phase, using a turbulent swirling flow. The state-of-the-art industrial cyclone designs are able to operate at elevated temperatures and moderate-to-high loading of solids, while at the same time meeting the required separation efficiency and having low investment and maintenance costs. This has led to the frequent use of cyclones as the only/initial stage of separation and cleaning processes rather than other industrial separators, e.g., bag filters, electrostatic separators, etc. Examples of cyclone separator applications in demanding industrial process conditions are high temperature gas–solid heat exchangers [1], e.g., in the cement industry; gasification and combustion of solid fuels [2–4]; coal pyrolysis [5]; and gas–solid separation in circulating fluidized beds (CFBs) [6].

With the main purpose of gaining improved insight into the flow physics and factors affecting the two important performance parameters of cyclones, i.e., pressure drop and separation efficiency, single–phase and gas–solid flows in pilot-scale cyclones have been extensively studied with the use of experimental methods (for example, see [7–10]). Accordingly, a number of simple semi-empirical/-theoretical models have been proposed to address the flow field and performance of cyclones (for example see [11–16]), and some of them are still being used in cyclone design and optimization.

In the past two decades, computational fluid dynamics (CFD) has been broadly applied to predict the separation efficiency and pressure drop of cyclone separators and to optimize cyclone designs. Compared to the models based on classical cyclone theory, three-dimensional CFD simulations have the advantage of taking into account the unsteadiness and asymmetry of cyclone flow. On the other hand, the presence of strong swirl and anisotropic turbulent flow as well as adverse pressure gradients in the cyclones has driven the CFD simulation studies to use more advanced turbulence models, e.g., the Reynolds stress transport model (RSTM) and large eddy simulation (LES), as well as higher-order discretization techniques, that are capable of capturing these specific flow physics. These models/techniques, however, are computationally demanding compared to the more commonly used models, e.g., k–ε. Furthermore, with the addition of extra physics to the CFD simulation of cyclones, e.g., presence of particles at high loadings and gas–solid heat transfer, additional complications will arise with respect to the solution's accuracy and stability.

In this paper, a brief introduction to the basic operation principles of cyclones at ambient and elevated temperatures is initially presented. Subsequently, the existing CFD simulation studies of cyclones are summarized and discussed based on the operating temperature, i.e., ambient and elevated temperature, and with a special focus on the studies with moderate-to-high loading of particles. The paper highlights specific process parameters that are able to be captured by the CFD simulations at these temperatures. Furthermore, the important sub-models utilized as well as specific challenges that may be encountered in such CFD simulations are addressed.

2. Fundamentals of Gas–Solid Cyclone Separators

The reverse-flow-type gas–solid cyclones are usually equipped with either tangential inlets, which are the main focus in this study, or axial inlets (for example, see [17,18]), usually referred to as swirl tubes [19]. A common configuration of a tangential inlet reverse–flow cyclone is schematically shown in Figure 1a. In this type of cyclone, the swirling gas flow pattern usually consists of a "double vortex": an outer vortex with a downward axial velocity and an upward-moving inner vortex. The tangential flow pattern, caused by the tangential inlet of the gas to the cyclone system, is referred to as a "Rankine vortex", comprised of a near-solid-body rotating flow at the core (i.e., tangential velocity is directly proportional to the radius) and a loss-free vortex flow (i.e., tangential velocity is inversely proportional to the radius) at the walls [19] (see Figure 1b). The outer vortex gradually loses its downward momentum in the bottom regions of the cyclone and changes its direction, i.e., "vortex end", to form the inner vortex. The inner vortex is led to the exit pipe, which is usually extended to the cyclone body to separate the inner vortex from the inlet velocity field. The axial distance from the "vortex end" location to the vortex finder is termed the "natural turning length" [19]. The double vortex structure is inherently unstable due to the presence of a radial pressure gradient imposed by the vortex itself [20]. This phenomenon is referred to as a precessing vortex core (PVC), which causes the location of the vortex axis to oscillate with a specific frequency. In some cyclones, in order to stabilize the vortex end position, a vortex stabilizer is installed either under or above the dust exit location (see Figure 1b), which leads to improved separation efficiency and desirable static pressure below the stabilizer [19].

The static pressure in the cyclone increases monotonically toward the walls to maintain the equilibrium of the rotating flow pattern. This pressure gradient is also present in the boundary layer of the cyclone walls, where the tangential velocity is small [19]. This leads to the presence of a secondary flow, namely inwardly-directed gas flow along the walls, at the cyclone roof, and the conical section, as depicted in Figure 1c.

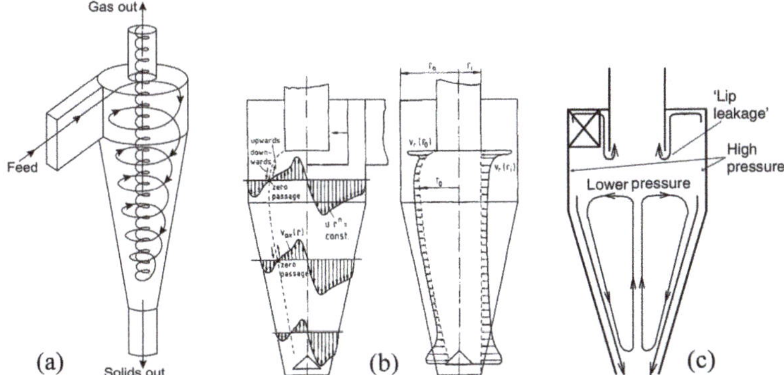

Figure 1. (**a**) Schematic drawing of a conical reverse-flow cyclone separator illustrating the basic operating principle and the presence of a double vortex inside the cyclone [21], reproduced with permission from G. Towler and R. Sinnott, Specification and design of solids-handling equipment, published by Elsevier, 2013. (**b**) Qualitative patterns of axial, tangential, and radial velocity components of the gas-flow field in cyclones (right) [22], reproduced with permission from M. Trefz and E. Muschelknautz, Extended cyclone theory for gas flows with high solids concentrations, published by John Wiley and Sons, 1993. (**c**) The secondary flow pattern caused by the swirl and pressure gradients in the cyclone [19], reproduced with permission from A. Hoffmann and L. Stein, Gas Cyclones and Swirl Tubes: Principles, Design and Operation; published by Springer Nature, 2007.

2.1. Performance Parameters of Cyclones

For the main purpose of gas–solid separation, two important design parameters of reversed-flow conical cyclones are separation efficiency and pressure drop. These parameters are affected by the gas–solid flow field inside the cyclone, which in turn is influenced by the cyclone's geometrical features as well as the operating conditions. In this section, these performance parameters are discussed and the influences of operating conditions, e.g., solids loading and gas temperature, on cyclone performance are explained.

2.1.1. Separation Efficiency

Cyclone separators are suitable for the separation of solid particles with a size range of 2–2000 microns, which are found in many industries, e.g., heavy industrial smoke, coal dust, cement dust, etc. [19] When solid particles are fed to a cyclone, they are affected by two main forces: the radially outward-directed centrifugal acceleration force, proportional to the cube of particle diameter, and the fluid drag force applied to the particles in the opposite direction; whereas the gravity force is reported to be of minor importance [23]. When the Stokes law applies for the drag force, which is often valid for solid particles in cyclones, the drag force is proportional to the particle diameter. This indicates the dominance of the centrifugal acceleration force for larger particles, leading to their improved separation. The larger particles experience the centrifugal effect as soon as the gas and solids experience the rotational flow at the inlet, and subsequently, they are pushed toward the walls and lose their momentum. Once the particles approach the wall, they start to move downward due to gravity and the drag force applied to them by the downward-directed gas flow. Finally, these particles are separated at the bottom of the cyclone.

Conversely, fine particles, e.g., 1–10 microns, have a greater tendency to follow the gas flow in a cyclone. These particles are poorly separated from the gas phase due to either following the bypass gas to the vortex finder (see Figure 1c) or entrainment to the inner vortex along the inner and outer vortex boundary. The turbulence dispersion of particles intensifies the re-entrainment of fine particles to the inner vortex followed by particle carry over to the vortex finder. On the other hand, the agglomeration of small particles as well as particle attrition can affect the separation efficiency. For example, in the reported grade efficiencies of a dilutely loaded pilot-scale cyclone, the separation efficiency of particles with a diameter of 0.8–1.0 μm is smaller than the separation efficiency of very fine particles, e.g., 0.3 μm, as well as of large particles [19]. This is known as "fish-hook" behavior [24] and is mainly due to agglomeration of small-sized particles to larger particles, leading to the improved grade efficiency of these particles. The mentioned trend is also observed in other experimental studies [25].

2.1.2. Pressure Loss

One of the main parameters considered in the design of industrial cyclone systems is the energy loss, usually termed the pressure loss. The smaller the pressure loss of the cyclone in a process, the cheaper the process cost. The overall pressure loss in a reversed-flow conical cyclone can be separated into three parts [19,20,26]:

1. Pressure loss at the inlet;
2. Pressure loss in the separation zone;
3. Pressure loss associated with the vortex finder.

The first contributor to the pressure loss is usually of minor importance. The pressure loss in the separation zone is mainly due to the frictional losses at the wall of the cyclone body. The vortex finder pressure loss is due to the dissipation of the swirl dynamic pressure in the vortex finder, which usually takes place without recovering the dynamic pressure into the static pressure. This pressure drop is the main contributor to the overall pressure drop for single-phase flow or dilutely loaded cyclones, and it is proportional to the square of the tangential velocity magnitude. The vortex finder pressure loss is usually affected by the frictional pressure loss, e.g., if the frictional pressure increases due to higher wall roughness or other operating conditions, the vortex will be weakened, leading to the reduction of the vortex finder pressure loss. This indicates that the contribution of different parts of pressure loss changes once the operating conditions of the cyclone change, e.g., an increase in the temperature or the solids loading, which will be further discussed in subsequent sections.

2.2. Effect of Solids Loading on Cyclone Performance

At low solid-to-gas mass loading ratios, e.g., 0.01 kg_s/kg_g, solid particles are clustered as thin strands on the cyclone walls and are transported in the downward direction in a spiral path, while at high mass loadings, e.g., 10 kg_s/kg_g, a major portion of (or the whole) wall surface area is covered with a layer of solid particles (also know as a dense strand in [27]), which is slipped directly into the solids discharge [22]. According to the existing experimental studies of cyclones, both overall and grade separation efficiencies are improved by increasing the inlet solids mass loading ratio [7,8,19]. Hoffmann and Stein [19] reported an overall improvement of the grade efficiency of almost all particle sizes for an increase in the inlet particle mass loading from 3.7×10^{-3} to 2.6×10^{-2} kg_s/kg_g. However, they have stated that the separation efficiency at high mass loading of particles is somewhat dependent on the inlet gas velocity of the cyclone, i.e., the swirl strength [28]. The overall separation efficiency versus inlet solids loading of selected experimental studies is presented in Figure 2. In this figure, the improvement of separation efficiency with the cyclone Reynolds number can be observed from the experimental data of [19] for a fixed value of kinematic response time of particles of $\tau_p = \frac{\rho_p d_p^2}{18\,\mu_g} = 0.1$ milliseconds. The improvement is more noticeable for lower mass loadings, i.e., below 0.01 kg_s/kg_g. The kinematic response time of particles used in the study of Fassani and Goldstein [29] is somewhat higher compared to the particles of other studies presented in Figure 2, leading to a very high efficiency of separation. For the rest of the studies, no conclusive argument can be stated regarding the effect of this parameter.

Despite available experimental studies on the effect of solids loading on the separation efficiency, there is no consensus regarding the exact mechanism of this improvement [19]. According to the concept of the critical mass loading ratio, ϕ_G, initially introduced by Muschelknautz and Brunner [8,13], when the inlet particle mass loading is higher than the critical mass loading, the share of particles corresponding to the extra amount of mass loading is ideally separated at the cyclone inlet, while the rest of the particles are separated based on a balance between the centrifugal forces and the turbulent dispersion, commonly referred to as "inner separation" [22]. The concept of Muschelknautz, however, suggests that the separation efficiency is independent of the mass loading for loadings smaller than ϕ_G, which is not consistent with the experimental data in the literature [28]. On the other hand, based on another concept introduced by Mothes and Löffler [30], the improvement in the separation efficiency at high solid mass loadings is attributed to the agglomeration of smaller particles, i.e., 2 μm diameter and smaller, to larger particles, i.e., 15 μm diameter. Hoffmann and Stein [19] interpreted the separation efficiency of cyclones at high mass loadings to be a combined effect of different contributions. For very fine particles, the agglomeration effect is more dominant, leading to a high efficiency of these particles, especially for a humid carrier gas [19]. Furthermore, they argued that the increased concentration of particles at the inlet can lead to reduced drag force, which improves the separation process.

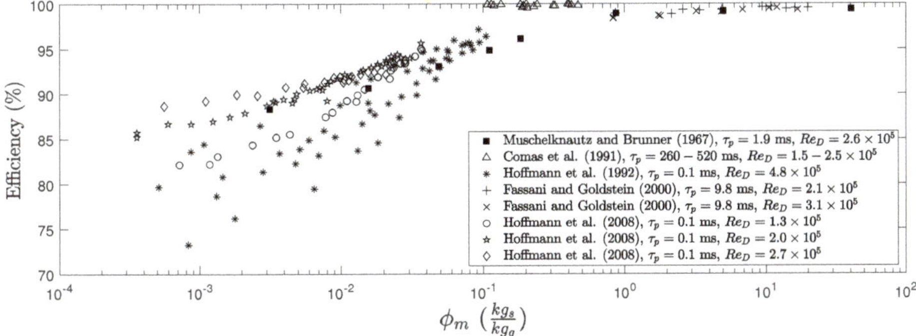

Figure 2. Overall cyclone separation efficiency versus mass loading ratio at the cyclone inlet for selected experimental data in the literature [7,8,19,29,31]. The Reynolds number is calculated based on the inlet velocity and the cyclone body diameter.

Baskakov et al. [32] and Muschelknautz and Brunner [8] have reported a decrease in the pressure drop up to a specific amount of solids loading, followed by an increase. A similar trend is also observed in other experimental [33] and CFD simulation studies [34,35]. This behavior is attributed to the increase of the frictional pressure loss due to the presence of particles near the walls. Although, at the same time as the friction velocity at the cyclone walls increases, the swirl intensity is weakened; consequently, the pressure loss associated with the vortex finder is decreased. Muschelknautz and Brunner [8] reported an 85% reduction in the tangential velocity peak by increasing the solids loading from 0 to around 20 kg_s/kg_g. At low mass loadings, the decrease in the vortex finder pressure loss is more pronounced, leading to the reduction of the overall pressure loss, whereas when the particle mass loading is sufficiently high, the vortex finder pressure loss becomes negligible due to a very weak tangential swirl and the frictional pressure drop still increases with the loading of particles [35]. As a consequence, the overall pressure drop starts to increase with the solids loading after reaching a minimum value.

A summary of selected experimental data on pressure drop versus solids loading in pilot-scale cyclones from the literature is presented in Figure 3. Despite the mentioned explanation, there are other studies in the literature that dispute the presence of a minimum pressure drop when the loading of particles changes. For example, in some studies [7,9,10,31], it is reported that by increasing the mass loading of particles in the cyclone, the pressure drop is reduced, although these studies are limited to small mass loadings, e.g., up to 0.5 kg_s/kg_g in [31], 0.1 kg_s/kg_g in [7], and 0.2 kg_s/kg_g in [9]. The mass loading ratio at which the minimum pressure drop occurs is reported to be around 0.25 (at an operating gas temperature of 250 °C) and 2.0 (at ambient temperature) in the studies of Baskakov et al. [32] and

Muschelknautz and Brunner [8], respectively. Furthermore, in some of the studies, it is reported that by increasing the inlet solids loading, the pressure drop increases [36] or remains at a nearly constant value lower than the pressure drop of the single-phase flow cyclone [29].

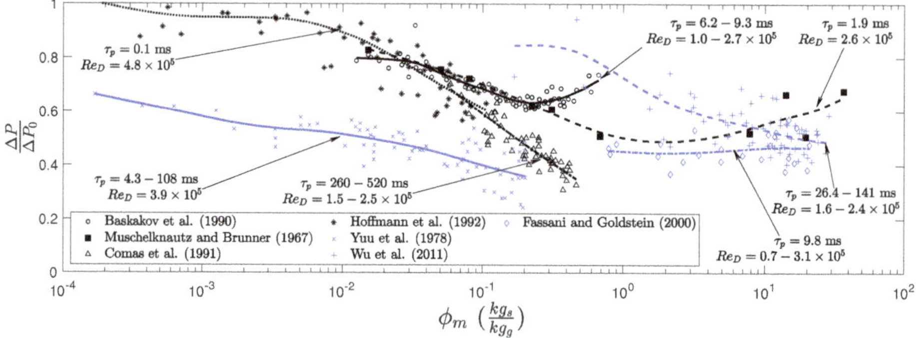

Figure 3. Experimental data of pressure drop (normalized with the pressure drop of a particle-free cyclone, ΔP_0) versus mass loading of pilot-scale cyclones from selected studies [7–9,29,31,32,37]. The lines are numerical fits to each set of experimental data. The Reynolds number is calculated based on the inlet velocity and the cyclone body diameter.

2.3. Effect of Operating Temperature on Cyclone Performance

The use of gas–solid cyclones is popular in high-temperature industrial processes, e.g., for drying, solidification, heating, flue gas cleaning purposes, etc. The high-temperature cyclone separators can be loaded with large amounts of particles, e.g., in gas–solid CFB cyclones, the solids loading can be on the order of 10–100 kg_s/kg_g [38]. As briefly mentioned earlier, the cyclone's performance is affected by its operating temperature. In this section, a brief introduction to the effect of cyclone gas temperature on separation efficiency, pressure drop, and gas–solid heat transfer is provided.

In general, as gas temperature rises, gas density and viscosity are, respectively, decreased and increased. At a fixed volumetric gas-flow rate to the cyclone, a direct consequence of this change in the fluid properties is the reduction of the cyclone's Reynolds number and the swirl intensity in the cyclone at elevated operating temperatures. As a result, the cyclone's pressure drop is reduced. This trend is reported in many experimental [39,40] and CFD studies (for example, see [41]) of cyclones operating at elevated temperatures.

For a fixed inlet gas velocity to the cyclone, as the temperature increases, the particle cut-size, i.e., the diameter at which the cyclone separation efficiency is equivalent to 50%, increases, indicating a negative influence of the operating temperature on the separation efficiency [39,40]. This behavior can be partly attributed to an intensified drag force applied to the particles by the carrier fluid as a consequence of increased gas viscosity at elevated temperatures. Furthermore, as stated before, the reduction in the swirl intensity leads to a weaker centrifugal effect on the particles compared to the intensified drag force, and as a result, the separation efficiency is reduced [40]. On the other hand, the gas density is inversely proportional to the temperature; but its effect on particle separation is less significant as only the difference between the gas and particle densities is the determining factor on the buoyancy force applied to the particles.

In the existing experimental studies in the literature, the heat transfer rate to the particles in gas–solid cyclone heat exchangers is positively affected by the loading of particles and the inlet gas velocity, while being inversely influenced by the particle size [1]. Improvement in the heat-transfer rate with solids loading is attributed to the reduced radial gas velocity in the cyclone as a result of stronger gas–solid momentum coupling. Consequently, the amount of gas bypassed to the vortex finder at the entrance of the cyclone is reduced, leading to the presence of a higher-temperature gas at the bottom regions of the cyclone and further improvement in the heat-transfer rate to the solids [1]. As the particle loading further increases, the rate of increase in the gas–solid heat transfer rate reduces [1,42] and approaches a certain value. Increased gas–solid heat transfer as a result of higher inlet gas velocity as well as the reduction of particle size can be simply explained by an improved driving factor for gas–solid heat transfer, i.e., more availability of the high temperature gas to transfer heat to the solids and improved gas–solid surface area for heat transfer, respectively.

One of the challenges that may exist at the cyclone's internal walls, especially for cyclones operating at elevated temperatures, is erosion due to wall–particle collision and/or corrosion due to the presence of aggressive species and agents in the gas, even in the presence of a strong refractory lining. It is more likely to have erosion on the cyclone wall in front of the gas inlet, caused by the very first impact that solids have with the cyclone wall, followed by a change in the direction of particle movement [43]. An immediate outcome of cyclone wall erosion is more rough surfaces on a cyclone's internal walls, affecting the pressure drop and separation efficiency of particles. A common solution for this challenge is "hardware" changes in the cyclone system during the maintenance period, e.g., installation of wear plates and replacement of eroded materials [19]. Furthermore, cyclone design modification is an option to improve the erosion behavior, e.g., adding a flat-disk vortex stabilizer close to the particle outlet reduces erosion [44]. Apart from erosion and corrosion, the deposition of particles on the cyclone walls can induce operational challenges in the long term. Deposit formation can take place on the outer surface of a vortex finder [45], dipleg [46], or on the internal walls of the cyclone body [47].

3. Approaches for the Numerical Modeling of Gas–Solid Systems

The use of numerical simulations is undoubtedly a valuable tool for understanding and improving industrial gas–solid systems. However, the presence of a very large span for length and time scales of both phases, which are influenced by each another, makes resolving all of the details of a gas–solid system through computational simulation a challenging task [48]. In the existing numerical methods, the gas–solid flow is resolved at different but limited ranges of length and time scales, and the smaller-scale details, if any, are modeled. A summary of the available approaches for a four-way coupled (to be explained later) gas–solid system is provided in Table 1. In the most fundamental approach, usually referred to as direct numerical simulation (DNS), the boundary layer over the particles surface is resolved. DNS can itself be categorized into two approaches: the resolved Eulerian–Lagrangian model, e.g., immersed boundary model, and the Lagrangian–Lagrangian approach using molecular dynamics, e.g., Lattice Boltzmann, applicable for simulations of gas–solid flows at extremely small scales of 0.01 and 0.001 m, respectively [48]. DNS is usually utilized for fundamental studies of gas–solid flows at small scales in order to develop sub-models, e.g., drag and Brownian motion models, for more simplified approaches.

In the Eulerian–Eulerian (E–E) approach, both phases are described as a continuum medium and solved in an Eulerian domain [49], while the gas–solid interactions are incorporated through a drag model and the particle–particle interactions through models for solid pressure and viscosity, e.g., the kinetic theory of granular flows (KTGF) [50]. A challenge in using the basic formulation of the E–E approach is the difficulty in taking into account poly-disperse particles in the model. However, there are several sub-models, such as the population balance model (PBM), that can be added to the basic E–E equations to account for different particle sizes, with the cost of higher computational overhead since separate momentum and continuity equations should be solved for each size bin [51–53]. In addition, some complex physical phenomena such as particle agglomeration and break-up can be modeled in combination with the PBM [54]. A limitation of the E–E approach is the validity of the continuum hypothesis for both phases, i.e., both phases are adequately present over the whole computational domain. This makes the model more appropriate for gas–solid systems with more homogeneous distributions of particles, mainly dense gas–solid systems such as fluidized beds.

In the (unresolved) Eulerian–Lagrangian (E–L) approach, also referred to as the discrete particle model (DPM), the carrier gas flow field is considered as a continuum medium and solved using an Eulerian formulation, while solid particles are tracked in a Lagrangian platform using Newton's second law. Compared to the resolved E–L model, categorized as DNS in this study, the boundary layer over the particle surface is not resolved in the unresolved E–L and the size of the computational grid is significantly larger than the particle size. One distinct advantage of this model compared to the E–E approach is the direct resolving of particle movement, i.e., fewer closure models are employed for the particle phase, which makes the model more accurate. Furthermore, poly-disperse systems can be readily modeled using the E–L method. The particle–particle interactions in the E–L model are usually incorporated as either hard-sphere or soft-sphere collision models. In the hard-sphere model, the collisions take place instantly and can be employed in both stochastic and deterministic collision approaches; thus it is suitable for dilute flows where binary collisions happen. Conversely, this model is not suitable for cases where particles go through multiple collisions with other particles simultaneously. The soft-sphere model, also referred to as the discrete element model (DEM), is a deterministic particle collision approach that takes into account overlapping particles during the collision and is able to incorporate multiple simultaneous collisions for a particle [55]. As the loading of particles in a gas–solid system increases, the E–L approach becomes computationally expensive, especially if the DEM model is utilized [55].

As an alternative, the particle–particle interactions can be modeled by using hybrid E–E and E–L approaches, such as the multi-phase particle in-cell (MP-PIC) method [56–58] and the dense discrete phase model (DDPM) [59]. The basic idea in the hybrid approach is to solve the carrier phase in an Eulerian frame with the inclusion of the particle phase volume fraction, taken from the Lagrangian tracking, in the conservation equations. Similar to the E–L method, the particles are tracked in a Lagrangian frame, while the particle–particle interactions are modeled using solid stress calculated from the Eulerian grid [58,59]. Since the interparticle interactions are modeled instead of being directly resolved, the computational overhead of the solution is reduced compared to the unresolved E–L approach, while the model is superior over the E–E model (e.g., straightforward handling of poly-dispersed particles).

Table 1. Summary of the available approaches for computational fluid dynamics (CFD) simulation of gas–solid flows with the inclusion of closure models to be considered in each approach. The table is inspired from [60].

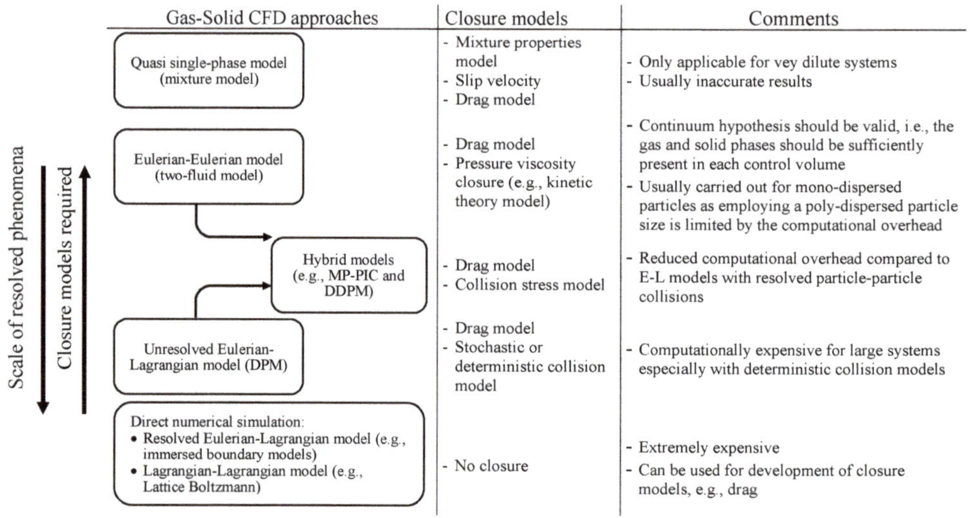

Gas-Solid CFD approaches	Closure models	Comments
Quasi single-phase model (mixture model)	- Mixture properties model - Slip velocity - Drag model	- Only applicable for vey dilute systems - Usually inaccurate results
Eulerian-Eulerian model (two-fluid model)	- Drag model - Pressure viscosity closure (e.g., kinetic theory model)	- Continuum hypothesis should be valid, i.e., the gas and solid phases should be sufficiently present in each control volume - Usually carried out for mono-dispersed particles as employing a poly-dispersed particle size is limited by the computational overhead
Hybrid models (e.g., MP-PIC and DDPM)	- Drag model - Collision stress model	- Reduced computational overhead compared to E-L models with resolved particle-particle collisions
Unresolved Eulerian-Lagrangian model (DPM)	- Drag model - Stochastic or deterministic collision model	- Computationally expensive for large systems especially with deterministic collision models
Direct numerical simulation: • Resolved Eulerian-Lagrangian model (e.g., immersed boundary models) • Lagrangian-Lagrangian model (e.g., Lattice Boltzmann)	- No closure	- Extremely expensive - Can be used for development of closure models, e.g., drag

4. CFD Simulation Studies of Gas–Solid Cyclones at Ambient Temperature

When solid particles are added to the gas flow in a cyclone separator, the mean and turbulent properties of the flow field may change. According to the classification of gas–solid interactions made by Elghobashi [61] (see Figure 4), for particle volume fractions lower than 10^{-6} (equivalent to the interparticle spacing of around $80d_p$ [20]), particles have a negligible influence on the turbulent flow field, and the gas–solid coupling is termed as one–way coupling. When the particle volume fraction reaches values higher than 10^{-6} but below 10^{-3} (equivalent to the interparticle spacing of around $8d_p$ [20]), the particles either attenuate or intensify the turbulence in the gas–solid flow, i.e., two-way coupled flow field, depending on the relative kinetic response time of particles to the characteristic flow time-scale [61]. In the context of gas–solid cyclone flows, attenuation is usually the case [20]. If the solids volume fraction is increased further, namely to the four-way coupling region, the interaction between particles becomes important too.

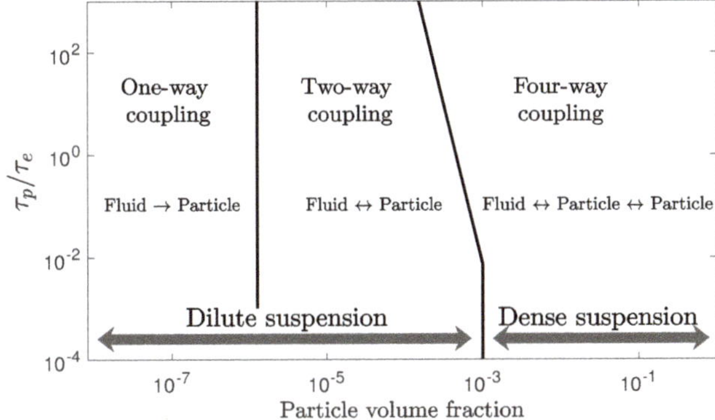

Figure 4. Map of gas–solid interaction regimes of particle-laden turbulent flows. τ_p and τ_e are the particle kinetic response time and time-scale of large eddies in a turbulent flow, respectively. Reproduced with permission from S. Elghobashi, On predicting particle-laden turbulent flows; published by Springer Nature, 1994 [61].

There are a vast number of CFD simulation studies of cyclone separators at ambient temperature, which are usually carried out for pilot-scale cyclone designs. In the current study and according to the above classification, these studies are separated into three groups of (I) single-phase gas flow; (II) one-way coupled gas–solid flow; and (III) two- and four-way coupled gas–solid flow in cyclones.

4.1. Group I: Single-Phase Flow CFD Simulations

As briefly mentioned earlier, the turbulent flow inside a particle-free conical cyclone is anisotropic and inherently unsteady due to the significant swirl and radial shear intensity as well as the adverse pressure gradient, which may induce recirculation regions. Accordingly, the turbulence model utilized in the CFD simulation of cyclones should be capable of correctly resolving or modeling the high curvature and intensive swirl flow and, at the same time, would be appropriate for modeling the adverse pressure gradient and recirculating flows [20].

As a result of being based on an isotropic turbulent flow assumption, the commonly used classical first-order turbulence models of the steady-state and unsteady Reynolds-averaged Navier–Stokes equations (URANS), i.e., zero-, one-, and two-equation models such as the k-ϵ model, are usually unable to accurately reproduce the flow field inside cyclones [62–65]. However, there are some attempts that have been carried out to improve the accuracy of the first-order turbulence models with the inclusion of anisotropy effects in their modified versions, e.g., the k-ϵ model based on renormalization group (RNG) theory [63,65], the k-ϵ curvature correction (cc) model [66], and the model proposed by Meier and

Mori [67]. These models provide improved accuracy in the prediction of single-phase flow in cyclones compared to the standard two-equation models. Alternatively, the second-order URANS closure models, e.g., the Reynolds stress transport model (RSTM) [66,68–70] and scale-resolving turbulence models, e.g., large eddy simulation (LES) [71,72], are able to capture the anisotropy effects in highly swirling flows and have been proven to be accurate in predicting the cyclone gas flow field. Among the two widely used variations of the differential RSTM in modeling the pressure-strain term, i.e., the Launder, Reece, and Rodi (LRR) model [73] and the Speziale, Sarkar, and Gatski (SSG) model [74], the latter provides superior accuracy [66]. With the recent advancements in computer capabilities, the use of LES for the prediction of the flow inside cyclones has become possible, but with a higher computational overhead cost. Using LES, it is possible to study cyclone flow physics in extensive detail and the accuracy of results is superior compared to the RSTM [75–77]. Nonetheless, the RSTM can provide reasonable predictions of the gas flow in cyclones using a relatively coarser computational grid compared to the ones used in LES studies, and at a much lower computational overhead than LES [76].

Apart from the turbulence model, there are important considerations related to the numerical aspects of CFD simulations of gas flow in cyclones. Due to the presence of strong velocity and pressure gradients, the use of high-order (i.e., second order and above) methods in the discretization of advection terms as well as the alignment of the computational cells with the flow direction (i.e., structured grids) are recommended in order to avoid unwanted numerical diffusion (for example, see [78]).

There are a large number of existing CFD studies focusing on single-phase flow characteristics in cyclones that are not fully listed here as it is not the main subject of this review. A summary of some of the previous single-phase flow studies, up to 2007, can be found in [20].

4.2. Group II: One-Way Coupled Gas–Solid Flow Simulations

In the second category of existing CFD studies of cyclones operating at ambient temperature, category (II), the one-way coupling method is employed for very dilute loading of particles (see Figure 4), and the E–L approach is commonly utilized. The particles are tracked either after the simulation has reached a steady-state solution, as a post-processing step [69,79–87], or in a transient way along the unsteady simulation [75,79,88–93]. After the completion of the particle-tracking procedure, the particle separation efficiencies are predicted and usually compared with the available experimental data. In general, it can be stated that there is a tendency for the particle cut-size diameter to be underpredicted in LES calculations [88,89,93] and overpredicted in RSTM calculations [20,93]. Furthermore, the separation efficiency is predicted more accurately when LES is used [93], due to a better prediction of the mean and fluctuating velocity fields. One must keep in mind that the accuracy of the predicted separation efficiency, however, depends on the validity of the basic assumptions for the simulations, e.g., the accumulation of particles near the cyclone walls may cast doubt on the validity of the one-way coupling assumption.

An important consideration in one-way coupled E–L simulations is the influence of turbulent dispersion, i.e., the movement of particles due to the presence of velocity fluctuations, especially for smaller-sized particles. If the resolved scales have comparable time-scalse as compared to the kinematic response time of particles, e.g., LES and DNS, the turbulent dispersion is resolved directly and no additional turbulent dispersion model is required [79,94], while the effect of sub-grid scale (SGS) structures in LES on particle movement can usually be neglected. For URANS simulations, however, it is common to apply a stochastic turbulent dispersion model [95]. Accurate prediction of the gas velocity fluctuations, i.e., rms velocities, is important for properly resolving particle dispersion due to turbulence. In some of the existing studies, it is reported that even though the mean velocities are properly captured by the RSTM, the rms fluctuations of the velocity field are underpredicted compared to the measurements [76,96], due to a failure in accurately predicting PVC behavior that gives rise to the rms velocities, especially in central regions of a cyclone [76]. Shukla et al. [96] reported an overprediction of separation efficiency for smaller-sized particles due to an underprediction of velocity fluctuations, as shown in Figure 5. On the other hand, LES is able to capture both mean and rms velocities properly, leading to accurate predictions of grade efficiency compared to the measurements [96].

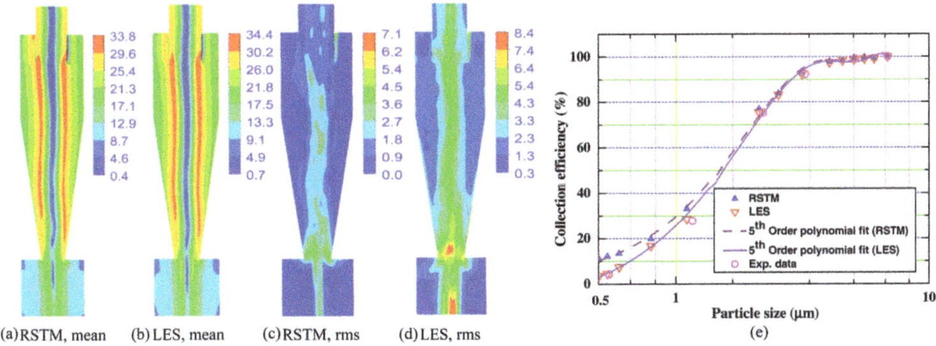

(a) RSTM, mean   (b) LES, mean   (c) RSTM, rms   (d) LES, rms   (e)

Figure 5. Comparison of predicted mean velocity profiles (**a,b**), rms velocities (**c,d**), and grade efficiencies (**e**) using the Reynolds stress transport model (RSTM) and large eddy simulation (LES) for a cyclone with a body diameter of 0.29 m and operating at ambient temperature and pressure [96]. The experimental data on separation efficiency are from [97]. Reproduced with permission from S. Shukla et al., The effect of modeling of velocity fluctuations on prediction of collection efficiency of cyclone separators; published by Elsevier, 2013.

4.3. Group III: Two- and Four-Way Coupled Gas–Solid Flow Simulations

Both the E–E and the E–L methods are employed for CFD simulation of two- and four-way coupled gas–solid flows. Generally speaking, and as discussed earlier, the volume fraction of particles in a cyclone separator is locally inhomogeneous, as the design of such equipment demands. Consequently, in the existing studies of cyclone separators, the E–L approach is preferred over the E–E approach.

4.3.1. E–L and Hybrid Model Simulations

As for the third category of CFD studies, category (III), compared to the previous categories, there are fewer E–L studies with two-way coupling [34,77,98–105] and only a limited number of studies with four-way coupling [35,106–109]. In almost all of the studies (except [99]), due to the lack of available experimental data for the cyclone flow loaded with particles, no comparison of the predicted modification of velocity profiles (compared to the single-phase flow) with the measurements is provided. Instead, the predicted overall/grade efficiencies [34,77,101,102] and/or pressure drops [34,35,98,101,102,106,107,109] are compared with the measurements for certain ranges of particle mass loading. A summary of the existing four-way coupled CFD studies of gas–solid cyclones operating at ambient temperature and with inlet mass loading ratios higher than 0.1 kg_s/kg_g is presented in Table 2.

Table 2. Details of the selected CFD studies of gas–solid flow in cyclones operating at ambient temperature with the four-way coupling method, in chronological order.

Author(s)	Cyclone Dimensions, $D \times H$ (m^2)	Cyclone Re Number [a]	Particle Diameter (µm)	Inlet Mass Loading (kg$_s$/kg$_g$)	CFD Solver	Turbulence Model/Turbulent Dispersion	Drag Model	Validation/Comments
Chu et al. [109]	0.2×0.8	272,000	2000 (mono-sized)	up to 2.5	ANSYS FLUENT (computational fluid dynamics–discrete element model, CFD–DEM, + user–defined functions, UDF)	Reynolds stress turbulent model, RSTM/not mentioned	friction and pressure gradient drags + particle rotation	Pressure drop is compared with the experiments for particle-free and particle-laden flows with solids loadings in the range of 0–2.5 kg$_s$/kg$_g$
Schneiderbauer et al. [110]	2.5×6.0	817,000	0.6–400 (size range)	0.22	ANSYS FLUENT 16 (hybrid Eulerian–Eulerian, E–E, and Eulerian–Lagrangian, E–L)	RSTM/not mentioned	heterogeneous model of [111]	Predicted grade and overall efficiencies are compared with the measurements. The implemented agglomeration model is reported to be crucial for proper prediction of grade efficiency, while the predicted overall efficiency is not influenced by presence of the agglomeration model.
Wei et al. [112]	0.3×1.1	113,000–263,000	2000 (mono-sized)	0.72–8.64	ANSYS FLUENT 15.0 coupled with EDEM 2.7 (CFD–DEM)	RSTM/not mentioned	Gidaspow [50]	Predicted pressure drops are compared with the experimental data for solids loadings of 0.72–8.64 kg$_s$/kg$_g$. Presence of solid strands and an ash top ring are reported.
Kozolub et al. [106]	0.2×0.78	75,300–130,000	2000 (size range)	0.61–2.9	ANSYS FLUENT 13.0 (dense discrete phase model, DDPM, based on kinetic theory of granular flow, KTGF)	RSTM/not mentioned	Wen–Yu [113]	Pressure drop is compared with the experimental data for particle-free and particle-laden flows with solids loadings in the range of 0–2.9 kg$_s$/kg$_g$. For the particle-laden flow, the trend of pressure drop change is well predicted while the values are somewhat overpredicted.
Sgrott and Sommerfeld [108]	0.29×1.16	280,000	0.5–60 (size range)	0.1	OpenFOAM 2.3.1 (CFD–DEM + agglomeration)	Large eddy simulation, LES/isotropic Langevin model	not mentioned particle rotation	For a particle-free flow, the predicted velocity profiles are compared with the experimental data of [114]. No validation is given for the particle-laden simulation case.
Zhou et al. [107]	0.29×1.16	30,000–188,000	2000–2800 (size range)	0.07–0.46	ANSYS FLUENT 6.3 (CFD–DEM)	RSTM/not mentioned	Gidaspow [50] particle rotation lift force	The predicted velocity profiles of a particle-free flow are compared with the experimental data of [114]. The pressure drop of the particle-free and particle-laden cases is compared with the measurements, while the difference between the particle-free and particle-laden pressure drops is not significant.
Hwang et al. [35]	0.2×0.8	272,000	2000 (mono-sized)	up to 20	ANSYS FLUENT 16.2 (DDPM–KTGF)	RSTM/discrete random walk, DRW	Wen–Yu [113]	The predicted pressure loss is compared with the experimental and numerical data of [109] for solid mass loadings up to 2.5.

[a] The Reynolds number is calculated based on the inlet velocity and the cyclone body diameter.

Effect of Solids Loading

As the solids loading ratio increases, the particle–particle interactions become important and neglecting these interactions in the cyclone CFD simulation may lead to poor predictions of cyclone performance. For example, Wasilewski and Duda [115] have reported an underprediction of both pressure drop and separation efficiency for two different pilot-scale cyclone geometries, though the overall trend of changes was well-predicted. The authors have claimed that this discrepancy is due to the simplification of the CFD model, neglecting particle–particle collisions and agglomeration. Both CFD–DEM [107–109] and hybrid E–E and E–L approaches [35,106] have been used to consider the effect of particle–particle interactions in existing cyclone separator simulation studies. In the CFD–DEM study of Zhou et al. [107], the effect of large (2000–2800 μm) and "ultra light" particles with a solid to gas density ratio of 33 is investigated for different loadings of particles up to 0.46 kg_s/kg_g. They have analyzed the significance of gas–solid and solid–solid forces in different regions of cyclones and concluded that the modification of the velocity field due to the presence of particles is hardly noticeable. On the other hand, based on CFD–DEM calculations, Chu et al. [109] reported a significant modification of the velocity field due to the presence of particles of nearly the same size but much heavier compared to the ones studied in [107]. According to this study, the particle–particle forces, which are much more dominant than the particle–gas forces, increase in magnitude with the loading of particles, with the most active region being inside the particle strands along the cyclone wall.

In most of the existing four-way coupled CFD simulations of cyclones using the hybrid methods [35,106], the focus is on the effect of solids loading on the performance of cyclones. Hwang et al. [35] carried out DDPM–KTGF simulations of a pilot-scale cyclone laden with 2000 μm diameter particles and reported a minimum in the predicted pressure drop versus solids loading. The comparison with available measurements, though, is carried out only for smaller solids loadings, i.e., loadings smaller than 2.5 kg_s/kg_g, while the minimum pressure drop takes place at solids loadings of around 6 kg_s/kg_g. Conversely, in the hybrid E–E and E–L study of Kozolub et al. [106], an increasing trend in the predicted pressure drop of a cyclone laden with 40–80 μm diameter particles is observed for all of the studied mass loadings up to 2.6 kg_s/kg_g. This trend is in line with the available measurements, but the values are somewhat underpredicted.

One of the important features that is captured by two- and four-way coupled CFD simulations of cyclones is the modulation of velocity field and turbulence statistics as compared to the particle-free flow. The swirl velocity is predicted to reduce substantially with the addition of particles to the flow and decrease further as the loading of particles increases [35,102,106,109,116]. The predicted reduction of tangential velocity magnitude as well as turbulent fluctuations by the addition of 0.05 and 0.1 kg_s/kg_g of particles to a pilot-scale cyclone is shown in Figure 6 based on the LES study of Derksen et al. [116]. The reduction of swirl is reported to be stronger in the loss-free vortex part of the swirl due to the higher concentration of particles in this region. On the other hand, the tangential velocity at regions close to the walls is still high (see Figure 6) due to the presence of particles and partial transfer of tangential momentum to the gas phase. The reduction of swirl can, in turn, reduce the effectiveness of centrifugal force on particle separation in the cyclone, making the particles be less concentrated near the walls. However, at the same time, the turbulence is attenuated by the presence of particles all over the cyclone body [116], leading to the reduction of turbulent dispersion; also, the amount of gas bypassed to the vortex finder at the entrance of the cyclone is reduced. The overall separation efficiency with respect to the solids loading is a balance between the significance of the above-mentioned effects.

The modification of cyclone efficiency with particle mass loading is not reported in the existing four-way coupled CFD studies as in most of the studies [35,107,109], the addition of large particles to the cyclone, e.g., 2000 μm in diameter, is investigated, leading to an ideal separation efficiency.

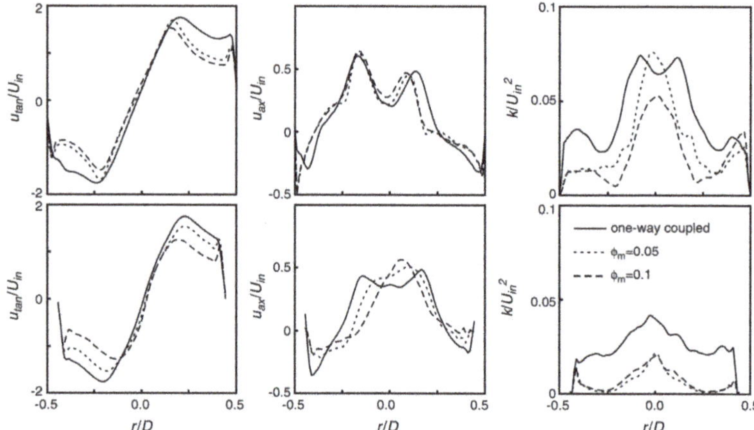

Figure 6. The predicted time-averaged values of axial and tangential velocities as well as the resolved turbulent kinetic energy of the gas for a pilot-scale cyclone with a body diameter of 0.29 m and operating with different mass loadings of particles using two-way coupled LES at 0.75D (top) and 2D (bottom) below the cyclone roof [116]. Reproduced with permission from J.J. Derksen, Simulation of mass-loading effects in gas-solid cyclone separators; published by Elsevier, 2006.

Agglomeration of Particles

As mentioned earlier, an important mechanism in solids separation in cyclones is particle agglomeration [19,30]. To model the particle agglomeration in a Lagrangian frame, it is common to assume either volume-equivalent (i.e., the new diameter has the same volume as the agglomerated particles) [117,118] or inertia-equivalent (i.e., the new diameter is calculated to maintain the inertia) [108,119,120] agglomeration. The agglomeration of particles in cyclones is studied in the CFD–DEM simulations of Sgrott and Sommerfeld [108] with a stochastic approach [121] to describe the particle–particle collisions. It is assumed that the only driving factor in interparticle adhesion is the van der Waals force. It has been shown that using the inertia-equivalent approach, the predicted agglomeration rate and the separation efficiency are slightly higher and lower, respectively, compared to the volume-equivalent method (see Figure 7). Furthermore, it is shown that for an inlet mass loading ratio of 0.1 kg_s / kg_g, the difference between the predicted velocity profiles considering two- and four-way coupling methods is small. However, when comparing the predicted grade efficiencies of two- and four-way coupled simulations, as shown in Figure 7, for very small particles, i.e., less than 2 μm in diameter, neglecting the agglomeration may lead to an underprediction of separation efficiency. This is in agreement with the agglomeration concept of Mothes and Löffler [30] regarding particle separation in cyclones, discussed in Section 2.2. The agglomeration effect is also investigated through one-way coupled CFD simulations of gas–solid cyclones [122].

4.3.2. E–E Simulations

E–E studies of gas–solid cyclone separators are very scarce, with the early investigations being allotted to the prediction of the change of cyclone hydrodynamics with the addition of particles [67,123]. In most of the existing E–E studies of gas–solid cyclones, the comparison with experimental data is carried out for cyclones operating with very low particle mass loadings [124–126], i.e., below 0.01 kg_s / kg_g, while the true advantage of using the E–E approach is for the high loading of particles. Costa et al. [126] implemented an E–E simulation of gas–solid flow in a cyclone with four-way coupling. At a solid mass loading of 0.009 kg_s / kg_g, the predicted grade efficiencies were improved by increasing the number of solid phases (corresponding to each size bin), but still overpredicted compared to the experimental data. Variations of this model are further utilized for optimization studies of gas–solid cyclones operating with either a very low mass loading of particles [127,128] or moderate loadings [129], i.e., 0.1 kg$_s$/kg$_g$, and higher.

For a particle mass loading of around 1 kg$_s$/kg$_g$, mixture models of cyclones are available [130,131], based on modeling the velocity slip between the phases (simplest approach mentioned in Table 1). The pressure drop estimated by these models is underpredicted [130,131], especially at high inlet solids loadings, i.e., 0.4 kg$_s$/kg$_g$, which, according to the authors, is due to the limitation of the granular and mixture models for densely loaded cyclones. Furthermore, the predicted separation efficiencies, compared to the measurements, are not conclusive [130]. In both studies, however, a reduction of swirl by increasing the mass loading is reported.

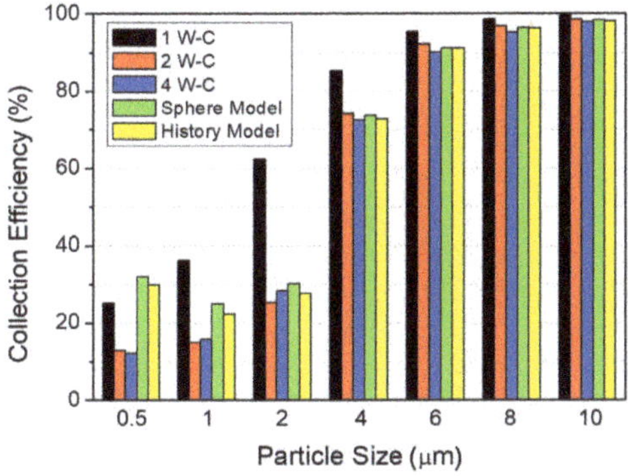

Figure 7. Grade efficiency predicted by the CFD simulation of Sgrott and Sommerfeld [108] for a pilot-scale cyclone loaded with particles with a diameter of 0.5–60 microns and a mass loading of 0.1 kg$_s$/kg$_g$. 1 W-C, 2 W-C, and 4 W-C refer to one-way coupling, two-way coupling, and four-way coupling methods (using CFD–DEM without agglomeration), respectively. Sphere and history models are volume-equivalent and inertia-equivalent approaches for agglomeration, respectively. Reproduced with permission from O.L. Sgrott and M. Sommerfeld, Influence of inter-particle collisions and agglomeration on cyclone performance and collection efficiency; published by John Wiley and Sons, 2018.

4.4. Summary

In this section, the existing CFD simulation studies of reversed-flow cyclone separators operating at ambient temperature are discussed and categorized based on the gas–solid coupling regime of the operating cyclone. For the single-phase gas flow in cyclones, the existing CFD simulation studies are able to accurately capture the velocity and pressure fields as compared to the available experimental data. According to these studies, the RSTM is favorable over other turbulence models with respect to accuracy and computational overhead. However, some modified versions of first-order turbulence models, e.g., k-ϵ curvature correction (cc) model [66], can still be used in the CFD simulations, especially when the solids loading is high and the swirl is less intense.

For dilute loading of particles and using the one-way coupling method, it is possible to properly capture the separation efficiency of particles in the cyclone as long as the one-way coupling condition is valid for the particle-laden flow. LES has proven to be more accurate in the prediction of grade efficiency, especially for small-sized particles, compared to RSTM, due to an improved prediction of velocity fluctuations and turbulent dispersion.

Among the available gas–solid simulation approaches listed in Table 1, the unresolved E–L and hybrid models are used more frequently, for dilute/medium and dense loading of particles in cyclones, respectively. For densely loaded cyclones, a few CFD studies focus on the effect of solids loading on performance, applying either DDPM–KTGF or CFD–DEM approaches, while only a handful of them investigate the loading of small-sized particles, i.e., 0.1–200 μm in diameter. In the existing CFD studies of densely loaded cyclones, comparison of the modified flow field with measurements is carried out very rarely, mainly due to challenges in the measurement methods for such systems. Furthermore, comparison/validation of the predicted pressure drops against measurements is usually carried out, rather than the separation efficiency.

5. CFD Simulation Studies of Gas–Solid Cyclones at Elevated Temperatures

In general, the available studies on the CFD simulation of cyclones operating at elevated temperatures are more scarce than the ones for cyclones operating at ambient temperature. Similar to the previous section, the CFD simulations of cyclone separators operating at elevated temperatures can be divided into three groups: single-phase flow, one-way coupled flow, and two/four-way coupled flow simulations. In this section, the three groups of studies are discussed with a focus on the process or physics that is investigated through the CFD simulation.

5.1. Group I: Single-Phase Flow CFD Simulations

Once the operating temperature increases, the physics of gas flow inside the cyclone become more complicated, mainly due to the heat transfer and subsequently variable gas density and viscosity in local regions, if the insulation is not perfect. In turn, since an extra transport equation for energy needs to be solved, the CFD calculation of gas flow becomes computationally more complex and demanding. However, in some of the existing CFD studies of cyclones operating at elevated temperatures, in order to avoid this complexity, the simulations are carried out at a fixed gas temperature, while the gas density and viscosity values are modified according to the operating temperature. Using this method, the predicted pressure drop is reported to be of acceptable accuracy compared to the measurements [131,132].

Similar to the CFD simulations of the cyclones at ambient temperature, accurate modeling of the turbulence is a must to achieve acceptable accuracy for high-temperature single-phase flow simulations of cyclones. However, as mentioned earlier, the increased viscosity as well as decreased density of the carrier gas at elevated temperatures leads to a reduction of swirl intensity and turbulence, indicating the possibility of using a wider range of available turbulence models while keeping the prediction accuracy reasonable. Gimbun et al. [133] have compared the predicted pressure drop of a pilot-scale cyclone with the experimental data of [39] using the RNG $k - \epsilon$ and RSTM. At lower temperatures, i.e., 300–800 K, RSTM is reported to provide a more accurate prediction of the pressure drop compared to the RNG $k - \epsilon$, while in the temperature range of 800–1200 K, the predicted pressure drops using RNG $k - \epsilon$ and RSTM are nearly the same and in agreement with the experimental data [133]. In the existing CFD studies of cyclones operating at elevated temperatures, both the $k - \epsilon$ [134] and RSTM [132] are used as turbulence models.

The tendency for weakened swirl and the modification of the internal velocity field, as an outcome of an elevated operating temperature, is reported in some of the existing CFD studies. According to the CFD simulation results of Shi et al. [132], an overall increase in the axial velocity profiles, weakened reverse flow at the center of the inner vortex, and weakened tangential velocity peaks (weakened swirl), are reported as the operating temperature of the studied pilot-scale cyclone increases from 20 °C to 800 °C. They reported that as the cyclone swirl intensity is weakened at elevated temperatures, the pressure drop is also reduced, and this reduction is dominantly affected by the gas density rather than its viscosity when the volumetric flow rate to the cyclone is kept constant [132]. The reduction of pressure drop with the gas temperature is reported in many of the existing CFD studies of cyclones operating at elevated temperatures [41,132,134–136]. An approach for the comprehensive analysis of local energy loss in cyclones is through evaluating entropy generation in CFD simulation, which has been carried out for cyclones operating at ambient [137,138] and elevated [135] temperatures. Duan et al. [135] have numerically investigated entropy generation in a cyclone system operating in a temperature range of 297–1123 K. According to this study, turbulence dissipation and wall friction are the main contributors to the energy loss. The ratio of contribution of both parameters is nearly unaffected by the gas temperature, whereas their magnitude decreases at higher operating temperatures due to the reduction in gas density. The maximum energy loss takes place in areas close to the vortex finder and the entrance of the dust bin.

5.2. Group II: One-Way Coupled Gas–Solid Flow Simulations

In the existing CFD studies of cyclones operating at elevated temperatures, the predicted separation efficiency is negatively affected by the temperature. Gimbun [41] has performed one-way coupled simulations of a pilot-scale cyclone using RSTM and a stochastic tracking of particles in a temperature range of 20–900 °C. For a fixed inlet gas velocity, the predicted grade separation efficiencies and particle cut-sizes were in proper agreement with the experimental data, and overall, the separation efficiency worsened with the temperature. The reduction in separation efficiency was stated to be a direct consequence of the weakened centrifugal effect. In the same line of research, Gimbun et al. [139] presented an accurate prediction of the particle cut-size for different cyclone types compared to experiments, at ambient and elevated temperatures. The reduction of separation efficiency is also reported in some of other studies (for example, see [134,136]).

One-way coupled heat transfer to solids particles fed in to a pilot-scale cyclone was studied by Mothilal and Pitchandi [140,141] using E–L simulation and RNG $k - \epsilon$ turbulence models and at an inlet air temperature of 200 °C. The predicted pressure drop was compared with the available experimental data for the particle-free case, while for the gas–solid cases, no validation was presented. By increasing the gas velocity at the cyclone inlet, the centrifugal effect on particles was improved and the predicted particle hold-up in the system was increased, especially for larger mass loadings of particles. This leads to a greater available surface area (for heat transfer) as well as a higher residence time of particles in the cyclone system.

5.3. Group III: Two- and Four-Way Coupled Gas–Solid Flow Simulations

In this group of studies, both E–L and E–E and hybrid approaches are applied for the CFD simulation of cyclones operating at elevated temperatures; these are discussed here with a focus on the studies that have investigated the performance of industrial-scale cyclones.

5.3.1. Cyclone Heat Exchangers

An important industrial application of cyclones is the solids preheating process used in, e.g., cement industry. The performance evaluation for full-scale cyclone heat exchangers is carried out in some of the existing studies and summarized in Table 3. Cristea and Conti [142,143] have applied a DDPM–KTGF approach to simulate the gas–solid flow in the preheater system (excluding the calciner) of a cement factory with an approximate height of 58 m. The gas–solid heat transfer as well as the calcination reaction are included in the CFD model [143]. As compared to the industrial measurements for the first-stage twin cyclones, the predicted pressure drop and separation efficiencies are in good agreement with the measurements, while the exit gas temperature is overpredicted. This discrepancy has been attributed to the complications that arise with the modeling of the high mass loading of particles in the upper-stage calciner [143]. Mikulčić et al. [144] have conducted a two-way coupled CFD simulation study of an industrial-scale cyclone belonging to the preheater system of a dry kiln process. The simulations aimed to assess the gas–solid heat transfer as well as the progress of the calcination process in the studied cyclone. The predicted pressure drop was in fair agreement with the measurements, while the outlet gas temperature was slightly overpredicted, and the authors did not provide any reasoning for this mismatch. In this CFD study, low gas temperature regions in the upper part of the cyclone and close to the walls were observed, due to the endothermic calcination reaction. Wasilewski [145] conducted a two-way coupled CFD simulation of a full-scale cyclone preheater and reported a reduction of the separation efficiency with increased temperatures and decreased particle loading. However, as compared to the measurements, an underprediction of the separation efficiencies is reported, which is claimed to be due to the limitation of the CFD model in taking into account the particle–particle interactions, i.e., collision and agglomeration. Presented in Figure 8 is their predicted trajectories and heat-transfer rate to laden particles of different diameters. The smaller-sized particles have a higher tendency to travel through the central regions of the cyclone body and escape from the vortex finder. In addition, they have a higher rate of heat exchange with the gas, especially at regions close to the inlet, compared to the larger-sized particles. In contrast, larger-sized particles have a higher tendency to travel in spiral clouds along the cyclone walls and toward the particle exit chamber.

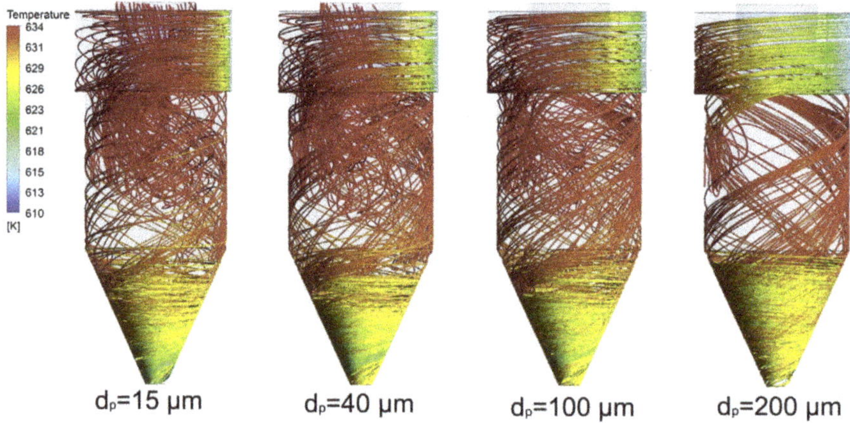

Figure 8. Predicted trajectories of particles of different diameters in an industrial-scale cyclone with a body diameter of 3.45 m. The trajectories are colored with particle temperature. The inlet gas and particle temperatures are 634 K and 611 K, respectively, and inlet solids mass loading is around 0.8 kg$_s$/kg$_g$ [145]. Reproduced with permission from M. Wasilewski, Analysis of the effects of temperature and the share of solid and gas phases on the process of separation in a cyclone suspension preheater; published by Elsevier, 2016.

Table 3. Selected CFD studies of industrial-scale cyclones operating at elevated temperatures, in chronological order.

Author(s)	Scale of Simulation	CFD Solver	Gas–Solid Model	Turbulence/Drag Models
Cristea and Conti [143]	Preheater system (≈ 58 m height)	ANSYS FLUENT 18.1	hybrid (DDPM–KTGF)	RSTM/Schiller and Naumann [146]
Mikulčić et al. [144]	Industrial cyclone (≈ 13 m height and 6 m diameter)	FIRE commercial solver	E–L (two–way coupled)	LES/not mentioned
Wasilewski [145]	Industrial cyclone (≈ 9 m height and 3.5 m diameter)	ANSYS FLUENT 14	E–L (two–way coupled)	RSTM/Schiller and Naumann [146]

There are also a few E–E simulation studies of cyclones at elevated temperatures. In the study of Vegini et al. [124], 2D axisymmetric simulations of full-scale cyclone separators connected in series and operating at elevated temperature were carried out. The selected turbulence model is a combination of the standard $k - \epsilon$ and Prandtl's longitudinal mixing model, which is claimed to be accurate enough for capturing the anisotropic effect of Reynolds stress in swirling turbulent flows. The numerical model was validated against the available grade efficiencies and pressure drops at ambient temperature and a dust load of around 0.004 kg$_s$/kg$_g$. In elevated temperature conditions, the predicted pressure drop values were generally overpredicted compared to the available long-term full-scale measurements. This overprediction was claimed to be due to the presence of false air in the system, and as a consequence, a reduction of swirl. The presence of false air is not considered in the simulations.

5.3.2. CFD Simulation of Cyclones as a Part of a Bigger System

There are a number studies that have carried out CFD simulations of elevated-temperature cyclones as a part of a bigger process system, e.g., a circulating fluidized bed (CFB) boiler. An important phenomenon that is explored in this type of study is the fluctuation and maldistribution of gas and solid flows at the inlet of parallel cyclones, which can, in turn, affect cyclone performance. The subject is investigated both experimentally [147–150] and numerically [151,152]. Zhang et al. [151] have reported temporal synchronized fluctuations of solid flux at the inlet of two parallel cyclones based on four-way coupled E–E simulations of a 150 *MW* CFB boiler operating at 917 °C. Jiang et al. [152] have carried out hybrid E–L and E–L (MP–PIC) CFD simulations of a scaled down circulating fluidized bed comprised of six parallel cyclones. They reported a non-uniformity in the average solid concentration downstream of the cyclone diplegs, based on both simulations and experimental data. However, some of the predicted solid concentrations were somewhat different from those of the measurements, with the highest relative error being around 28%.

5.4. Summary

The existing CFD studies of cyclones operating at elevated temperatures were discussed in this section, with a focus on the ones investigating cyclone performance parameters, i.e., pressure drop, separation efficiency, and gas–solid heat transfer. The gas-flow simulation studies reveal that CFD simulations are capable of capturing the reduced swirl intensity and reduced pressure drop as a result of elevated temperature. Furthermore, in some of the studies, the first-order turbulence models are satisfactorily used to calculate the gas flow field. The reduction in separation efficiency of dilutely loaded cyclones, i.e., one-way coupled studies, due to the increase in temperature, is captured. A number of CFD studies of industrial-scale cyclones are also described. However, it is common for the full-scale measurements of these systems not to exist or to be limited, e.g., to only the pressure drop of the cyclone for a specific operating condition, indicating the lack of complete validation of the accuracy of current CFD studies.

6. Outlook

In this review, the capability of computational fluid dynamics in predicting the gas–solid flow field and performance of cyclone separators, operating at dilute–high loading of particles and ambient and elevated temperatures, is discussed. Summaries of ambient- and elevated-temperature CFD studies are provided in Sections 4.4 and 5.4. In general, it can be stated that the CFD simulation can be used as a powerful tool to successfully predict the flow field and performance of cyclone separators, i.e., pressure drop and separation efficiency, with the E–L and hybrid methods being the most frequently used approaches. However, as the mass loading and the operating temperature increase, the validation of the internal flow field, which is important for a better understanding of the separation process, becomes challenging due to the lack of experimental data. For these cyclones, the validation is so far limited to the comparison of the pressure drop and seldom the measured separation efficiency.

The drag models that are usually applied in two- and four-way coupled simulations are well-known homogeneous models, e.g., Wen–Yu [113] and Gidaspow [50], while heterogeneous models, e.g., the energy-minimization-multi-scale (EMMS) approach [153], are rarely used (for example, see [111]). In the authors' opinion, heterogeneous drag models can be more suitable for CFD simulation of heavily loaded cyclones (e.g., inlet mass loading on the order of 1–100 kg_s/kg_g), as it is likely to have cluster formation in the internal regions, apart from regions close to the cyclone walls, especially for the elevated operating temperatures with weakened centrifugal effect.

Author Contributions: Conceptualization, M.N. and H.W.; investigation, M.N.; writing–original draft preparation, M.N.; writing–review and editing, M.N.; supervision, H.W., B.L., Y.T., W.W., K.D.-J.; project administration, H.W.; funding acquisition, H.W., K.D.-J. All authors have read and agreed to the published version of the manuscript.

Funding: This project is supported by the Sino-Danish Centre for Education and Research; and the ProBu - Process Technology for Sustainable Building Materials Production - project (grant number: 8055-00014B) funded by Innovation Fund Denmark, FLSmidth A/S, Rockwool International A/S and Technical University of Denmark.

Conflicts of Interest: The authors declare no conflict of interest.

Abbreviations

The following abbreviations are used in this manuscript:

CFB	Circulating fluidized bed
CFD	Computational fluid dynamics
DDPM	Dense discrete phase model
DEM	Discrete element method
DRW	Discrete random walk
E–E	Eulerian–Eulerian
E–L	Eulerian–Lagrangian
LES	Large eddy simulation
LRR	Launder, Reece,and Rodi model [73] (variation of the RSTM)
KTGF	Kinetic theory of granular flows
PBM	Population balance model
PVC	Precessing vortex core
RSTM	Reynolds stress transport model
SGS	Subgrid-scale
SSG	Speziale, Sarkar, and Gatski model [74] (variation of the RSTM)
UDF	User-defined function
URANS	Unsteady Reynolds-averaged Navier–Stokes

References

1. Jain, A.; Mohanty, B.; Pitchumani, B.; Rajan, K. Studies on Gas-Solid Heat Transfer in Cyclone Heat Exchanger. *J. Heat Transf.* **2006**, *128*, 761. [CrossRef]
2. He, P.; Luo, S.; Cheng, G.; Xiao, B.; Cai, L.; Wang, J. Gasification of biomass char with air-steam in a cyclone furnace. *Renew. Energy* **2012**, *37*, 398–402. [CrossRef]
3. Barnhart, J.; Laurendeau, N. Pulverized coal combustion and gasification in a cyclone reactor. 1. Experiment. *Ind. Eng. Chem. Process Des. Dev.* **1982**, *21*, 671–680. [CrossRef]
4. Zhen, F. A biomass pyrolysis gasifier applicable to rural China. *Fuel Sci. Technol. Int.* **1993**, *11*, 1025–1035. [CrossRef]
5. Green, N.W.; Duraiswamy, K.; Lumpkin, R.E. Pyrolysis with Cyclone Burner. U.S. Patent 4102773A, 25 July 1978.
6. Muschelknautz, U.; Muschelknautz, E. *Separation Efficiency of Recirculating Cyclones in Circulating Fluidized Bed Combustions*; VGB PowerTech: Essen, Germany, 1999; Volume 79, pp. 48–53.
7. Hoffmann, A.; van Santen, A.; Allen, R.; Clift, R. Effects of geometry and solid loading on the performance of gas cyclones. *Powder Technol.* **1992**, *70*, 83–91. [CrossRef]
8. Muschelknautz, E.; Brunner, K. Untersuchungen an Zyklonen. *Chemie Ing. Tech.* **1967**, *39*, 531–538. [CrossRef]
9. Yuu, S.; Jotaki, T.; Tomita, Y.; Yoshida, K. The reduction of pressure drop due to dust loading in a conventional cyclone. *Chem. Eng. Sci.* **1978**, *33*, 1573–1580. [CrossRef]
10. Kang, S.; Kwon, T.; Kim, S. Hydrodynamic characteristics of cyclone reactors. *Powder Technol.* **1989**, *58*, 211–220. [CrossRef]
11. Alexander, R. Fundamentals of cyclone design and operation. *Proc. Australas Inst. Min. Met.* **1949**, *152*, 202.
12. Barth, W. Berechnung und Auslegung von Zyklonabscheidern auf Grund neuerer Untersuchungen. *Brennst-Waerme-Kraft* **1956**, *8*, 1–9.
13. Muschelknautz, E. Die Berechnung von Zyklonabscheidern für Gase. *Chemie Ing. Tech.* **1972**, *44*, 63–71. [CrossRef]
14. Meißner, P.; Löffler, F. Zur Berechnung des Strömungsfeldes im Zyklonabscheider. *Chemie Ing. Tech.* **1978**, *50*, 471. [CrossRef]
15. Reydon, R.; Gauvin, W. Theoretical and experimental studies of confined vortex flow. *Can. J. Chem. Eng.* **1981**, *59*, 14–23. [CrossRef]

16. Muschelknautz, E.; Krambrock, W. Aerodynamische Beiwerte des Zyklonabscheiders aufgrund neuer und verbesserter Messungen. *Chem. Ing. Tech.* **1970**, *42*, 247–255. [CrossRef]

17. Hsiao, T.; Chen, D.; Greenberg, P.; Street, K. Effect of geometric configuration on the collection efficiency of axial flow cyclones. *J. Aerosol Sci.* **2011**, *42*, 78–86. [CrossRef]

18. Xiong, Z.; Ji, Z.; Wu, X. Development of a cyclone separator with high efficiency and low pressure drop in axial inlet cyclones. *Powder Technol.* **2014**, *253*, 644–649. [CrossRef]

19. Hoffmann, A.; Stein, L. *Gas Cyclones and Swirl Tubes*; Springer: Berlin/Heidelberg, Germany, 2007; pp. 1–422. [CrossRef]

20. Cortés, C.; Gil, A. Modeling the gas and particle flow inside cyclone separators. *Prog. Energy Combust. Sci.* **2007**, *33*, 409–452. [CrossRef]

21. Towler, G.; Sinnott, R. Specification and design of solids-handling equipment. *Chem. Eng. Des.* **2013**, 937–1046. [CrossRef]

22. Trefz, M.; Muschelknautz, E. Extended cyclone theory for gas flows with high solids concentrations. *Chem. Eng. Technol.* **1993**, *16*, 153–160. [CrossRef]

23. ter Linden, A. Investigations into Cyclone Dust Collectors. *Proc. Inst. Mech. Eng.* **1949**, *160*, 233–251. [CrossRef]

24. Majumder, A.; Shah, H.; Shukla, P.; Barnwal, J. Effect of operating variables on shape of "fish-hook" curves in cyclones. *Miner. Eng.* **2007**, *20*, 204–206. [CrossRef]

25. Haig, C.; Hursthouse, A.; McIlwain, S.; Sykes, D. The effect of particle agglomeration and attrition on the separation efficiency of a Stairmand cyclone. *Powder Technol.* **2014**, *258*, 110–124. [CrossRef]

26. Barth, W. Design and Layout of the Cyclone Separator on the Basis of New Investigations. *Brennstow-Wäerme-Kraft (BWK)* **1956**, *8*, 1–9.

27. Hugi, E.; Reh, L. Focus on solids strand formation improves separation performance of highly loaded circulating fluidized bed recycle cyclones. *Chem. Eng. Process. Process Intensif.* **2000**, *39*, 263–273. [CrossRef]

28. Hoffmann, A.; Arends, H.; Sie, H. An experimental investigation elucidating the nature of the effect of solids loading on cyclone performance. *Filtr. Sep.* **1991**, *28*, 188–193. [CrossRef]

29. Fassani, F.; Goldstein, L. A study of the effect of high inlet solids loading on a cyclone separator pressure drop and collection efficiency. *Powder Technol.* **2000**, *107*, 60–65. [CrossRef]

30. Mothes, H.; Löffler, F. Motion and deposition of particles in cyclones. *Ger. Chem. Eng.* **1985**, *8*, 223–233.

31. Comas, M.; Comas, J.; Chetrit, C.; Casal, J. Cyclone pressure drop and efficiency with and without an inlet vane. *Powder Technol.* **1991**, *66*, 143–148. [CrossRef]

32. Baskakov, A.; Dolgov, V.; Goldobin, Y. Aerodynamics and heat transfer in cyclones with particle-laden gas flow. *Exp. Therm. Fluid Sci.* **1990**, *3*, 597–602. [CrossRef]

33. Huang, Y.; Mo, X.; Yang, H.; Zhang, M.; Lv, J. Effects of Cyclone Structures on the Pressure Drop Across Different Sections in Cyclone Under Gas-Solid Flow. In *Clean Coal Technology and Sustainable Development*; Springer: Singapore, 2016; pp. 301–307. [CrossRef]

34. Ashry, E.; Abdelrazek, A.; Elshorbagy, K. Numerical and experimental study on the effect of solid particle sphericity on cyclone pressure drop. *Sep. Sci. Technol.* **2018**, *53*, 2500–2516. [CrossRef]

35. Hwang, I.; Jeong, H.; Hwang, J. Numerical simulation of a dense flow cyclone using the kinetic theory of granular flow in a dense discrete phase model. *Powder Technol.* **2019**, *356*, 129–138. [CrossRef]

36. Li, S.; Yang, H.; Wu, Y.; Zhang, H. An Improved Cyclone Pressure Drop Model at High Inlet Solid Concentrations. *Chem. Eng. Technol.* **2011**, *34*, 1507–1513. [CrossRef]

37. Wu, X.; Liu, J.; Xu, X.; Xiao, Y. Modeling and experimental validation on pressure drop in a reverse-flow cyclone separator at high inlet solid loading. *J. Therm. Sci.* **2011**, *20*, 343–348. [CrossRef]

38. Dewil, R.; Baeyens, J.; Caerts, B. CFB cyclones at high temperature: Operational results and design assessment. *Particuology* **2008**, *6*, 149–156. [CrossRef]

39. Bohnet, M. Influence of the gas temperature on the separation efficiency of aerocyclones. *Chem. Eng. Process. Process Intensif.* **1995**, *34*, 151–156. [CrossRef]

40. Patterson, P.; Munz, R. Cyclone collection efficiencies at very high temperatures. *Can. J. Chem. Eng.* **1989**, *67*, 321–328. [CrossRef]

41. Gimbun, J. CFD Simulation of Aerocyclone Hydrodynamics and Performance at Extreme Temperature. *Eng. Appl. Comput. Fluid Mech.* **2008**, *2*, 22–29. [CrossRef]

42. Szekely, J.; Carr, R. Heat transfer in a cyclone. *Chem. Eng. Sci.* **1966**, *21*, 1119–1132. [CrossRef]

43. Danyluk, S.; Shack, W.; Park, J.; Mamoun, M. The erosion of a Type 310 stainless steel cyclone from a coal gasification pilot plant. *Wear* **1980**, *63*, 95–104. [CrossRef]
44. Chen, Y.; Nieskens, M.; Karri, S.; Knowlton, T. Developments in cyclone technology improve FCC unit reliability. *Pet. Technol. Q.* **2010**, *15*, 65–71.
45. Kim, S.; Lee, J.; Kim, C.; Koh, J.; Kim, G.; Choi, S. Characteristics of Deposits Formed in Cyclones in Commercial RFCC Reactor. *Ind. Eng. Chem. Res.* **2012**, *51*, 10238–10246. [CrossRef]
46. Kim, S.; Lee, J.; Koh, J.; Kim, G.; Choi, S.; Yoo, I. Formation and Characterization of Deposits in Cyclone Dipleg of a Commercial Residue Fluid Catalytic Cracking Reactor. *Ind. Eng. Chem. Res.* **2012**, *51*, 14279–14288. [CrossRef]
47. Yoshida, H.; Fukui, K.; Yoshida, K.; Shinoda, E. Particle separation by Iinoya's type gas cyclone. *Powder Technol.* **2001**, *118*, 16–23. [CrossRef]
48. van der Hoef, M.; van Sint Annaland, M.; Deen, N.; Kuipers, J. Numerical Simulation of Dense Gas-Solid Fluidized Beds: A Multiscale Modeling Strategy. *Annu. Rev. Fluid Mech.* **2008**, *40*, 47–70. [CrossRef]
49. Fan, L.; Zhu, C. *Principles of Gas-Solid Flows*; Cambridge Series in Chemical Engineering; Cambridge University Press: Cambridge, UK, 1998.
50. Gidaspow, D. *Multiphase Flow and Fluidization: Continuum and Kinetic Theory Descriptions*; Academic Press: Cambridge, MA, USA, 1994; p. 467.
51. Zeneli, M.; Nikolopoulos, A.; Nikolopoulos, N.; Grammelis, P.; Karellas, S.; Kakaras, E. Simulation of the reacting flow within a pilot scale calciner by means of a three phase TFM model. *Fuel Process. Technol.* **2017**, *162*, 105–125. [CrossRef]
52. Mathiesen, V.; Solberg, T.; Hjertager, B. Predictions of gas-particle flow with an Eulerian model including a realistic particle size distribution. *Powder Technol.* **2000**, *112*, 34–45. [CrossRef]
53. Rizk, M.A. Mathematical modeling of densely loaded, particle-laden turbulent fLows. *Automization Sprays* **1993**, *3*, 1–27.
54. Ibsen, C.; Helland, E.; Hjertager, B.; Solberg, T.; Tadrist, L.; Occelli, R. Comparison of multifluid and discrete particle modelling in numerical predictions of gas particle flow in circulating fluidised beds. *Powder Technol.* **2004**, *149*, 29–41. [CrossRef]
55. Cundall, P.; Strack, O. A discrete numerical model for granular assemblies. *Géotechnique* **1979**, *29*, 47–65. [CrossRef]
56. Andrews, M.; O'Rourke, P. The multiphase particle-in-cell (MP-PIC) method for dense particulate flows. *Int. J. Multiph. Flow* **1996**, *22*, 379–402. [CrossRef]
57. Snider, D.; O'Rourke, P. The multiphase particle-in-cell (MP-PIC) method for dense particle flow. *Int. J. Multiph. Flow* **2010**, *22*, 277–314, [CrossRef]
58. Andrews, M.; Baillergeau, D.; O'Rourke, P. A Lagrangian particle method for the simulation of dense particulate flows. *Am. Soc. Mech. Eng. Fluids Eng. Div.* **1994**, *199*, 37–46.
59. Popoff, B.; Braun, M. A Lagrangian Approach to Dense Particulate Flows. In Proceedings of the ICMF 2007 6th International Conference on Multiphase Flow, Leipzig, Germany, 9–13 July 2007; p. 11.
60. van der Hoef, M.; Ye, M.; van Sint Annaland, M.; Andrews, A.; Sundaresan, S.; Kuipers, J. Multiscale modeling of gas-fluidized beds. *Adv. Chem. Eng.* **2006**, *31*, 65–149. [CrossRef]
61. Elghobashi, S. On predicting particle-laden turbulent flows. *Appl. Sci. Res.* **1994**, *52*, 309–329. [CrossRef]
62. Hamdy, O.; Bassily, M.; El-Batsh, H.; Mekhail, T. Numerical study of the effect of changing the cyclone cone length on the gas flow field. *Appl. Math. Model.* **2017**, *46*, 81–97. [CrossRef]
63. Hoekstra, A.; Derksen, J.; Van Den Akker, H. An experimental and numerical study of turbulent swirling flow in gas cyclones. *Chem. Eng. Sci.* **1999**, *54*, 2055–2065. [CrossRef]
64. Vaitiekūnas, P.; Jakštonienė, I. Analysis of numerical modelling of turbulence in a conical reverse-flow cyclone. *J. Environ. Eng. Landsc. Manag.* **2010**, *18*, 321–328. [CrossRef]
65. Kaya, F.; Karagoz, I. Performance analysis of numerical schemes in highly swirling turbulent flows in cyclones. *Curr. Sci.* **2008**, *94*, 1273–1278.
66. Grotjans, H. Application of higher order turbulence models to cyclone flows. *VDI Berichte* **1999**, *1511*, 175–182.
67. Meier, H.; Mori, M. Anisotropic behavior of the Reynolds stress in gas and gas-solid flows in cyclones. *Powder Technol.* **1999**, *101*, 108–119. [CrossRef]

68. Qian, F.; Zhang, J.; Zhang, M. Effects of the prolonged vertical tube on the separation performance of a cyclone. *J. Hazard. Mater.* **2006**, *136*, 822–829. [CrossRef] [PubMed]

69. Wang, B.; Xu, D.; Chu, K.; Yu, A. Numerical study of gas-solid flow in a cyclone separator. *Appl. Math. Model.* **2006**, *30*, 1326–1342. [CrossRef]

70. Afolabi, L.; Aroussi, A.; Isa, N. Numerical modelling of the carrier gas phase in a laboratory-scale coal classifier model. *Fuel Process. Technol.* **2011**, *92*, 556–562. [CrossRef]

71. Derksen, J.; Van den Akker, H. Simulation of vortex core precession in a reverse-flow cyclone. *AIChE J.* **2000**, *46*, 1317–1331. [CrossRef]

72. Derksen, J. Simulations of confined turbulent vortex flow. *Comput. Fluids* **2005**, *34*, 301–318. [CrossRef]

73. Launder, B.; Reece, G.; Rodi, W. Progress in the development of a Reynolds-stress turbulence closure. *J. Fluid Mech.* **1975**, *68*, 537. [CrossRef]

74. Speziale, C.; Sarkar, S.; Gatski, T. Modelling the pressure-strain correlation of turbulence: An invariant dynamical systems approach. *J. Fluid Mech.* **1991**, *227*, 245–272. [CrossRef]

75. Elsayed, K.; Lacor, C. Analysis and Optimisation of Cyclone Separators Geometry Using RANS and LES Methodologies. In *Turbulence and Interactions. Notes on Numerical Fluid Mechanics and Multidisciplinary Design*; Deville, M., Estivalezes, J.L., Gleize, V., Lê, T.H., Terracol, M., Vincent, S., Eds.; Springer, Berlin/Heidelberg, Germany, 2014; pp. 65–74. [CrossRef]

76. Gronald, G.; Derksen, J. Simulating turbulent swirling flow in a gas cyclone: A comparison of various modeling approaches. *Powder Technol.* **2011**, *205*, 160–171. [CrossRef]

77. Jang, K.; Lee, G.; Huh, K. Evaluation of the turbulence models for gas flow and particle transport in URANS and LES of a cyclone separator. *Comput. Fluids* **2018**, *172*, 274–283. [CrossRef]

78. Shukla, S.; Shukla, P.; Ghosh, P. Evaluation of numerical schemes using different simulation methods for the continuous phase modeling of cyclone separators. *Adv. Powder Technol.* **2011**, *22*, 209–219. [CrossRef]

79. Derksen, J. Separation performance predictions of a Stairmand high-efficiency cyclone. *AIChE J.* **2003**, *49*, 1359–1371. [CrossRef]

80. Kępa, A. The influence of a plate vortex limiter on cyclone separator. *Sep. Sci. Technol.* **2016**, *51*, 1–13. [CrossRef]

81. Song, C.; Pei, B.; Jiang, M.; Wang, B.; Xu, D.; Chen, Y. Numerical analysis of forces exerted on particles in cyclone separators. *Powder Technol.* **2016**, *294*, 437–448. [CrossRef]

82. Guo, Q.; Wang, Q.; Lu, X.; Ji, P.; Kang, Y. Numerical Simulation for a CFB Boiler's Cyclone Separator with Structure Optimizations. In *Clean Coal Technology and Sustainable Development*; Springer: Singapore, 2016; pp. 257–263. [CrossRef]

83. Haig, C.; Hursthouse, A.; Sykes, D.; Mcilwain, S. The rapid development of small scale cyclones - numerical modelling versus empirical models. *Appl. Math. Model.* **2016**, *40*, 6082–6104. [CrossRef]

84. Amirante, R.; Tamburrano, P. Tangential inlet cyclone separators with low solid loading. *Eng. Comput.* **2016**, *33*, 2090–2116. [CrossRef]

85. Sun, X.; Zhang, Z.; Chen, D. Numerical modeling of miniature cyclone. *Powder Technol.* **2017**, *320*, 325–339. [CrossRef]

86. Zhang, G.; Chen, G.; Yan, X. Evaluation and improvement of particle collection efficiency and pressure drop of cyclones by redistribution of dustbins. *Chem. Eng. Res. Des.* **2018**, *139*, 52–61. [CrossRef]

87. Mazyan, W.; Ahmadi, A.; Brinkerhoff, J.; Ahmed, H.; Hoorfar, M. Enhancement of cyclone solid particle separation performance based on geometrical modification: Numerical analysis. *Sep. Purif. Technol.* **2018**, *191*, 276–285. [CrossRef]

88. Elsayed, K.; Lacor, C. Numerical modeling of the flow field and performance in cyclones of different cone-tip diameters. *Comput. Fluids* **2011**, *51*, 48–59. [CrossRef]

89. de Souza, F.; de Vasconcelos Salvo, R.; de Moro Martins, D. Large Eddy Simulation of the gas-particle flow in cyclone separators. *Sep. Purif. Technol.* **2012**, *94*, 61–70. [CrossRef]

90. Misiulia, D.; Andersson, A.; Lundström, T. Computational Investigation of an Industrial Cyclone Separator with Helical-Roof Inlet. *Chem. Eng. Technol.* **2015**, *38*, 1425–1434. [CrossRef]

91. Akiyama, O.; Kato, C. Numerical Investigations of Unsteady Flows and Particle Behavior in a Cyclone Separator. *J. Fluids Eng.* **2017**, *139*, 091302. [CrossRef]

92. Yohana, E.; Tauviqirrahman, M.; Putra, A.R.; Diana, A.E.; Choi, K.H. Numerical analysis on the effect of the vortex finder diameter and the length of vortex limiter on the flow field and particle collection in a new cyclone separator. *Cogent Eng.* **2018**, *5*. [CrossRef]

93. Wasilewski, M.; Anweiler, S.; Masiukiewicz, M. Characterization of multiphase gas-solid flow and accuracy of turbulence models for lower stage cyclones used in suspension preheaters. *Chin. J. Chem. Eng.* **2018**. [CrossRef]

94. Lessani, B.; Nakhaei, M. Large-eddy simulation of particle-laden turbulent flow with heat transfer. *Int. J. Heat Mass Transf.* **2013**, *67*, 974–983. [CrossRef]

95. Berlemont, A.; Desjonqueres, P.; Gouesbet, G. Particle lagrangian simulation in turbulent flows. *Int. J. Multiph. Flow* **1990**, *16*, 19–34. [CrossRef]

96. Shukla, S.; Shukla, P.; Ghosh, P. The effect of modeling of velocity fluctuations on prediction of collection efficiency of cyclone separators. *Appl. Math. Model.* **2013**, *37*, 5774–5789. [CrossRef]

97. Zhao, B. Development of a new method for evaluating cyclone efficiency. *Chem. Eng. Process. Process Intensif.* **2005**, *44*, 447–451. [CrossRef]

98. Zhou, F.; Sun, G.; Han, X.; Zhang, Y.; Bi, W. Experimental and CFD study on effects of spiral guide vanes on cyclone performance. *Adv. Powder Technol.* **2018**, *29*, 3394–3403. [CrossRef]

99. Derksen, J.; van den Akker, H.; Sundaresan, S. Two-way coupled large-eddy simulations of the gas-solid flow in cyclone separators. *AIChE J.* **2008**, *54*, 872–885. [CrossRef]

100. Li, G.; Lu, Y. Cyclone separation in a supercritical water circulating fluidized bed reactor for coal/biomass gasification: Structural design and numerical analysis. *Particuology* **2018**, *39*, 55–67. [CrossRef]

101. Huang, A.; Ito, K.; Fukasawa, T.; Fukui, K.; Kuo, H. Effects of particle mass loading on the hydrodynamics and separation efficiency of a cyclone separator. *J. Taiwan Inst. Chem. Eng.* **2018**, *90*, 61–67. [CrossRef]

102. Liu, S.; Guo, Y.; Sun, G.; Weng, L. Flow Simulation and Performance Comparison of Rough-Cut Cyclones with Different Structures in Dilute Phase. *Chem. Eng. Technol.* **2015**, *38*, 1809–1816. [CrossRef]

103. Tamjid, S.; Hashemabadi, S.; Shirvani, M. Prediction of catalyst attrition in a regeneration cyclone. *Filtr. Sep.* **2010**, *47*, 29–33. [CrossRef]

104. Cui, J.; Chen, X.; Gong, X.; Yu, G. Numerical Study of Gas-Solid Flow in a Radial-Inlet Structure Cyclone Separator. *Ind. Eng. Chem. Res.* **2010**, *49*, 5450–5460. [CrossRef]

105. Wan, G.; Sun, G.; Xue, X.; Shi, M. Solids concentration simulation of different size particles in a cyclone separator. *Powder Technol.* **2008**, *183*, 94–104. [CrossRef]

106. Kozołub, P.; Klimanek, A.; Białecki, R.; Adamczyk, W. Numerical simulation of a dense solid particle flow inside a cyclone separator using the hybrid Euler-Lagrange approach. *Particuology* **2017**, *31*, 170–180. [CrossRef]

107. Zhou, H.; Hu, Z.; Zhang, Q.; Wang, Q.; Lv, X. Numerical study on gas-solid flow characteristics of ultra-light particles in a cyclone separator. *Powder Technol.* **2019**, *344*, 784–796. [CrossRef]

108. Sgrott, O.; Sommerfeld, M. Influence of inter-particle collisions and agglomeration on cyclone performance and collection efficiency. *Can. J. Chem. Eng.* **2019**, *97*, 511–522. [CrossRef]

109. Chu, K.; Wang, B.; Xu, D.; Chen, Y.; Yu, A. CFD-DEM simulation of the gas-solid flow in a cyclone separator. *Chem. Eng. Sci.* **2011**, *66*, 834–847. [CrossRef]

110. Schneiderbauer, S.; Haider, M.; Hauzenberger, F.; Pirker, S. A Lagrangian-Eulerian hybrid model for the simulation of industrial-scale gas-solid cyclones. *Powder Technol.* **2016**, *304*, 229–240. [CrossRef]

111. Schneiderbauer, S.; Pirker, S. Filtered and heterogeneity-based subgrid modifications for gas-solid drag and solid stresses in bubbling fluidized beds. *AIChE J.* **2014**, *60*, 839–854. [CrossRef]

112. Wei, J.; Zhang, H.; Wang, Y.; Wen, Z.; Yao, B.; Dong, J. The gas-solid flow characteristics of cyclones. *Powder Technol.* **2017**, *308*, 178–192. [CrossRef]

113. Wen, C.; Yu, Y. Mechanics of fluidization. *Chem. Eng. Progress, Symp. Ser.* **1966**, *62*, 100–111.

114. Hoekstra, A. Gas fLow Field and Collection Efficiency of Cyclone Separators. Ph.D. Thesis, Delft University of Technology, Delft, The Netherlands, 2000.

115. Wasilewski, M.; Duda, J. Multicriteria optimisation of first-stage cyclones in the clinker burning system by means of numerical modelling and experimental research. *Powder Technol.* **2016**, *289*, 143–158. [CrossRef]

116. Derksen, J.; Sundaresan, S.; van den Akker, H. Simulation of mass-loading effects in gas-solid cyclone separators. *Powder Technol.* **2006**, *163*, 59–68. [CrossRef]

117. Ho, C.; Sommerfeld, M. Modelling of micro-particle agglomeration in turbulent flows. *Chem. Eng. Sci.* **2002**, *57*, 3073–3084. [CrossRef]
118. Ho, C.; Sommerfeld, M. Numerische Berechnung der Staubabscheidung im Gaszyklon unter Berücksichtigung der Partikelagglomeration. *Chem. Ing. Tech.* **2005**, *77*, 282–290. [CrossRef]
119. Breuer, M.; Almohammed, N. Modeling and simulation of particle agglomeration in turbulent flows using a hard-sphere model with deterministic collision detection and enhanced structure models. *Int. J. Multiph. Flow* **2015**, *73*, 171–206. [CrossRef]
120. Almohammed, N.; Breuer, M. Modeling and simulation of agglomeration in turbulent particle-laden flows: A comparison between energy-based and momentum-based agglomeration models. *Powder Technol.* **2016**, *294*, 373–402. [CrossRef]
121. Sommerfeld, M. Validation of a stochastic Lagrangian modelling approach for inter-particle collisions in homogeneous isotropic turbulence. *Int. J. Multiph. Flow* **2001**, *27*, 1829–1858. [CrossRef]
122. Gronald, G.; Staudinger, G. Investigations on agglomeration in gas cyclones. In Proceedings of the World Congress on Particle Technology, Tsukuba, Japan, 23–27 Aprril 2006.
123. Meier, H.; Mori, M. Gas-solid flow in cyclones: The Eulerian-Eulerian approach. *Comput. Chem. Eng.* **1998**, *22*.
124. Vegini, A.; Meier, H.; Iess, J.; Mori, M. Computational Fluid Dynamics (CFD) Analysis of Cyclone Separators Connected in Series. *Ind. Eng. Chem. Res.* **2008**, *47*, 192–200. [CrossRef]
125. Meier, H.; Vegini, A.; Mori, M. Four-Phase Eulerian-Eulerian Model for Prediction of Multiphase Flow in Cyclones. *J. Comput. Multiph. Flows* **2011**, *3*, 93–105. [CrossRef]
126. Costa, K.; Sgrott, O.; Decker, R.; Reinehr, E.; Martignoni, W.; Meier, H. Effects of phases' numbers and solid-solid interactions on the numerical simulations of cyclones. *Chem. Eng. Trans.* **2013**, *32*, 1567–1572. [CrossRef]
127. Sgrott, O.; Noriler, D.; Wiggers, V.; Meier, H. Cyclone optimization by COMPLEX method and CFD simulation. *Powder Technol.* **2015**, *277*, 11–21. [CrossRef]
128. Sgrott, O.; Costa, K.; Noriler, D.; Wiggers, V.; Martignoni, W.; Meier, H. Cyclones' project optimization by combination of an inequality constrained problem and computational fluid dynamics techniques (CFD). *Chem. Eng. Trans.* **2013**, *32*, 2011–2016. [CrossRef]
129. Luciano, R.; Silva, B.; Rosa, L.; Meier, H. Multi-objective optimization of cyclone separators in series based on computational fluid dynamics. *Powder Technol.* **2018**, *325*, 452–466. [CrossRef]
130. Qian, F.; Huang, Z.; Chen, G.; Zhang, M. Numerical study of the separation characteristics in a cyclone of different inlet particle concentrations. *Comput. Chem. Eng.* **2007**, *31*, 1111–1122. [CrossRef]
131. Kharoua, N.; Khezzar, L.; Nemouchi, Z. Study of the Pressure Drop and Flow Field in Standard Gas Cyclone Models Using the Granular Model. *Int. J. Chem. Eng.* **2011**, *2011*, 1–11. [CrossRef]
132. Shi, L.; Bayless, D.; Kremer, G.; Stuart, B. CFD simulation of the influence of temperature and pressure on the flow pattern in cyclones. *Ind. Eng. Chem. Res.* **2006**, *45*, 7667–7672. [CrossRef]
133. Gimbun, J.; Chuah, T.; Fakhru'l-Razi, A.; Choong, T.S. The influence of temperature and inlet velocity on cyclone pressure drop: A CFD study. *Chem. Eng. Process. Process Intensif.* **2005**, *44*, 7–12. [CrossRef]
134. Shin, M.; Kim, H.; Jang, D.; Chung, J.; Bohnet, M. A numerical and experimental study on a high efficiency cyclone dust separator for high temperature and pressurized environments. *Appl. Therm. Eng.* **2005**, *25*, 1821–1835. [CrossRef]
135. Duan, L.; Wu, X.; Zhongli, J. The influence of temperature on the entropy generation of cyclone separator. In Proceedings of the 11th International Conference on Fluidized Bed Technology, Beijing, China, 1 May 2014; pp. 923–928.
136. Siadaty, M.; Kheradmand, S.; Ghadiri, F. Study of inlet temperature effect on single and double inlets cyclone performance. *Adv. Powder Technol.* **2017**, *28*, 1459–1473. [CrossRef]
137. Duan, L.; Wu, X.; Ji, Z.; Fang, Q. Entropy generation analysis on cyclone separators with different exit pipe diameters and inlet dimensions. *Chem. Eng. Sci.* **2015**, *138*, 622–633. [CrossRef]
138. Duan, L.; Wu, X.; Ji, Z.; Xiong, Z.; Zhuang, J. The flow pattern and entropy generation in an axial inlet cyclone with reflux cone and gaps in the vortex finder. *Powder Technol.* **2016**, *303*, 192–202. [CrossRef]
139. Gimbun, J.; Chuah, T.; Choong, T.; Fakhru'l-Razi, A. A CFD Study on the Prediction of Cyclone Collection Efficiency. *Int. J. Comput. Methods Eng. Sci. Mech.* **2005**, *6*, 161–168. [CrossRef]

140. Mothilal, T.; Pitchandi, K. Influence of inlet velocity of air and solid particle feed rate on holdup mass and heat transfer characteristics in cyclone heat exchanger. *J. Mech. Sci. Technol.* **2015**, *29*, 4509–4518. [CrossRef]

141. Mothilal, T.; Pitchandi, K. Computational fluid dynamic analysis on the effect of particles density and body diameter in a tangential inlet cyclone heat exchanger. *Therm. Sci.* **2017**, *21*, 2883–2895. [CrossRef]

142. Cristea, E.; Conti, P. Coupled 3-D CFD-DDPM numerical simulation of turbulent swirling gas-particle flow within cyclone suspension preheater of cement kilns. Proceedings of the ASME 2016 Fluids Engineering Division Summer Meeting collocated with the ASME 2016 Heat Transfer Summer Conference and the ASME 2016 14th International Conference on Nanochannels, Microchannels, and Minichannels, Washington, DC, USA, 10–2013;14 July 2016. [CrossRef]

143. Cristea, E.; Conti, P. Hybrid Eulerian Multiphase-Dense Discrete Phase Model Approach for Numerical Simulation of Dense Particle-Laden Turbulent Flows Within Vertical Multi-Stage Cyclone Heat Exchanger. In Proceedings of the ASME 2018 5th Joint US-European Fluids Engineering Division Summer Meeting, Montreal, QC, Canada, 15–20 July 2018. [CrossRef]

144. Mikulčić, H.; Vujanović, M.; Ashhab, M.; Duić, N. Large eddy simulation of a two-phase reacting swirl flow inside a cement cyclone. *Energy* **2014**, *75*, 89–96. [CrossRef]

145. Wasilewski, M. Analysis of the effects of temperature and the share of solid and gas phases on the process of separation in a cyclone suspension preheater. *Sep. Purif. Technol.* **2016**, *168*, 114–123. [CrossRef]

146. Schiller, L.; Naumann, A. Uber die grundlegenden berechungen bei der schwerkraftaufbereitung. *Vereines Dtsch. Ingenieure* **1933**, *77*, 318–320.

147. Masnadi, M.; Grace, J.; Elyasi, S.; Bi, X. Distribution of multi-phase gas-solid flow across identical parallel cyclones: Modeling and experimental study. *Sep. Purif. Technol.* **2010**, *72*, 48–55. [CrossRef]

148. Grace, J. Maldistribution of flow through parallel cyclones in circulating fluidized beds. In Proceedings of the 9th International Conference on Circulating Fluidized Beds, CFB 2008, in Conjunction with the 4th International VGB Workshop on Operating Experience with Fluidized Bed Firing Systems, Hamburg, Germany, 13–16 May 2008; Code 106582.

149. Kim, T.; Choi, J.; Shun, D.; Kim, S.; Kim, S.; Grace, J. Wear of water walls in a commercial circulating fluidized bed combustor with two gas exits. *Powder Technol.* **2007**, *178*, 143–150. [CrossRef]

150. Zhou, X.; Cheng, L.; Wang, Q.; Luo, Z.; Cen, K. Non-uniform distribution of gas-solid flow through six parallel cyclones in a CFB system: An experimental study. *Particuology* **2012**, *10*, 170–175. [CrossRef]

151. Zhang, N.; Lu, B.; Wang, W.; Li, J. 3D CFD simulation of hydrodynamics of a 150MWe circulating fluidized bed boiler. *Chem. Eng. J.* **2010**, *162*, 821–828. [CrossRef]

152. Jiang, Y.; Qiu, G.; Wang, H. Modelling and experimental investigation of the full-loop gas-solid flow in a circulating fluidized bed with six cyclone separators. *Chem. Eng. Sci.* **2014**, *109*, 85–97. [CrossRef]

153. Wang, W.; Li, J. Simulation of gas-solid two-phase flow by a multi-scale CFD approach of the EMMS model to the sub-grid level. *Chem. Eng. Sci.* **2007**, *62*, 208–231. [CrossRef]

Article

Comparison of Surface Tension Models for the Volume of Fluid Method

Kurian J. Vachaparambil * and Kristian Etienne Einarsrud

Department of Materials Science and Engineering, Norwegian University of Science and Technology (NTNU), 7491 Trondheim, Norway
* Correspondence: kurian.j.vachaparambil@ntnu.no; Tel.: +47-45517425

Received: 24 July 2019; Accepted: 13 August 2019; Published: 15 August 2019

Abstract: With the increasing use of Computational Fluid Dynamics to investigate multiphase flow scenarios, modelling surface tension effects has been a topic of active research. A well known associated problem is the generation of spurious velocities (or currents), arising due to inaccuracies in calculations of the surface tension force. These spurious currents cause nonphysical flows which can adversely affect the predictive capability of these simulations. In this paper, we implement the Continuum Surface Force (CSF), Smoothed CSF and Sharp Surface Force (SSF) models in OpenFOAM. The models were validated for various multiphase flow scenarios for Capillary numbers of 10^{-3}–10. All the surface tension models provide reasonable agreement with benchmarking data for rising bubble simulations. Both CSF and SSF models successfully predicted the capillary rise between two parallel plates, but Smoothed CSF could not provide reliable results. The evolution of spurious current were studied for millimetre-sized stationary bubbles. The results shows that SSF and CSF models generate the least and most spurious currents, respectively. We also show that maximum time step, mesh resolution and the under-relaxation factor used in the simulations affect the magnitude of spurious currents.

Keywords: surface tension modelling; VOF; rising bubbles; capillary rise

---

1. Introduction

For a comprehensive understanding of flow physics in multiphase systems, which is ubiquitous in both nature and technological processes, consideration of surface tension is important. For instance, the break down of a fluid jet into droplets is used to form droplets in inkjets [1] and lab-on-chip devices [2] while the thinning and breakdown dynamics of non-Newtonian fluid filaments is critical in its application in jetting [3,4]. Flow scenarios such as underground water flows [5], oil recovery [6] and paper-based microfluidics [7] are examples of flow through porous media where dominance of surface tension may produce a capillary rise. The detachment diameter of the bubble [8,9] and shape of rising bubble [10] during bubble evolution in champagne, boiling and electrochemical gas evolution is dependent on surface tension, as is the droplet size produced during atomisation of fuels [11], spraying [12,13] and growth of a bubble in confined geometries [14]. The effect of surface tension is also important in events such as nucleation of bubbles [15,16] and droplets [17].

Due to the importance and complex nature of multiphase flows, numerical simulations, especially computational fluid dynamics (CFD), are commonly used to study and understand these processes. The CFD strategies used to model multiphase flows can broadly be divided into four categories: Euler–Euler (EE), Euler–Lagrange (EL), interface tracking and capturing methods. The EE approach assumes that phases are interpenetrating, which is efficient when modelling large-scale industrial processes [18,19], while EL tracks the dispersed phases individually, which can be computationally expensive [20,21]. As both EE and EL approaches do not resolve the complete interactions between the phases, they require so called "closure laws" (see [18–21]). Interface tracking methods, such as the moving mesh method, use a separate boundary-fitted moving mesh for each phase [22]. Although interface tracking methods are quite accurate, they are typically used to model bubble or droplets with mild-moderate deformations [22,23] but to handle complex interface deformations these methods require a global or local re-meshing [24]. Interface capturing methods use a fixed grid with functions to capture the interface such as the Volume Of Fluid method (VOF) [25], level-set [26] and diffuse interface methods [27]. Other methods available in the literature employ a hybrid interface tracking-capturing approach, such as the immersed boundary [28] and front tracking method [29]. Due to its ability to conserve mass (both level-set [30] and phase-field [31] models have difficulties in conserving mass), robustness and ability to produce reasonably sharp interfaces VOF is very popular in multiphase simulations [32–57] and implemented in both commercial (ANSYS Fluent® and Flow-3D®) and open source (OpenFOAM®) CFD packages.

Due to the popularity of open source CFD packages, this paper predominantly delves into the VOF formulation and reported development in interFoam, which is the VOF-based solver available in OpenFOAM®. In the VOF method, a scalar function representing the volume fraction of phases in the computational cells is advected. The advection of the volume fraction equation is done using specific discretisation schemes, such as the interface compression method [58], to prevent the excessive smearing of the interface thickness. Apart from interface compression method, recent work has explored reconstruction of interface using techniques such as the isoAdvector method [59,60] and piece wise linear interface calculation (PLIC) algorithm [61]. Although the VOF approach in theory produces a sharp interface, the "real" VOF, which is implemented in solvers such as interFoam, produces a non-sharp interface, which stretches over a few computational cells. This non-sharp nature of the volume fraction leads to errors in the calculated curvature which generates spurious currents that is quantified in the work by Harvie et al. [62], appearing as vortices around the interface (see [63,64]). The presence of these spurious currents introduces non-physical velocities near the interface, which can increase the interfacial mass transfer while modelling condensation [32] and evaporation [57] scenarios and adversely effects the accuracy of simulations. In the literature, spurious currents in VOF methods can be reduced by the following approaches:

- force balance, which is achieved by discretising the surface tension and pressure forces at the same location [65];
- accurate calculation of the curvature (see Table 1); and
- choosing the appropriate time step for the solver (see [63]).

Table 1. An overview of improved curvature calculations and surface tension models developed for VOF method.

Publication	Remarks
Brackbill et al. [66]	Introduced the Continuum Surface Force (CSF) and density scaled CSF models. These methods are very common due to its relatively straightforward implementation in a VOF framework.
Ubbink [67]	Proposed using a smoothed volume fraction to calculate curvature (referred to as "Smoothed CSF" in this paper). Using a smoothed volume fraction to compute the curvature instead of non-smoothed volume fraction in CSF model reduced spurious current up to one order of magnitude [56]. This method has been used in modelling condensing bubbles [32] and droplets in microfluidic devices [56]. A similar smoothening of α_1 was proposed by Heyns and Oxtoby [68].
Sussman and Puckett [69]	Developed a fully coupled level-set VOF (CLSVOF) method which combines the mass conservativeness of VOF method with smoothness of the level-set function to reduce spurious currents. The CLSVOF method has been used to applications such as splashing droplets [45], flow through microfluidic devices [46], wave breaking [47] and droplet evaporation [43]. Another variant of coupled level set approach is the simple coupled level-set VOF (S-CLSVOF) proposed by Albadawi et al. [70].
Raessi et al. [71]	Proposed a method to calculate κ based on advected normals. The spurious currents were lower than CSF (within the same order) while modelling cases such as stationary bubble, rising bubble and Rayleigh–Taylor instability [71].
Renardy and Renardy [72]	Introduced parabolic reconstruction of surface tension (PROST) algorithm which uses a least-squares fit of the interface to a quadratic surface. The spurious current produced by the algorithm is lower by two orders of magnitude compared to CSF [72]. The model was used to simulate droplet deformation including breakup [48,72].
Cifani et al. [61]	Implemented piecewise linear interface calculation (PLIC) algorithm (proposed by Youngs [73]) to reconstruct the interface in interFoam and managed to reduce spurious currents.
Pilliod and Puckett [74]	Developed an efficient least squares volume-of-fluid interface reconstruction algorithm (ELIVRA) which reconstructs the interface using a least square method to fit the interface to a linear surface.
Popinet [75]	Proposed calculating curvature using height functions. Use of height functions have reduced spurious current (slightly outperformed PROST algorithm [75]) and has been shown to model flow in microchannels [49], rising bubble [34,44] and other multiphase flows [50].
Raeini et al. [76]	Introduced a sharp surface force formulation to calculate the capillary force, which is then filtered to reduce the spurious currents (known as FSF model). Neglecting the filtering terms in the FSF model provides a sharp surface formulation of surface tension known as SSF, which is described in [42]. The SSF has been reported to be reduce the spurious currents by two to three orders in comparison to CSF [42]. The FSF model has been reported to provide periodic bursts in the velocity fields but lower spurious current than SSF [42]. The approach has been used to model bubbles in microfluidic devices [51] and flow through porous media [52].
Denner et al. [77]	Proposed the use of artificial viscosity model, which applies artificial shear stress in the tangential direction to interface, to dampen the spurious currents. The model has been used to model rising bubble and capillary instability of a water jet [77].
Lafaurie et al. [78]	Proposed an alternative to the CSF model, known as the Continuum Surface Stress (CSS) model, determines surface tension as divergence of stress tensor without relying on complex curvature calculations. Due to imbalance in the surface tension and pressure, CSS model can also produce spurious currents [35] which has reported to be in the same order as CSF [72]. CSS model has been used to model static droplets and rising bubbles [35], but it does not provide reliable results for falling films [41].

To analyse the force balance (described in [65]), Deshpande et al. [63] evaluated interFoam and showed that there is no imbalance in the surface tension and pressure forces due to inconsistent discretisation. However, the iterative process, which is used to solve pressure equation, introduces an imbalance which is related to the user defined tolerance level of the solution [63]. An overview of literature that provides an improved estimate of the interface curvature and surface tension modelling approaches is provided in Table 1. The improved representation of the interface (which aids in accurate calculation of the interface curvature) is provided bycoupled level-set VOF (CLSVOF) method, height functions and interface reconstruction algorithms (like piecewise linear interface calculation (PLIC), parabolic reconstruction of surface tension (PROST) and efficient least squares volume-of-fluid interface reconstruction (ELIVRA) algorithms), whereas the other methods discussed in Table 1 provide alternative approaches to model surface tension. To ensure that spurious currents do not grow over time, a stability condition, proposed by Brackbill et al. [66], for explicit treatment of surface tension is

$$\Delta t < \sqrt{\frac{\rho_{avg}(\Delta x)^3}{2\pi\sigma}}, \tag{1}$$

where Δx, σ and ρ_{avg} are grid spacing, surface tension and average of density of both phases, respectively. As proposed by Galusinski and Vigneaux [79], a comprehensive constraint on the time step must consider the effect of both density and viscosity which can be expressed as

$$\Delta t \leq \frac{1}{2}\left(C_2\tau_\mu + \sqrt{(C_2\tau_\mu)^2 + 4C_1\tau_\rho^2}\right), \tag{2}$$

where τ_μ and τ_ρ are time scales which are defined as $\mu_{avg}\Delta x/\sigma$ and $\sqrt{\rho_{avg}(\Delta x)^3/\sigma}$, where μ_{avg} is the average dynamic viscosity between the phases, respectively. An evaluation of interFoam, by Deshpande et al. [63], proposed that time step should satisfy

$$\Delta t \leq \max\left(C_2\tau_\mu, 10C_1\tau_\rho\right), \tag{3}$$

along with the time step constraint discussed in Equation (2). Deshpande et al. [63] also calculated the values of C_1 and C_2 for interFoam to be equal to 0.01 and 10, respectively. Further details of the numerical methods used to model surface tension is available in the recent review work by Popinet [80].

In the literature, comparison between surface tension models is often done for a specific of flow phenomenon and at times a static scenario is used to quantify the spurious currents. Some examples of flow phenomena used to compare surface tension models are rising bubbles whose diameters are in the order of few millimetres [33–35], translating and rotating bubbles [64], oscillating droplets or bubbles [34], stagnant bubbles or droplets [34,35,39,64], Rayleigh–Taylor instability [37,38], Taylor bubbles [64], falling films [41], droplet splashing [38,39], capillary rise [42] and bubble evolution [37,40]. These typically compare the CSF model with height functions [33,34,64], PROST [37], PLIC [42], CLSVOF and its variants [37–40,64], FSF and SSF [42], and CSS [35,41] models. Although the flow scenarios that are used to compare surface tension models are diverse, they can be broadly categorised based on the dominance of surface tension in the flow using the Capillary number (Ca), which is defined as the ratio of viscous to surface tension forces in the system. Flow phenomena such as capillary rise and stationary bubbles are examples of low values of Ca whereas flows with larger values of Ca include rising bubbles and falling films.

During processes such as gas evolution during electrochemical reactions and boiling, nucleated bubbles grow by mass transfer across the interface [15,16] or coalescence [8], but once the bubble detaches it may deform as it rises up and/or interacts with other bubbles [53]. Other complex processes, such as splashing, involve droplet spreading on a surface which is accompanied by formation of secondary smaller droplets at the rim [81]. To reliably model these processes, surface tension models must be able to accurately handle flow scenarios with both small and large capillary numbers.

In literature, the work by Hoang et al. [56] implemented the Smoothed CSF approach to model the steady motion of bubbles in a straight two-dimensional channel, the formation of bubbles in two- and three-dimensional T-junctions, and the breakup of droplets in three-dimensional T-junctions. A study by Heyns and Oxtoby [68] implemented a selection of surface tension modelling approaches (e.g., the CSF, a variant of Smoothed CSF and a force-balanced higher-resolution artificial compressive formulation) to model a stationary bubble. To the best of the authors' knowledge, a recent study by Yamamoto et al. [36] is the only one of its kind where different surface tension models (i.e., S-CLSVOF, density scaled S-CLSVOF and CSF) are compared based on a variety of processes with various capillary numbers (e.g., rising bubbles, capillary rise, capillary wave and thermocapillary flows).

In this study, we implemented three different surface tension models, namely CSF [66], Smoothed CSF [67] and SSF [76], in interFoam available in OpenFOAM 6. To investigate the capability of the surface tension models to handle various flow scenarios, we used two benchmark cases:

- two-dimensional rising bubbles (proposed by Hysing et al. [54], Klostermann et al. [55]); and
- two-dimensional capillary rise.

These two benchmark cases were selected due to their relevance in a variety of processes. To compare the spurious currents generated by the surface tension models, a stationary bubble was simulated. For practical applications, the time step constraint can substantially increase the computational time, thus the temporal development of the spurious currents with the surface tension models were also examined. Furthermore, the evolution of spurious currents with mesh resolution and under-relaxation factor used for the simulations was also investigated. In the interest of knowledge dissemination, the solvers and the test cases (implemented in OpenFOAM 6) discussed in the paper are available in the Supplementary Materials.

2. Numerical Model

2.1. Governing Equations

The VOF approach (developed by Hirt and Nichols [25]) denotes the individual phases using a scalar function called volume fraction, represented as

$$\alpha_1(\vec{x}, t) = \begin{cases} 0 \text{ (within Phase 2 or gas)} \\ 0 < \alpha_1 < 1 \text{ (at the interface)} \\ 1 \text{ (within Phase 1 or liquid)}, \end{cases} \tag{4}$$

where α_1 is the volume fraction of liquid. The fluid properties such as density (ρ) and viscosity (μ) in a control volume are calculated as

$$\chi = \chi_1 \alpha_1 + \chi_2 \alpha_2 \text{ where } \chi \in [\rho, \mu], \tag{5}$$

where χ_1 and χ_2 represent the fluid property of liquid and gas phase, respectively.

Considering the fluids to be incompressible, isothermal and immiscible, the VOF approach solves a single set of continuity and momentum equation for the whole domain. The continuity equation is written as

$$\nabla \cdot \vec{U} = 0, \tag{6}$$

where $\vec{U}$ is the fluid velocity. The momentum equation is

$$\frac{\partial \rho \vec{U}}{\partial t} + \nabla \cdot (\rho \vec{U} \vec{U}) = -\nabla p + \rho \vec{g} + \nabla \cdot \mu \left(\nabla \vec{U} + \nabla \vec{U}^T \right) + \vec{F}_{st}, \tag{7}$$

where last term represents the surface tension forces, the second last term is the viscous term, $\vec{g}$ is the acceleration due to gravity and p is the pressure. Advection of the volume fraction of liquid (α_1) is implemented in interFoam as

$$\frac{\partial \alpha_1}{\partial t} + \nabla \cdot (\alpha_1 \vec{U}) + \nabla \cdot (\alpha_1 (1 - \alpha_1) \vec{U}_r) = 0, \tag{8}$$

where the third term is an artificial compression term used to sharpen the interface [58,61]. The artificial compression term uses a relative velocity ($\vec{U}_r$) defined as

$$\vec{U}_r = C_\alpha \left| \frac{\phi}{|S_f|} \right| \vec{n}_f, \tag{9}$$

where ϕ, S_f, C_α and $\vec{n}_f$ are the velocity flux, face surface area, an adjustable compression factor and unit normal vector to the interface, respectively. In the literature, C_α can be set between 0 and 4, where C_α equal to zero and one correspond to the case of no and moderate compression, respectively [56]. The increase in the value of C_α sharpens the interface but increases the magnitude of spurious currents (see [51,56]). To model practical flow scenarios using interFoam, the value of C_α is generally set to unity [32,63].

2.2. Surface Tension Models

This section introduces the three surface tension models: CSF, Smoothed CSF and SSF.

2.2.1. The Continuum Surface Force (CSF) Model

Proposed by Brackbill et al. [66], the CSF model provides a volumetric representation of surface tension, represented as

$$\vec{F}_{st} = \sigma \kappa \nabla \alpha_1, \tag{10}$$

where σ is the surface tension and κ is the curvature, defined in Equation (13). The unit normal vector at the interface is calculated as

$$\hat{n} = \frac{\nabla \alpha_1}{|\nabla \alpha_1| + \delta}, \tag{11}$$

where δ is a small non-zero term to ensure that the denominator does not become zero. δ is calculated as $10^{-8} / \left(\frac{\sum_N V_i}{N} \right)^{1/3}$, where N is the number of computational cells and $\sum_N V_i$ provides the sum of the volumes of individual cells (represented by i). Once $\hat{n}$ is calculated, it is corrected to account for wall adhesion through

$$\hat{n} = \hat{n}_w \cos\theta + \hat{t}_w \cos\theta \tag{12}$$

where θ is the contact angle of the gas–liquid interface at the walls (measured in the liquid phase), and $\hat{n}_w$ and $\hat{t}_w$ are unit vectors that are normal and tangential to the wall, respectively [82]. The curvature of the interface is then calculated as

$$\kappa = -\nabla \cdot \hat{n}. \tag{13}$$

2.2.2. The Smoothed CSF Model

The Smoothed CSF model (by Ubbink [67]) proposed modifying CSF by modifying the calculation of curvature of interface by using a smoothed volume fraction of liquid (α_1).

The smoothed volume fraction field is calculated using a smoother proposed by Lafaurie et al. [78], which has been implemented in the literature [32,56] and is represented as

$$\widetilde{\alpha_1} = \frac{\sum_{f=1}^{N} < \alpha_1 >_{c \to f} S_f}{\sum_{f=1}^{N} S_f}, \tag{14}$$

where the indices c and f are the cell and face centre indices, respectively. $< \alpha_1 >_{c \rightarrow f}$ represents the interpolation of α_1 from cell to face centre. The smoothening of volume fraction, done using Equation (14), is applied twice to obtain a smooth volume fraction field, which is used in Equation (15). Implementation of Equation (14) in interFoam is done using the subroutine developed in the work by [56]. Based on the smoothed volume fraction field, the unit normal to the interface is calculated as

$$\tilde{n} = \frac{\nabla \tilde{\alpha_1}}{|\nabla \tilde{\alpha_1}| + \delta},$$

(15)

which is then corrected for wall adhesion (based on Equation (12)). The curvature of the interface is then calculated as

$$\tilde{\kappa} = -\nabla \cdot \tilde{n}.$$

(16)

The surface tension can be represented using the modified curvature ($\tilde{\kappa}$ in Equation (16)), which can be represented as

$$\vec{F}_{st} = \sigma \tilde{\kappa} \nabla \alpha_1.$$

(17)

2.2.3. The Sharp Surface Force (SSF) Model

In the SSF model, proposed by Raeini et al. [76], smoothened and sharpened volume fraction fields are used to calculate curvature and gradient of of volume fraction.

The smoothened volume fraction (α_s) is calculated based on interpolating the cell-centred values of α_1 to the cell faces using a three consecutive smoothening steps described using Equations (18a)–(18c)

$$\alpha_{s1} = C \left\langle < \alpha_1 >_{c \rightarrow f} \right\rangle_{f \rightarrow c} + \left(1 - C \right) \alpha_1,$$

(18a)

$$\alpha_{s2} = C \left\langle < \alpha_{s1} >_{c \rightarrow f} \right\rangle_{f \rightarrow c} + \left(1 - C \right) \alpha_{s1},$$

(18b)

$$\alpha_s = C \left\langle < \alpha_{s2} >_{c \rightarrow f} \right\rangle_{f \rightarrow c} + \left(1 - C \right) \alpha_{s2},$$

(18c)

where C is set equal to 0.5. The unit normal to the interface is then calculated as

$$\hat{n}_s = \frac{\nabla \alpha_s}{|\nabla \alpha_s| + \delta},$$

(19)

which is then corrected for wall adhesion (based on Equation (12)). The curvature (κ_s) is calculated using Equation (19) as

$$\kappa_s = -\nabla \cdot \hat{n}_s.$$

(20)

The interface curvature is smoothed by using a three step procedure, which can be broadly summarised into Equations (21a), (21c), and (21d). The first step involves smoothening the curvature calculated in Equation (20) as

$$\kappa_{f1} = \left(2\sqrt{\alpha_c(1 - \alpha_c)} \right) \kappa_s + \left(1 - 2\sqrt{\alpha_c(1 - \alpha_c)} \right) \kappa_s^*$$

(21a)

where α_c is defined as $\min(1, \max(\alpha_1, 0))$ and

$$\kappa_s^* = \frac{\left\langle < w\kappa_s >_{c \rightarrow f} \right\rangle_{f \rightarrow c}}{\left\langle < w >_{c \rightarrow f} \right\rangle_{f \rightarrow c}}, \quad w = \sqrt{\alpha_c(1 - \alpha_c) + 10^{-3}}.$$

(21b)

The second step further smoothens the curvature (calculated in Equation (21a)) as

$$\kappa_{f2} = \left(2\sqrt{\alpha_c(1-\alpha_c)}\right)\kappa_s + \left(1 - 2\sqrt{\alpha_c(1-\alpha_c)}\right)\kappa_{s2}^*, \text{ where } \kappa_{s2}^* = \frac{\left\langle <w\kappa_{f1}>_{c\to f}\right\rangle_{f\to c}}{\left\langle <w>_{c\to f}\right\rangle_{f\to c}}. \qquad (21c)$$

The final step calculates the the final curvature as

$$\kappa_{final} = \frac{<w\kappa_{f2}>_{c\to f}}{<w>_{c\to f}}. \qquad (21d)$$

The surface tension is then given as

$$\vec{F}_{st} = \sigma\kappa_{final}\nabla\alpha_{sh}, \qquad (22)$$

where α_{sh} is a sharpened volume fraction of liquid defined in Equation (23).

$$\alpha_{sh} = \frac{1}{1-C_{sh}}\left[\min\left(\max\left(\alpha_1, \frac{C_{sh}}{2}\right), 1 - \frac{C_{sh}}{2}\right) - \frac{C_{sh}}{2}\right], \qquad (23)$$

where C_{sh} is a sharpening coefficient. A value of C_{sh}=0 reduces α_{sh} to α_1, whereas C_{sh}=1 provides sharp representation of the interface (which is numerically unstable). We used C_{sh}=0.98 for static cases and C_{sh}=0.5 for dynamic cases.

3. Solver Settings

To simplify the treatment of pressure boundary condition and density change across the interface, interFoam uses p_{rgh} which is defined as $p - \rho\vec{g}\cdot\vec{x}$, where $\rho\vec{g}\cdot\vec{x}$ is the hydrostatic component of pressure [58]. The volume fraction evolution equation (Equation (8)) is solved using the Multidimensional Universal Limiter with Explicit Solution (MULES) algorithm, which preserves the boundedness of volume fraction [61,63]. Once volume fraction is solved, the continuity equation (Equation (6)) and momentum equation (Equation (7)) are solved using the Pressure Implicit with Splitting of Operator (PISO) algorithm [83]. In PISO, a predicted velocity is updated using a pressure correction procedure to advance velocity and pressure fields in time [58,63]. To understand the implementation and solution algorithm of the governing equations (Equations (6)–(8)) in interFoam, please refer to the work by Rusche [58] or Deshpande et al. [63]. The discretisation schemes, solvers and others parameters used to solve the governing equations for all the simulations discussed in this paper are presented in Tables 2–4, respectively. Under-relaxation factors, if set to less than unity, cause damping of the solution, which can lead to longer computational time for the solution reach to a steady state value. In flow scenarios where there is no steady state solution, using an under-relaxation factor can lead to erroneous results due to under-prediction of the flow variables. We used an under-relaxation factor in the solver equal to one for dynamic cases and 0.9 for static cases. The effect of using an under relaxation factor of one on static cases is also investigated.

Table 2. Discretisation schemes.

Modeling Term	Keyword	Scheme	Remarks
Time derivatives	ddtSchemes	Euler	First order implicit method (see [84])
Divergence term	$\nabla \cdot (\rho \vec{U} \vec{U})$	vanLeerV	Modified vanLeer for vector fields (see [84])
	$\nabla \cdot (\vec{U} \alpha_1)$	vanLeer	See [85]
	$\nabla \cdot (\vec{U}_c \alpha_1 (1 - \alpha_1))$	interfaceCompression	See [63]
Gradient term	gradSchemes	linear	Operator with ∇ (see [84])
Laplacian term	laplacianSchemes	linear corrected	Operator with ∇^2 (see [84])
Others	snGradSchemes	corrected	Surface normal gradients (see [84])
	interpolationSchemes	linear	Interpolates values (see [84])

Table 3. Solvers used for the discretised equation.

Equation	Linear Solver	Smoother/Preconditioner	Tolerance
Pressure correction equation	PCG	DIC	10^{-20} (based on [63])
Momentum equation	smoothSolver	symGaussSeidel	10^{-12}
Volume fraction equation	smoothSolver	symGaussSeidel	10^{-12}

Table 4. Other parameters used in solving the discretised equations.

Parameter	Value	Notes
nAlphaCorr	2	Number of α_1 correction [55]; typically set equal to 1 or 2 for time-dependent flows [86].
nAlphaSubCycles	1	Represents the number of sub-cycles within α_1 equation [84].
cAlpha (C_α)	1	Used for interface compression in Equation (9).
MULESCorr	yes	Switches on semi-implicit MULES [87].
nLimiterIter	3	Number of MULES iterations over the limiter [87].
momentumPredictor	no	Controls solving of the momentum predictor; typically set to 'no' for multiphase and low Reynolds number flows [84].
minIter	1	Minimum number of iterations used in momentum calculation.
nOuterCorrectors	1	PISO algorithm is selected by setting this parameter equal to unity (in PIMPLE algorithm) [84].
nCorrectors	3	The number of times the PISO algorithm solves the pressure and momentum equation in each step; usually set to 2 or 3 [84].
nNonOrthogonalCorrectors	0	Used when meshes are non-orthogonal [84].

4. Validation: Benchmark Test Cases

4.1. Two Dimensional Rising Bubbles

Due to the computational overhead of modelling a three-dimensional rising bubble, we model the buoyancy driven motion of a single bubble as proposed by Hysing et al. [54], Klostermann et al. [55]. The work by Hysing et al. [54] reported benchmarking data such as the bubble shape, rising velocity and circularity for two cases. These benchmarking data are produced based on numerical simulations using codes such as TP2D, FreeLIFE and MoonNMD [54]. In the work by Klostermann et al. [55], the benchmark proposed by Hysing et al. [54] was used to evaluate the VOF solver in OpenFOAM®️ (i.e., interFoam) for various meshes.

The computational domain used for the simulation is a rectangle of dimensions 1 m $\times$ 2 m where the bubble of diameter 0.5 m was initialised such that the centre of the bubble is at a distance of 0.5 m from the bottom and side walls. As mesh convergence could not be achieved perfectly in previous works [36,55], we used a uniform grid 160×320 for the simulations, corresponding to the fine mesh used in [54]. The pressure boundary conditions used in the simulations were zero gradient on the side and bottom walls, and a Dirchlet condition (equal to zero) at the top wall. The volume fraction of fluid used a zero gradient boundary condition on all walls. The velocity boundary conditions used for the simulations were no slip on top and bottom walls, but slip condition was implemented for the side walls. The fluid properties associated with the test cases, which are abbreviated as TC1 and TC2, are tabulated in Table 5. The maximum Courant number used by the solver was set equal to 0.01 and maximum time step permitted was based on Equations (2) and (3). The test cases are distinguished based on Reynolds (Re), Eötvös (Eo) and Capillary (Ca) numbers, which are defined as

$$Re = \frac{U_g L}{\nu_1}, Eo = \frac{\rho_1 U_g^2 L}{\sigma}, Ca = \frac{Eo}{Re} \tag{24}$$

with L and U_g being the characteristic length scale (equal to 0.5 m) and characteristic velocity (defined as $\sqrt{|\vec{g}|L}$), respectively. The bubble shape was obtained at $\alpha_1 = 0.5$ and rising velocity was calculated based on bubble volume averaged vertical component of the velocity vector [54,55]. For validation, we used the the data reported by Klostermann et al. [55] and Hysing et al. [54] (for the predictions by the FreeLIFE solver, which is referred to as 'Benchmark' in this paper) for a uniform grid of 160×320.

Table 5. Physical parameters used for the rising bubble simulations (see [54]).

Cases	ρ_1(kg/m^3)	ρ_2(kg/m^3)	ν_1(m^2/s)	ν_2(m^2/s)	σ(N/m)	$\vec{g}$(m/s^2)*	Re	Eo	Ca
TC1	1000	100	10^{-2}	10^{-2}	24.5	(0 −0.98 0)	35	10	0.286
TC2	1000	1	10^{-2}	10^{-1}	1.96	(0 −0.98 0)	35	125	3.571

*$\vec{g}$ is the reduced gravity as described in [54].

The first test case, TC1, corresponds to the case where surface tension effects are dominant [55]. The temporal evolution of the bubble as predicted by the various surface tension models is compared in Figure 1. Due to the stronger surface tension effects, the interface deforms into an ellipsoidal bubble (see Figure 2). The bubble shape (at t = 3 s) predicted by CSF model provides a slightly better agreement to the benchmark data compared to the other surface tension models. The surface tension models also tend to underpredict the position of the bubble at t = 3 s. This underprediction could be attributed to the lower rising velocity (see Figure 3), which has also been reported in previous studies using OpenFOAM [36,54,55]. Although bubble shape and rising velocity provide an overview of the capability of the surface tension models, the quantification of the errors associated with the models was based on the maximum rising velocity (V_{max}) and the time at which the V_{max} occurred (tabulated in Table 6). The benchmarking data show that SSF model provides a better agreement to the data reported by Hysing et al. [54] (absolute error is less than 2%) and Klostermann et al. [55] (absolute error is nearly 1.5%) in comparison to the other models.

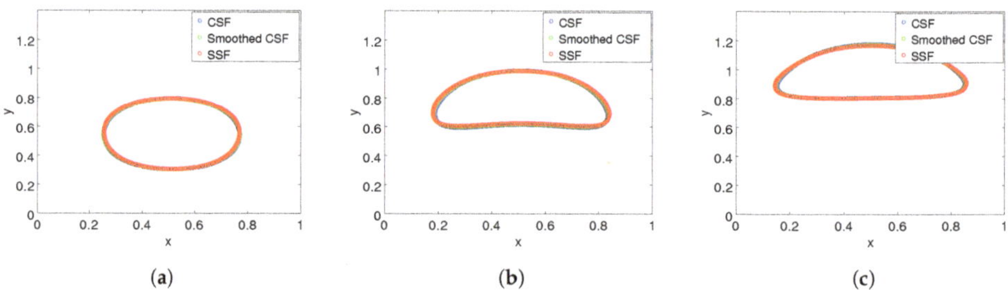

(a)   (b)   (c)

Figure 1. Temporal evolution of the bubble for TC1: (**a**) t = 0.5 s; (**b**) t = 1.5 s; and (**c**) t = 2.5 s.

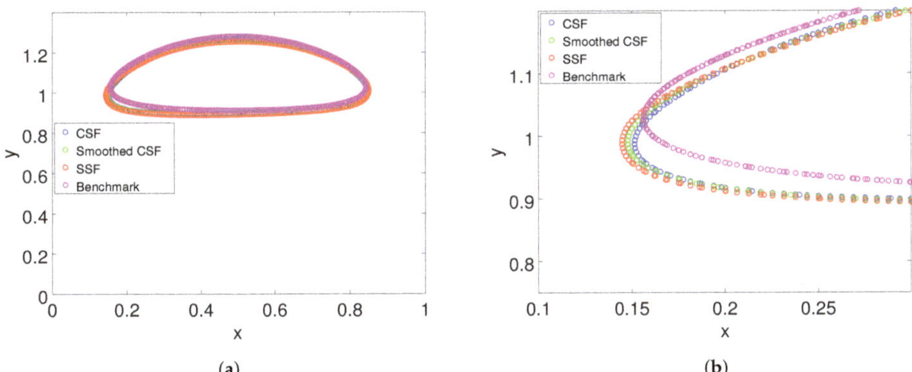

Figure 2. Validation Bubble shape for TC1 at $t = 3$ s: (**a**) bubble morphology; and (**b**) detailed.

Table 6. Benchmark quantities for TC1.

Parameter	CSF	Smoothed CSF	SSF	[55]	Benchmark ([54])
V_{max}	0.2366	0.2375	0.2386	0.2365	0.2419
$t(V_{max})$	0.9632	0.9491	0.9104	0.9219	0.9270

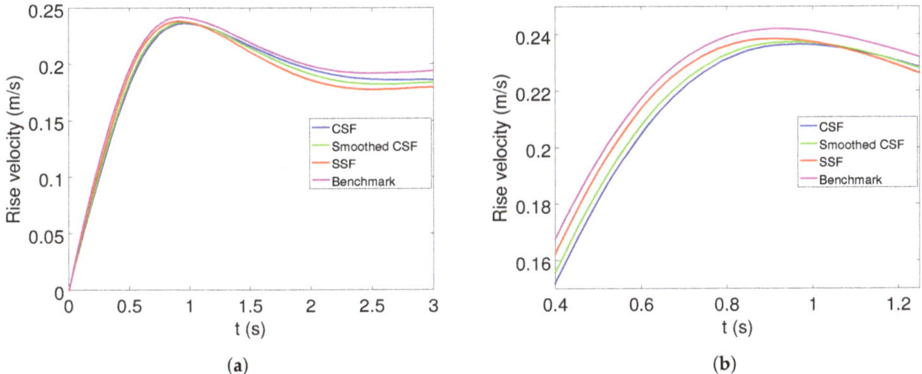

Figure 3. Validation Bubble rising velocity for TC1: (**a**) temporal changes of bubble rising velocity; and (**b**) detailed.

The other test case, TC2, corresponds to a case where the surface tension effects are lower [55]. This results in larger deformation of interface as the bubble evolves (see Figure 4) and eventually forms a skirted bubble that has thin filaments that breaks down into smaller droplets (see Figure 5). Comparing the surface tension models to the benchmark for final bubble shape shows that the models agree quite well (see Figure 5) but there is a difference between the models with respect to the prediction of the skirted part of the bubble (see Figure 5b). Figure 6 shows that the surface tension models in comparison to the benchmark data under-predicts the rise velocity. Comparing with the benchmark, the SSF model provides the closest agreement for V_{max1} (absolute error is nearly 3.5% [54] and less than 0.1% for [55]) and $t(V_{max1})$ (absolute error is nearly 3–3.5% for both [54,55]) (see Table 7). On the other hand, CSF model agrees with the benchmarking data for V_{max2} (absolute error is nearly 5.7% [54] and 0% for [55]) and $t(V_{max2})$ (absolute error is nearly 0.6% for both [54,55]) (see Table 7).

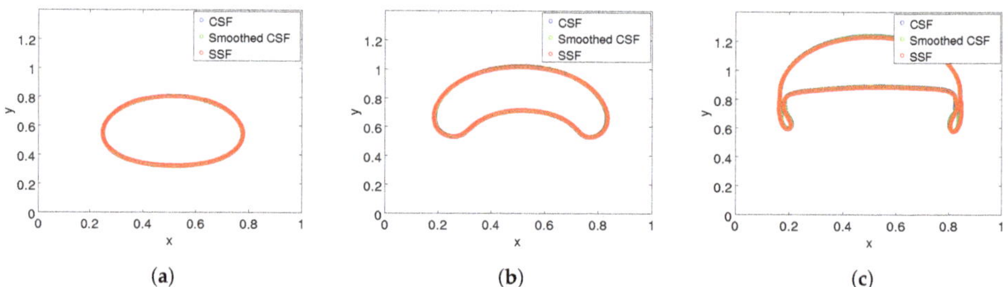

Figure 4. Temporal evolution of the bubble for TC2: (**a**) $t = 0.5$ s; (**b**) $t = 1.5$ s; and (**c**) $t = 2.5$ s.

Table 7. Benchmark quantities for TC2.

Parameter	CSF	Smoothed CSF	SSF	[55]	Benchmark ([54])
V_{max1}	0.2434	0.2429	0.2427	0.2431	0.2514
$t(V_{max1})$	0.7663	0.7637	0.7502	0.7250	0.7281
V_{max2}	0.2302	0.2290	0.2260	0.2302	0.2440
$t(V_{max2})$	1.9721	1.9700	1.9729	1.9594	1.9844

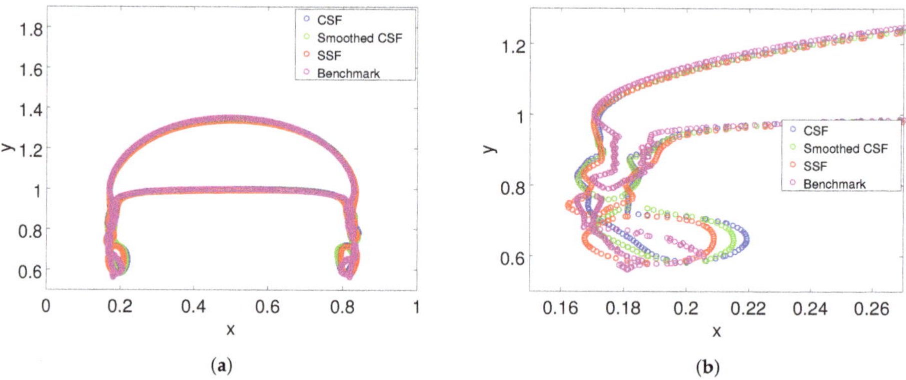

Figure 5. Validation Bubble shape for TC2 at $t = 3$ s: (**a**) bubble morphology; and (**b**) detailed.

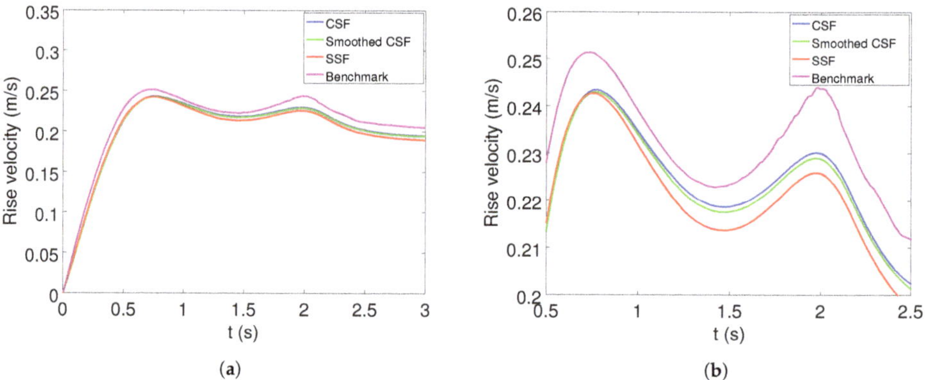

Figure 6. Validation Bubble rising velocity for TC2: (**a**) temporal changes of bubble rising velocity; and (**b**) detailed.

In the previous work by Klostermann et al. [55], the spurious currents were reported to be the reason for the error between the benchmark ([54]) and their simulations (for both TC1 and TC2). Thus, the differences in the predictions, for the rising bubble simulations, between the three surface tension models considered in this paper and their departure from the benchmark can also be attributed to spurious currents generated by these models (which is discussed below). For TC2, the larger variation between the surface tension models after the first peak in the transient evolution of the rise velocity (see Figure 6) can be attributed to the differences in the shapes of filament or satellite droplets (based on the work of Yamamoto et al. [36]). Interestingly, there are also some differences in the predictions by the CSF model (for both TC1 and TC2) and the data reported by Klostermann et al. [55], which could be attributed to the difference in the solver settings (e.g., the discretisation schemes, linear solvers and number of iterations) and/or the variations within the different versions of OpenFOAM. The influence of the discretisation schemes on the predicting the flow variables has been previously investigated in [88,89] but further investigation into the effects of other solver settings (e.g., the choice of linear solver and number of iterations) on the solution is required to quantify its effect. As OpenFOAM gets updated, some of the functionalities and/or the algorithms are modified, for example, the artificial interface compression term used in advection of α_1 (defined in Equation (9)) is computed differently in the older versions of the software (see [55]). To the best knowledge of the authors, no study has reported a comparison of the performance of various versions of OpenFOAM for specific flow scenarios. These settings, especially discretisation schemes and interface compression algorithms, would effect the generation and evolution of spurious currents, which could be the potential source of the discrepancy between our simulations and the data reported in literature.

4.2. Two-Dimensional Capillary Rise

The rise of liquid through a narrow tube or between two parallel plates, which occurs as a consequence of the wetting of the walls by the liquid, is known as capillary rise. As the liquid rises, it reaches a point of equilibrium when the vertical component of the force exerted by surface tension is balanced by the gravitational force acting on the risen liquid column. This equilibrium point (for liquid rising between two vertical parallel plates) is denoted using a height (h_b), which can be analytically calculated as

$$h_b = \frac{2\sigma\cos\theta}{\Delta\rho|\vec{g}|a},$$

(25)

where $\Delta\rho$ is the difference between densities of liquid and gas, and a is the distance between the plates [90].

To study capillary rise, we used a rectangular domain of dimensions 1 mm $\times$ 20 mm, where a (defined in Equation (25)) is equal to 1 mm, with a uniform mesh of 20 $\times$ 400. This mesh resolution provided the most accurate prediction of capillary rise for the same computational domain while using CSF model in the previous work by Yamamoto et al. [36]. The boundary conditions for velocity field imposes a no slip boundary condition for the walls and pressure based condition (applied to both inlet and outlet) that computes inlet velocity based on the patch-face normal component of the internal-cell velocity and outflow using the zero gradient condition. The volume fraction field uses a zero gradient condition at walls (with a contact angle of 45°) and outlet, along with a Dirchlet condition (equal to one) at inlet. The boundary condition for pressure uses a Dirichlet condition (equal to zero) at inlet and outlet whereas the walls use a Neumann boundary condition. The materials properties used for the simulations are described in Table 8. The initial volume fraction of liquid in the domain is set such that the liquid–gas interface is at a height of 8 mm from the bottom surface. The maximum time step (which satisfies both Equations (2) and (3)) and maximum Courant number were set equal to 3.5 μs and 0.1, respectively.

Table 8. Physical parameters used for the capillary rise simulations.

ρ_1(kg/m^3)	ρ_2(kg/m^3)	ν_1(m^2/s)	ν_2(m^2/s)	σ(N/m)	$\vec{g}$(m/s^2)*	Ca
1000	1	10^{-6}	1.48×10^{-5}	0.07	$(0-10\,0)$	0.0014

*This value of $\vec{g}$ is used to study capillary rise by Yamamoto et al. [36].

Once the interface position stabilised (see Figure 7), the capillary height $h_{b,calc}$ was calculated approximately from the volume fraction field as

$$h_{b,calc} = \frac{\int_S \alpha_1 dS}{a}, \tag{26}$$

where the numerator is the area occupied by the liquid in the computational domain [36]. The capillary rise height calculated from the simulations is compared to the analytically derived h_b (which was determined to be 9.9 mm using Equation (25)) in Table 9.

Table 9. Errors associated with the surface tension models on prediction of capillary rise.

Surface Tension Model	$h_{b,calc}$ (mm)	$E(h) = (h_{b,calc} - h_b)/h_b$
CSF	9.16	-0.076
Smoothed CSF	Capillary height did not stabilise during simulations (see Figure 7)	
SSF	9.26	-0.065

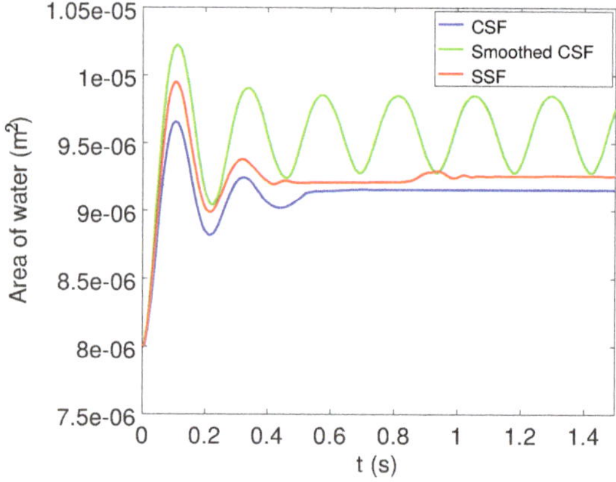

Figure 7. Evolution of the water column during capillary rise.

Table 9 shows that SSF model provides a better prediction of the capillary rise height compared to CSF model. A previous work by Yamamoto et al. [36] reported an error of 7.7% for a capillary rise model using the CSF model. Interestingly, the Smoothed CSF model could not provide a reliable capillary rise prediction due to the oscillation of the water column (see Figure 7). This discrepancy can be explained based on the evolution of the spurious currents (U_{sc} defined in Equation (27)), which are plotted in Figure 8. The magnitude of spurious currents (U_{sc}) generated in the simulations was computed at each time step as

$$U_{sc} = \max(|\vec{U}|). \tag{27}$$

The periodic growth and decay of the spurious currents in the Smoothed CSF model (see Figure 8) results in the unrealistic motion of the interface whereas the CSF model which has much larger magnitude of spurious currents is much more periodic (see Figure 8), which reduces the net motion of the liquid–gas interface. Compared to CSF and Smoothed CSF models, the spurious current evolution in the SSF model is lowered by nearly two orders of magnitude (see Figure 8).

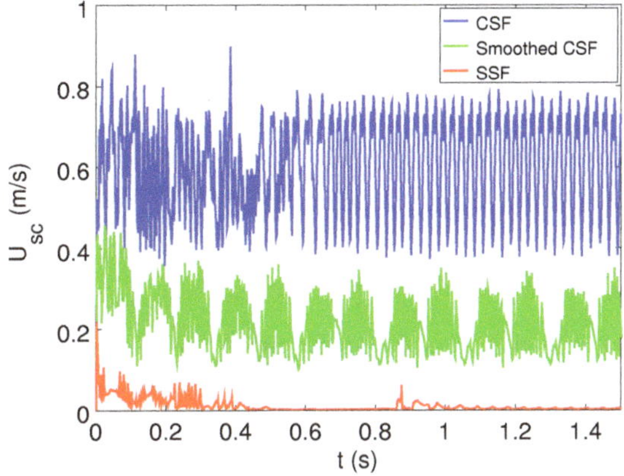

Figure 8. Evolution of spurious currents during the capillary rise simulations. It is worth pointing out that the figure is plotted using data extracted at every 500th point from the dataset obtained from simulations in order to reduce the rendering time of the image but care has been taken to showcase the larger temporal variations of U_{sc}.

5. Analysis: Spurious Current

To study the spurious currents generated during the simulations, we simulated a stationary bubble where the effect of gravity was neglected. A bubble of diameter $2R$ was set at the centre of a square domain of dimensions $4R \times 4R$. The properties of the two phases and other physical parameters used for the simulations described in this section are tabulated in Table 10. For these simulations, the boundaries were assigned the Dirichlet condition, equal to 101325 Pa, for pressure and zero gradient condition for both α_1 and $\vec{U}$. The simulations were run until an end time of 0.05 s to ensure that initial transients (if any) were eliminated with maximum time step calculated based on Equations (2) and (3) along with maximum Courant number of 0.1.

Table 10. Physical parameters used for the simulations in the analysis of spurious current.

$\rho_1 (kg/m^3)$	$\rho_2 (kg/m^3)$	$\nu_1 (m^2/s)$	$\nu_2 (m^2/s)$	$\sigma (N/m)$	$\vec{g} (m/s^2)$
1000	1	10^{-6}	1.48×10^{-5}	0.07	(0 0 0)

The accuracy of the surface tension models was calculated based on the following parameters: Laplace pressure, magnitude of spurious currents and mass imbalance. For a two-dimensional bubble, the Laplace pressure can be calculated using the Young–Laplace equation as

$$\Delta p_c' = \frac{\sigma}{R}. \tag{28}$$

The Laplace pressure inside the bubble was calculated from the simulation as

$$\Delta p_c = \frac{\int_V \alpha_2 p dV}{\int_V \alpha_2 dV} - p_0, \tag{29}$$

where p_0 is the operating pressure (which was equal to 101325 Pa). The mean error associated with the Laplace pressure calculated by the various surface tension models was determined as

$$\overline{E}(\Delta p_c) = \frac{\overline{\Delta p_c} - \Delta p_c'}{\Delta p_c'}, \tag{30}$$

where the overbar represents the time averaged variables.

5.1. Stagnant Bubble of Few Millimetres

In this test case, we modelled a bubble with a radius of 2.5 mm using fluid properties described in Table 10 and under-relaxation factor of 0.9. The computations were performed using a uniform structured grid. The total number of mesh elements and maximum time step (which satisfies both Equations (2) and (3)) used in the simulations are described in Table 11.

Table 11. Details of mesh and the associated maximum time step calculated based on Equations (2) and (3) used for stationary bubble simulations.

Mesh	Mesh Resolution (mm^2)	Total Number of Cells	$R/\delta x$*	Maximum Time Step (s)
M0	0.5×0.5	400	5	9×10^{-5}
M1	0.25×0.25	1600	10	3×10^{-5}
M2	0.125×0.125	6400	20	1×10^{-5}
M3	0.083×0.083	14400	30	6×10^{-6}

*$R/\delta x$ is the ratio of the radius of the bubble and the cell size.

To understand how spurious currents occur with various surface tension models, U_{sc} is plotted at $t = 0.05$ s for the grid described by M3 in Figure 9. In the surface tension models considered in this study, the spurious currents occur around the interface but their magnitudes are much larger in the bubble than outside. To quantify the spurious currents from the simulations, the magnitude of spurious currents and capillary pressure are tabulated in Table 12. The spurious currents generated by the surface tension models tends to reduce with finer meshes for both SSF and Smoothed CSF. On the other hand, the increase in spurious current for CSF can be explained based on the dependence on the mesh size (Δx) is given by

$$C_{\Delta x} \sim \sqrt{\frac{\sigma}{\rho \Delta x}}, \tag{31}$$

where $C_{\Delta x}$ is the magnitude of the spurious velocities (studied for CSF model [63,66]). Equation (31) indicates that smaller mesh sizes result in larger values of spurious currents for CSF model. As shown in Table 12, the Laplace pressure predicted by the surface tension models does not perfectly match $\Delta p'_c$ but both Smoothed CSF and SSF provides a better prediction in comparison to CSF.

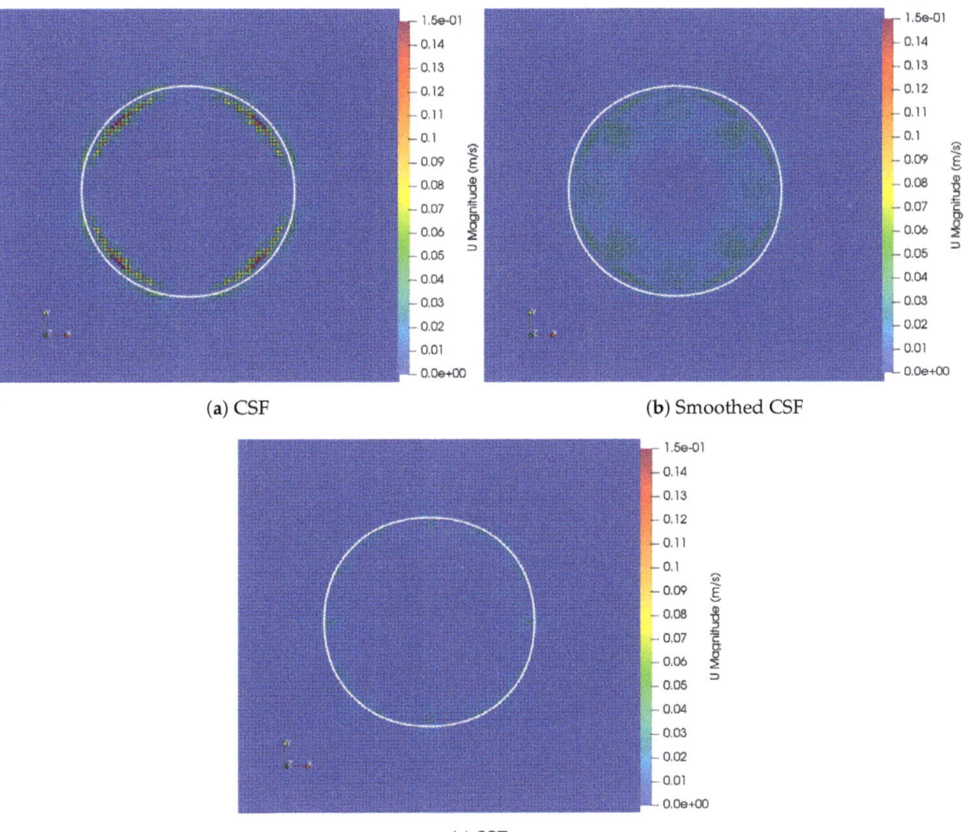

(**a**) CSF

(**b**) Smoothed CSF

(**c**) SSF

Figure 9. Comparison of spurious current generated by surface tension models at $t = 0.05$ s using M3 mesh. The gas–liquid interface in the domain is represented using a contour (in white) that is plotted at $\alpha_1 = 0.5$.

Table 12. Comparison of spurious currents based on mesh and surface tension models (using an under-relaxation factor of 0.9).

Surface Tension Model	Mesh	$\overline{U_{sc}}$	$Ca = \dfrac{\rho_1 v_1 \overline{U_{sc}}}{\sigma}$	$\overline{\Delta p_c}$	$\overline{E}(\Delta p_c)$	Mass Imbalance
CSF	M0	0.133	0.002	22.29	-0.20	0
	M1	0.171	0.002	23.03	-0.18	0
	M2	0.174	0.002	24.06	-0.14	0
	M3	0.189	0.003	24.77	-0.12	0
Smoothed CSF	M0	0.096	0.001	24.12	-0.14	0
	M1	0.088	0.001	25.14	-0.10	0
	M2	0.062	0.001	25.19	-0.10	0
	M3	0.049	0.001	26.09	-0.07	0
SSF	M0	0.045	0.001	23.95	-0.14	0
	M1	0.087	0.001	25.12	-0.10	0
	M2	0.036	0.001	25.88	-0.08	0
	M3	0.041	0.001	25.55	-0.09	0

5.2. Effect of Time Step

The two time step constraints were from Brackbill et al. [66] (Equation (1)) and Deshpande et al. [63] (Equations (2) and (3)). To study the effect of time step constraint, the simulations used a bubble of 2.5 mm with the M3 mesh (see Table 11) and fluid properties described in Table 10 using an under-relaxation factor of 0.9. The maximum time steps (Δt) used for the simulations are 25 μs (based on [66]) and 6 μs (based on [63]).

The temporal evolution of U_{sc} is compared for the surface tension models in Figure 10. Using the time step dictated by Deshpande et al. [63], the spurious currents generated by the CSF model are reduced by less than half in comparison to when time step constraint proposed by Brackbill et al. [66] was used. The other models show an absolute difference in the mean spurious current of nearly 7% and 6%, respectively, for the time step constraints (see Table 13).

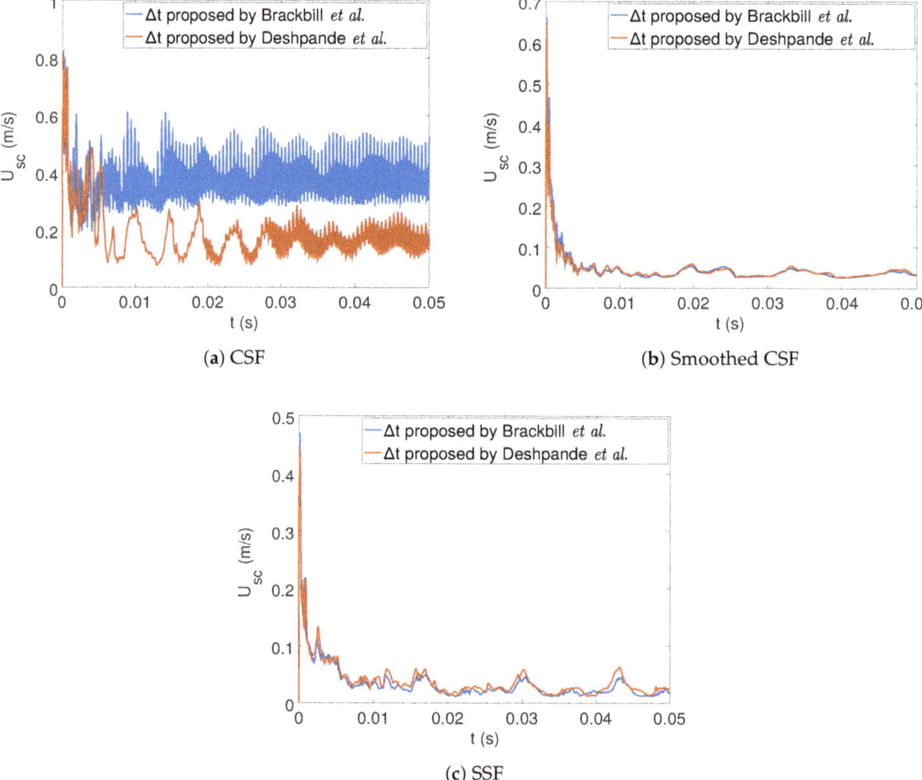

(a) CSF

(b) Smoothed CSF

(c) SSF

Figure 10. Evolution of spurious currents for various surface tension models.

Table 13. Comparison of spurious currents for the time stepping constraints based on M3 mesh and surface tension models (while using an under-relaxation factor of 0.9).

Surface Tension Model	$\overline{U_{sc}}$ based on Brackbill et al. [66]	$\overline{U_{sc}}$ based on Deshpande et al. [63]
CSF	0.395950	0.189170
Smoothed CSF	0.052188	0.048619
SSF	0.038550	0.040984

5.3. Effect of Under-Relaxation Factor

To understand the effect of under-relaxation factor, we considered a case which used an under-relaxation factor of unity for modelling the stationary bubble of 2.5 mm with M3 mesh. Table 14 provides a summary of the spurious current and the Laplace pressure in the bubble. Comparison of the results from under-relaxation factor of 0.9 (see Table 12) and 1 (see Table 14) shows that spurious currents generated by Smoothed CSF model is substantially larger when using a larger under-relaxation factor (nearly twice). The SSF model provides the least amount of spurious currents for both the under-relaxation factors and the CSF model generates larger spurious currents with larger mesh density (as described by Equation (31)). It is also worth pointing out that the evolution of spurious currents for the time step constraints provide marginally higher spurious currents for CSF model (0.1% using the time step constraint by Equation (1)) but the Smoothed CSF and SSF models show a spurious current reduction by nearly 10% and 11%, respectively (see Table 15). Based on the evolution of spurious currents based on time step constraint, the SSF model generates the least spurious current when compared to Smoothed CSF and CSF models.

Table 14. Comparison of spurious currents based on mesh and surface tension models (using no under-relaxation and time step dictated by Deshpande et al. [63]).

Surface Tension Model	Mesh	$\overline{U_{sc}}$	$Ca = \frac{\rho_1 \nu_1 \overline{U_{sc}}}{\sigma}$	$\overline{\Delta p_c}$	$\overline{E}(\Delta p_c)$	Mass Imbalance
	M0	0.158	0.002	22.27	−0.20	0
CSF	M1	0.279	0.004	23.09	−0.18	0
	M2	0.510	0.007	24.34	−0.13	0
	M3	0.723	0.010	24.47	−0.13	0
	M0	0.154	0.002	24.02	−0.14	0
Smoothed CSF	M1	0.122	0.002	24.97	−0.11	0
	M2	0.104	0.001	25.21	−0.10	0
	M3	0.075	0.001	26.03	−0.07	0
	M0	0.042	0.001	24.07	−0.14	0
SSF	M1	0.065	0.001	24.86	−0.11	0
	M2	0.033	0.000	26.04	−0.07	0
	M3	0.036	0.001	25.64	−0.08	0

Table 15. Comparison of spurious currents for the time stepping constraints based on M3 mesh and surface tension models (while providing no under-relaxation to the flow variables).

Surface Tension Model	$\overline{U_{sc}}$ based on Brackbill et al. [66]	$\overline{U_{sc}}$ based on Deshpande et al. [63]
CSF	0.722930	0.723390
Smoothed CSF	0.082800	0.075458
SSF	0.040016	0.035903

6. Conclusions

In the study, we successfully implemented CSF, Smoothed CSF and SSF models in OpenFOAM and compared them based on their ability to simulate a two-dimensional stationary bubble, rising bubbles and capillary rise. The flow scenarios modelled corresponds to a variety of capillary numbers (in the order of 10^{-3}, 0.1 and 1), which is relevant in various industrial processes. The numerical simulations show that:

- For a stationary bubble with a 2.5 mm radius, CSF and SSF models generate the most and least amount of spurious currents, respectively. For the finest mesh used, Smoothed CSF and SSF models reduce spurious currents by nearly one-tenth and one-twentieth of the CSF model (when no under-relaxation factor is used), respectively. When using a lower under-relaxation factor (for the finest mesh), Smoothed CSF and SSF models reduce the spurious currents by approximately one-fourth of the CSF model.

- The time step constraints proposed by Brackbill et al. [66] and Deshpande et al. [63] show that spurious currents generated by the CSF is significantly reduced while using a lower under-relaxation factor. In Smoothed CSF and SSF models, when using the same under-relaxation factor, the time step constraint slightly reduces the spurious currents by 6–7%. Interestingly, when no under-relaxation is used, the CSF model generates marginally larger (nearly 0.1%) spurious currents with the time step constraint proposed by Deshpande et al. [63], but other models show a reduction in spurious current by less than 10%.
- The Laplace pressure in the bubbles predicted by Smoothed CSF and SSF is more accurate with an error of 7–9% for the higher mesh densities than CSF model with negligible imbalance in mass of the phases.
- Although using a lower under-relaxation factor reduces the spurious currents and predicts the Laplace pressure in the stationary bubble (for all the surface tension models considered) quite reasonably, it can adversely effect the accuracy of dynamic cases such as rising bubbles by underestimating the flow variables.
- Using a higher mesh density results in larger spurious currents for CSF model but they are reduced for both Smoothed CSF and SSF models for the static case considered.
- The effect of mesh resolution was studied only for the stationary bubble in this work. For the case of rising bubbles, previous works [36,55], using the CSF model, reported challenges in achieving a mesh independent solution. Similarly, for capillary rise using the CSF model, Yamamoto et al. [36] reported an increasing error when using a finer mesh. The meshes used in this paper correspond to the finest grid (used in FreeLIFE solver) implemented by Hysing et al. [54] and the grid that provided a most accurate model for capillary rise in the work by Yamamoto et al. [36]. We expect similar effects of mesh resolution for both Smoothed CSF and SSF models for dynamic cases, as they are variants of the same formulation. The quantification of these errors will be treated in future work.
- Rising bubbles were successfully modelled using the surface tension models and validated based on the final bubble shape and rising velocities proposed by Hysing et al. [54] and Klostermann et al. [55].
- Modelling the capillary rise with SSF was shown to provide a more accurate representation than the CSF model. Interestingly, the Smoothed CSF could not reliably simulate capillary rise due to spurious currents.

Although the surface tension models considered in this study did not eliminate spurious currents entirely, the comparison provides insights into the limitations of these models. Based on the simulations done in this study, the SSF model seems to provide a versatile surface tension formulation that generates small spurious currents and provides a more accurate representation of various processes in comparison to the standard CSF model.

Supplementary Materials: The following are available online at https://www.mdpi.com/2227-9717/7/8/542/s1.

Author Contributions: K.J.V. built the models, ran the simulations, post-processed and analysed the results, and wrote the manuscript. K.E.E. contributed contributed in supervising, reviewing of the results and revising manuscript.

Funding: We would also like to thank the Department of Material Science and Engineering, NTNU, for funding this research.

Acknowledgments: The authors would like to thank the OpenFOAM community (both developers and contributors). We would also like to thank the reviewers whose comments improved the manuscript.

References

1. Basaran, O.A.; Gao, H.; Bhat, P.P. Nonstandard inkjets. *Annu. Rev. Fluid Mech.* **2013**, *45*, 85–113. [CrossRef]
2. Clark, I.C.; Abate, A.R. Microfluidic bead encapsulation above 20 kHz with triggered drop formation. *Lab Chip* **2018**, *18*, 3598–3605. [CrossRef] [PubMed]
3. Jomy Vachaparambil, K. An Analytical and Numerical Study of Droplet Formation and Break-off for Jetting of Dense Suspensions. Master's Thesis, KTH, School of Engineering Sciences (SCI), Mechanics, Stockholm, Sweden, 2016.
4. Valette, R.; Hachem, E.; Khalloufi, M.; Pereira, A.; Mackley, M.; Butler, S. The effect of viscosity, yield stress, and surface tension on the deformation and breakup profiles of fluid filaments stretched at very high velocities. *J. Non-Newton. Fluid Mech.* **2019**, *263*, 130–139. [CrossRef]
5. Høst-Madsen, J.; Jensen, K.H. Laboratory and numerical investigations of immiscible multiphase flow in soil. *J. Hydrol.* **1992**, *135*, 13–52. [CrossRef]
6. Muggeridge, A.; Cockin, A.; Webb, K.; Frampton, H.; Collins, I.; Moulds, T.; Salino, P. Recovery rates, enhanced oil recovery and technological limits. *Philos. Trans. Royal Soc. A Math. Phys. Eng. Sci.* **2014**, *372*, 20120320. [CrossRef] [PubMed]
7. Osborn, J.L.; Lutz, B.; Fu, E.; Kauffman, P.; Stevens, D.Y.; Yager, P. Microfluidics without pumps: reinventing the T-sensor and H-filter in paper networks. *Lab Chip* **2010**, *10*, 2659–2665. [CrossRef] [PubMed]
8. Brussieux, C.; Viers, P.; Roustan, H.; Rakib, M. Controlled electrochemical gas bubble release from electrodes entirely and partially covered with hydrophobic materials. *Electrochimica Acta* **2011**, *56*, 7194–7201. [CrossRef]
9. Guan, P.; Jia, L.; Yin, L.; Tan, Z. Effect of bubble contact diameter on bubble departure size in flow boiling. *Exp. Heat Transf.* **2016**, *29*, 37–52. [CrossRef]
10. Clift, R.; Grace, J.; Weber, M. *Bubbles, Drops, and Particles*; Dover Civil and Mechanical Engineering Series; Dover Publications: Mineola, NY, USA, 2005.
11. Vuorinen, V.A.; Hillamo, H.; Kaario, O.; Nuutinen, M.; Larmi, M.; Fuchs, L. Effect of droplet size and atomization on spray formation: A priori study using large-eddy simulation. *Flow Turbul. Combust.* **2011**, *86*, 533–561. [CrossRef]
12. Hilz, E.; Vermeer, A.W. Spray drift review: The extent to which a formulation can contribute to spray drift reduction. *Crop Prot.* **2013**, *44*, 75–83. [CrossRef]
13. Bouffard, J.; Kaster, M.; Dumont, H. Influence of process variable and physicochemical properties on the granulation mMechanism of mannitol in a fluid bed top spray granulator. *Drug Dev. Ind. Pharm.* **2005**, *31*, 923–933. [CrossRef] [PubMed]
14. Gallino, G.; Gallaire, F.; Lauga, E.; Michelin, S. Physics of Bubble-Propelled Microrockets. *Adv. Funct. Mater.* **2018**, *28*, 1800686. [CrossRef]
15. Vachaparambil, K.J.; Einarsrud, K.E. Analysis of bubble nucleation mechanisms in supersaturated solutions: A macroscopic perspective. *Meet. Abstr.* **2018**, *MA2018-01*, 1366.
16. Vachaparambil, K.J.; Einarsrud, K.E. Explanation of bubble nucleation mechanisms: A gradient theory approach. *J. Electrochem. Soc.* **2018**, *165*, E504–E512. [CrossRef]
17. Xu, W.; Lan, Z.; Peng, B.; Wen, R.; Ma, X. Heterogeneous nucleation capability of conical microstructures for water droplets. *RSC Adv.* **2015**, *5*, 812–818. [CrossRef]
18. Alexiadis, A.; Dudukovic, M.; Ramachandran, P.; Cornell, A.; Wanngård, J.; Bokkers, A. Liquid–gas flow patterns in a narrow electrochemical channel. *Chem. Eng. Sci.* **2011**, *66*, 2252–2260. [CrossRef]
19. Gupta, A.; Roy, S. Euler–Euler simulation of bubbly flow in a rectangular bubble column: Experimental validation with Radioactive Particle Tracking. *Chem. Eng. J.* **2013**, *225*, 818–836. [CrossRef]
20. Shams, E.; Finn, J.; Apte, S.V. A numerical scheme for Euler–Lagrange simulation of bubbly flows in complex systems. *Int. J. Numer. Methods Fluids* **2011**, *67*, 1865–1898. [CrossRef]
21. Legendre, D.; Zevenhoven, R. A numerical Euler–Lagrange method for bubble tower CO2 dissolution modeling. *Chem. Eng. Res. Des.* **2016**, *111*, 49–62. [CrossRef]
22. Tuković, Ž.; Jasak, H. A moving mesh finite volume interface tracking method for surface tension dominated interfacial fluid flow. *Comput. Fluids* **2012**, *55*, 70–84. [CrossRef]
23. Charin, A.; Tuković, Ž.; Jasak, H.; Silva, L.; Lage, P. A moving mesh interface tracking method for simulation of liquid–liquid systems. *J. Comput. Phys.* **2017**, *334*, 419–441. [CrossRef]

24. Menon, S.; Mooney, K.G.; Stapf, K.; Schmidt, D.P. Parallel adaptive simplical re-meshing for deforming domain CFD computations. *J. Comput. Phys.* **2015**, *298*, 62–78. [CrossRef]
25. Hirt, C.; Nichols, B. Volume of fluid (VOF) method for the dynamics of free boundaries. *J. Comput. Phys.* **1981**, *39*, 201–225. [CrossRef]
26. Chang, Y.; Hou, T.; Merriman, B.; Osher, S. A Level Set Formulation of Eulerian Interface Capturing Methods for Incompressible Fluid Flows. *J. Comput. Phys.* **1996**, *124*, 449–464. [CrossRef]
27. Jacqmin, D. Calculation of Two-Phase Navier–Stokes Flows Using Phase-Field Modeling. *J. Comput. Phys.* **1999**, *155*, 96–127. [CrossRef]
28. Peskin, C.S. Flow patterns around heart valves: A numerical method. *J. Comput. Phys.* **1972**, *10*, 252–271. [CrossRef]
29. Pivello, M.; Villar, M.; Serfaty, R.; Roma, A.; Silveira-Neto, A. A fully adaptive front tracking method for the simulation of two phase flows. *Int. J. Multiph. Flow* **2014**, *58*, 72–82. [CrossRef]
30. Cubos-Ramírez, J.; Ramírez-Cruz, J.; Salinas-Vázquez, M.; Vicente-Rodríguez, W.; Martinez-Espinosa, E.; Lagarza-Cortes, C. Efficient two-phase mass-conserving level set method for simulation of incompressible turbulent free surface flows with large density ratio. *Comput. Fluids* **2016**, *136*, 212–227. [CrossRef]
31. Yue, P.; Zhou, C.; Feng, J.J. Spontaneous shrinkage of drops and mass conservation in phase-field simulations. *J. Comput. Phys.* **2007**, *223*, 1–9. [CrossRef]
32. Samkhaniani, N.; Ansari, M. Numerical simulation of bubble condensation using CF-VOF. *Prog. Nucl. Energy* **2016**, *89*, 120–131. [CrossRef]
33. Baltussen, M.; Kuipers, J.; Deen, N. A critical comparison of surface tension models for the volume of fluid method. *Chem. Eng. Sci.* **2014**, *109*, 65–74. [CrossRef]
34. Patel, H.; Kuipers, J.; Peters, E. Computing interface curvature from volume fractions: A hybrid approach. *Comput. Fluids* **2018**, *161*, 74–88. [CrossRef]
35. Jafari, A.; Shirani, E.; Ashgriz, N. An improved three-dimensional model for interface pressure calculations in free-surface flows. *Int. J. Comput. Fluid Dyn.* **2007**, *21*, 87–97. [CrossRef]
36. Yamamoto, T.; Okano, Y.; Dost, S. Validation of the S-CLSVOF method with the density-scaled balanced continuum surface force model in multiphase systems coupled with thermocapillary flows. *Int. J. Numer. Methods Fluids* **2017**, *83*, 223–244. [CrossRef]
37. Gerlach, D.; Tomar, G.; Biswas, G.; Durst, F. Comparison of volume-of-fluid methods for surface tension-dominant two-phase flows. *Int. J. Heat Mass Transf.* **2006**, *49*, 740–754. [CrossRef]
38. Yokoi, K. A practical numerical framework for free surface flows based on CLSVOF method, multi-moment methods and density-scaled CSF model: Numerical simulations of droplet splashing. *J. Comput. Phys.* **2013**, *232*, 252–271. [CrossRef]
39. Yokoi, K. A density-scaled continuum surface force model within a balanced force formulation. *J. Comput. Phys.* **2014**, *278*, 221–228. [CrossRef]
40. Ningegowda, B.; Premachandran, B. A coupled level set and volume of fluid method with multi-directional advection algorithms for two-phase flows with and without phase change. *Int. J. Heat Mass Transf.* **2014**, *79*, 532–550. [CrossRef]
41. Albert, C.; Raach, H.; Bothe, D. Influence of surface tension models on the hydrodynamics of wavy laminar falling films in volume of fluid-simulations. *Int. J. Multiph. Flow* **2012**, *43*, 66–71. [CrossRef]
42. Pavuluri, S.; Maes, J.; Doster, F. Spontaneous imbibition in a microchannel: analytical solution and assessment of volume of fluid formulations. *Microfluid. Nanofluid.* **2018**, *22*, 90. [CrossRef]
43. Gumulya, M.; Utikar, R.P.; Pareek, V.; Mead-Hunter, R.; Mitra, S.; Evans, G.M. Evaporation of a droplet on a heated spherical particle. *Chem. Eng. J.* **2015**, *278*, 309–319. [CrossRef]
44. Gumulya, M.; Joshi, J.B.; Utikar, R.P.; Evans, G.M.; Pareek, V. Bubbles in viscous liquids: Time dependent behaviour and wake characteristics. *Chem. Eng. Sci.* **2016**, *144*, 298–309. [CrossRef]
45. Li, D.; Duan, X. Numerical analysis of droplet impact and heat transfer on an inclined wet surface. *Int. J. Heat Mass Transf.* **2019**, *128*, 459–468. [CrossRef]
46. Sontti, S.G.; Atta, A. Formation characteristics of Taylor bubbles in power-law liquids flowing through a microfluidic co-flow device. *J. Ind. Eng. Chem.* **2018**, *65*, 82–94. [CrossRef]
47. Wang, Z.; Yang, J.; Koo, B.; Stern, F. A coupled level set and volume-of-fluid method for sharp interface simulation of plunging breaking waves. *Int. J. Multiph. Flow* **2009**, *35*, 227–246. [CrossRef]

48. Khismatullin, D.; Renardy, Y.; Renardy, M. Development and implementation of VOF-PROST for 3D viscoelastic liquid–liquid simulations. *J. Non-Newton. Fluid Mech.* **2006**, *140*, 120–131. [CrossRef]

49. Guo, Z.; Fletcher, D.F.; Haynes, B.S. Implementation of a height function method to alleviate spurious currents in CFD modelling of annular flow in microchannels. *Appl. Math. Model.* **2015**, *39*, 4665–4686. [CrossRef]

50. Fuster, D.; Agbaglah, G.; Josserand, C.; Popinet, S.; Zaleski, S. Numerical simulation of droplets, bubbles and waves: State of the art. *Fluid Dyn. Res.* **2009**, *41*, 065001. [CrossRef]

51. Aboukhedr, M.; Georgoulas, A.; Marengo, M.; Gavaises, M.; Vogiatzaki, K. Simulation of micro-flow dynamics at low capillary numbers using adaptive interface compression. *Comput. Fluids* **2018**, *165*, 13–32. [CrossRef]

52. Raeini, A.Q.; Blunt, M.J.; Bijeljic, B. Direct simulations of two-phase flow on micro-CT images of porous media and upscaling of pore-scale forces. *Adv. Water Resour.* **2014**, *74*, 116–126. [CrossRef]

53. Liu, Q.; Palm, B. Numerical study of bubbles rising and merging during convective boiling in micro-channels. *Appl. Therm. Eng.* **2016**, *99*, 1141–1151. [CrossRef]

54. Hysing, S.; Turek, S.; Kuzmin, D.; Parolini, N.; Burman, E.; Ganesan, S.; Tobiska, L. Quantitative benchmark computations of two-dimensional bubble dynamics. *Int. J. Numer. Methods Fluids* **2009**, *60*, 1259–1288. [CrossRef]

55. Klostermann, J.; Schaake, K.; Schwarze, R. Numerical simulation of a single rising bubble by VOF with surface compression. *Int. J. Numer. Methods Fluids* **2013**, *71*, 960–982. [CrossRef]

56. Hoang, D.A.; van Steijn, V.; Portela, L.M.; Kreutzer, M.T.; Kleijn, C.R. Benchmark numerical simulations of segmented two-phase flows in microchannels using the Volume of Fluid method. *Comput. Fluids* **2013**, *86*, 28–36. [CrossRef]

57. Hardt, S.; Wondra, F. Evaporation model for interfacial flows based on a continuum-field representation of the source terms. *J. Comput. Phys.* **2008**, *227*, 5871–5895. [CrossRef]

58. Rusche, H. Computational Fluid Dynamics of Dispersed Two-Phase Flows at High Phase Fractions. Ph.D. Thesis, Imperial College London (University of London), London, UK, 2003.

59. Roenby, J.; Bredmose, H.; Jasak, H. A computational method for sharp interface advection. *Royal Soc. Open Sci.* **2016**, *3*, 160405. [CrossRef] [PubMed]

60. Scheufler, H.; Roenby, J. Accurate and efficient surface reconstruction from volume fraction data on general meshes. *J. Comput. Phys.* **2019**, *383*, 1–23. [CrossRef]

61. Cifani, P.; Michalek, W.; Priems, G.; Kuerten, J.; van der Geld, C.; Geurts, B. A comparison between the surface compression method and an interface reconstruction method for the VOF approach. *Comput. Fluids* **2016**, *136*, 421–435. [CrossRef]

62. Harvie, D.; Davidson, M.; Rudman, M. An analysis of parasitic current generation in volume of fluid simulations. *Appl. Math. Model.* **2006**, *30*, 1056–1066. [CrossRef]

63. Deshpande, S.S.; Anumolu, L.; Trujillo, M.F. Evaluating the performance of the two-phase flow solver interFoam. *Comput. Sci. Discov.* **2012**, *5*, 014016. [CrossRef]

64. Abadie, T.; Aubin, J.; Legendre, D. On the combined effects of surface tension force calculation and interface advection on spurious currents within Volume of Fluid and Level Set frameworks. *J. Comput. Phys.* **2015**, *297*, 611–636. [CrossRef]

65. Francois, M.M.; Cummins, S.J.; Dendy, E.D.; Kothe, D.B.; Sicilian, J.M.; Williams, M.W. A balanced-force algorithm for continuous and sharp interfacial surface tension models within a volume tracking framework. *J. Comput. Phys.* **2006**, *213*, 141–173. [CrossRef]

66. Brackbill, J.; Kothe, D.; Zemach, C. A continuum method for modeling surface tension. *J. Comput. Phys.* **1992**, *100*, 335–354. [CrossRef]

67. Ubbink, O. Numerical Prediction of Two Fluid Systems with Sharp Interfaces. Ph.D Thesis, Imperial College London (University of London), London, UK, 1997.

68. Heyns, J.A.; Oxtoby, O.F. Modelling surface tension dominated multiphase flows using the VOF approach. In Proceedings of 6th European Conference on Computational Fluid Dynamics, Barcelona, Spain, 20–25 July 2014.

69. Sussman, M.; Puckett, E.G. A coupled level set and volume-of-fluid method for computing 3D and axisymmetric incompressible two-phase flows. *J. Comput. Phys.* **2000**, *162*, 301–337. [CrossRef]

70. Albadawi, A.; Donoghue, D.; Robinson, A.; Murray, D.; Delauré, Y. Influence of surface tension implementation in volume of fluid and coupled volume of fluid with level set methods for bubble growth and detachment. *Int. J. Multiph. Flow* **2013**, *53*, 11–28. [CrossRef]

71. Raessi, M.; Mostaghimi, J.; Bussmann, M. A volume-of-fluid interfacial flow solver with advected normals. *Comput. Fluids* **2010**, *39*, 1401–1410. [CrossRef]

72. Renardy, Y.; Renardy, M. PROST: A Parabolic Reconstruction of Surface Tension for the Volume-of-Fluid Method. *J. Comput. Phys.* **2002**, *183*, 400–421. [CrossRef]

73. Youngs, D. Time-dependent multi-material flow with large fluid distortion. In *Numerical Methods for Fluid Dynamics*; Morton, K.W., Baines, M.J., Eds.; Academic Press: New York, USA, 1982, pp. 273–285.

74. Pilliod, J.E.; Puckett, E.G. Second-order accurate volume-of-fluid algorithms for tracking material interfaces. *J. Comput. Phys.* **2004**, *199*, 465–502. [CrossRef]

75. Popinet, S. An accurate adaptive solver for surface-tension-driven interfacial flows. *J. Comput. Phys.* **2009**, *228*, 5838–5866. [CrossRef]

76. Raeini, A.Q.; Blunt, M.J.; Bijeljic, B. Modelling two-phase flow in porous media at the pore scale using the volume-of-fluid method. *J. Comput. Phys.* **2012**, *231*, 5653–5668. [CrossRef]

77. Denner, F.; Evrard, F.; Serfaty, R.; van Wachem, B.G. Artificial viscosity model to mitigate numerical artefacts at fluid interfaces with surface tension. *Comput. Fluids* **2017**, *143*, 59–72. [CrossRef]

78. Lafaurie, B.; Nardone, C.; Scardovelli, R.; Zaleski, S.; Zanetti, G. Modelling merging and fragmentation in multiphase flows with SURFER. *J. Comput. Phys.* **1994**, *113*, 134–147. [CrossRef]

79. Galusinski, C.; Vigneaux, P. On stability condition for bifluid flows with surface tension: Application to microfluidics. *J. Comput. Phys.* **2008**, *227*, 6140–6164. [CrossRef]

80. Popinet, S. Numerical models of surface tension. *Annu. Rev. Fluid Mech.* **2018**, *50*, 49–75. [CrossRef]

81. Wang, Y.; Dandekar, R.; Bustos, N.; Poulain, S.; Bourouiba, L. Universal rim thickness in unsteady sheet fragmentation. *Phys. Rev. Lett.* **2018**, *120*, 204503. [CrossRef] [PubMed]

82. Saha, A.A.; Mitra, S.K. Effect of dynamic contact angle in a volume of fluid (VOF) model for a microfluidic capillary flow. *J. Colloid Interface Sci.* **2009**, *339*, 461–480. [CrossRef] [PubMed]

83. Issa, R. Solution of the implicitly discretised fluid flow equations by operator-splitting. *J. Comput. Phys.* **1986**, *62*, 40–65. [CrossRef]

84. Greenshields, C.J. *OpenFOAM User Guide*; Version 6; The OpenFOAM Foundation: London, UK.

85. Van Leer, B. Towards the ultimate conservative difference scheme. II. Monotonicity and conservation combined in a second-order scheme. *J. Comput. Phys.* **1974**, *14*, 361–370. [CrossRef]

86. Maki, K. Ship Resistance Simulations with OpenFOAM. In Proceedings of the 6th OpenFOAM Workshop: The Pennsylvania State University, State College, PA, USA, 13–16 June 2011.

87. Greenshields, C. OpenFOAM 2.3.0: Multiphase Modelling. 2014. Available online: https://openfoam.org/release/2-3-0/multiphase/ (accessed on 12 August 2019).

88. Song, B.; Liu, G.R.; Lam, K.Y.; Amano, R.S. On a higher-order bounded discretization scheme. *Int. J. Numer. Methods Fluids* **2000**, *32*, 881–897. [CrossRef]

89. Min, T.; Yoo, J.Y.; Choi, H. Effect of spatial discretization schemes on numerical solutions of viscoelastic fluid flows. *J. Non-Newton. Fluid Mech.* **2001**, *100*, 27–47. [CrossRef]

90. Bullard, J.W.; Garboczi, E.J. Capillary rise between planar surfaces. *Phys. Rev. E* **2009**, *79*, 011604. [CrossRef] [PubMed]

 *processes*

Correction

Correction: Vachaparambil, K.J. Comparison of Surface Tension Models for the Volume of Fluid Method. *Processes* 2019, 7, 542

Kurian J. Vachaparambil * and Kristian Etienne Einarsrud *

Department of Materials Science and Engineering, Norwegian University of Science and Technology (NTNU), 7491 Trondheim, Norway
* Correspondence: kurian.j.vachaparambil@ntnu.no (K.J.V.); kristian.e.einarsrud@ntnu.no (K.E.E.)

Received: 21 January 2020; Accepted: 22 January 2020; Published: 25 January 2020

Corrections:

In Equations (2) and (3), τ_μ and τ_ρ should be defined as $\mu_{avg}\Delta x/\sigma$ and $\sqrt{\rho_{avg}(\Delta x)^3/\sigma}$, respectively.

In Equation (9), $\vec{n}_f$ is the unit normal vector to the interface and $\vec{S}_f$ is the face surface area.

In Table 8 and Table 10, the kinematic viscosity of gas or phase 2 should be equal to 1.48×10^{-5} m^2/s, as provided in the simulation case files available in the Supplementary Material.

The results reported in [1] are not affected by these typographical errors.

Conflicts of Interest: The authors declare no conflict of interest.

Reference

1. Vachaparambil, K.J.; Einarsrud, K.E. Comparison of Surface Tension Models for the Volume of Fluid Method. *Processes* **2019**, *7*, 542. [CrossRef]

 *processes*

Article

Application of CFD to Analyze the Hydrodynamic Behaviour of a Bioreactor with a Double Impeller

Mohammadreza Ebrahimi [1,2], Melih Tamer [2], Ricardo Martinez Villegas [2], Andrew Chiappetta [2] and Farhad Ein-Mozaffari [1,*]

[1] Department of Chemical Engineering, Ryerson University, 350 Victoria Street, Toronto, ON M5B 2K3, Canada; mebrahimi@ryerson.ca

[2] Manufacturing Technology, Sanofi Pasteur Canada, 1755 Steeles Avenue West, North York, ON M2R 3T4, Canada; Melih.Tamer@sanofi.com (M.T.); Ricardo.Villegas@sanofi.com (R.M.V.); Andrew.Chiappetta@sanofi.com (A.C.)

* Correspondence: fmozaffa@ryerson.ca; Tel.: +1-416-979-5000 (ext. 4251)

Received: 18 August 2019; Accepted: 27 September 2019; Published: 3 October 2019

Abstract: Stirred bioreactors are commonly used unit operations in the pharmaceutical industry. In this study, computational fluid dynamics (CFD) was used in order to analyze the influence of the impeller configuration (Segment–Segment and Segment–Rushton impeller configurations) and the impeller rotational speed (an operational parameter) on the hydrodynamic behaviour and mixing performance of a bioreactor equipped with a double impeller. A relatively close agreement between the power values obtained from the CFD model and those measured experimentally was observed. Various parameters such as velocity profiles, stress generated by impellers due to the turbulence and velocity gradient, flow number, and mixing time were used to compare the CFD simulations. It was observed that the impeller's RPM could change the intensity of the interaction between the impellers when a Segment–Rushton impeller was used. In general, increasing the RPM led to an increase in total power and the stress acting on the cells and to a shorter mixing time. At a constant RPM, the Segment–Rushton impeller configuration had higher total power and stress acting on cells compared to the Segment–Segment impeller configuration. At lower RPM values (i.e., 50 and 100), the Segment–Segment impeller provided a shorter mixing time. Conversely, at the highest RPM (i.e., 150) the Segment–Rushton impeller had a shorter mixing time compared to the Segment–Segment impeller; this was attributed to the high level of turbulence generated with the former impeller configuration at high RPM.

Keywords: Stirred fermenter; dual-impeller; Segment impeller; CFD; Optimization

1. Introduction

Stirred fermenters (stirred bioreactors) are widely applied in the pharmaceutical industry to produce pharmaceutical compounds. The use of these unit operations has become the leading solution for production of microbial cells at an industrial scale [1]. Despite the common use of stirred fermenters, the understanding of mixing and hydrodynamic characteristics of these systems is still limited and the optimization of mixing conditions in this equipment has remained a challenging task. In general, a stirred fermenter needs to be designed and operated in order to achieve the following:

i. To achieve a uniform/homogenous mixing; the uniform mixing guarantees a homogeneous cell suspension and the concentration of nutrients and oxygen (if the process is aerobic) in the culture medium. This consequently increases the fermenter yield [2].

ii.   To minimize the mechanical and hydrodynamic stress intensity generated by impeller(s) on cells; high intensity of shear and normal stresses in a stirred fermenter is usually undesirable as it could damage the cells and change the cell morphology.

Hydrodynamic and mixing characteristics of stirred fermenters depend greatly on the geometrical parameters (e.g., impeller's type, size, and position as well as the number of impellers) and operating conditions (e.g., impeller's rotational speed, the medium composition, and microbial cell properties). Various types of impellers can be used in stirred bioreactors to enhance mixing homogeneity, to reduce mixing time and power consumption [3], and to control the shear conditions and hydrodynamic forces experienced by cells [4–6]. The type of impeller is decided upon based on the bioprocess application and mechanical properties of the cells. Conventionally, the Rushton impeller has been widely used in aerated stirred bioreactors due to its high turbulence and gas dispersion capabilities as well as its high gas–liquid mass transfer [4,6]. However, it is known that this impeller has a high power number: it creates a non-homogenous turbulence energy dissipation distribution (high values in the vicinity of the impeller and low values in the other zone of the mixing tank) as well as compartmentalization with axial flow barriers [6]. One type of impeller that has recently drawn attention in stirred bioreactors is referred to as a Segment or Elephant Ear (EE) (because of its blade shape). It typically has three wide blades, and its main characteristics are high solidity and low shear intensity along the impeller blade [7]. This impeller is particularly considered as a suitable candidate for cultivation of shear-sensitive microorganisms [4,8]. Venkat and Chalmers [9] characterized the flow pattern around this impeller using Particle Tracking Velocimetry experimental techniques. Zhu et al. [7] studied the performance of EE upward-pumping (EEU) and EE downward-pumping (EED) under both aerated and unaerated conditions experimentally. In the unaerated experiments, the power number of EED was larger than the power number of EEU, and the maximum values of turbulence kinetic energy (TKE) and flow numbers were similar for both EEU and EED. The authors concluded that upward-pumping mode showed some advantages over downward-pumping for aerated cases. Simmons et al. [10] compared the power and flow characteristics (flow pattern and mixing time) of an EE impeller with two other axial impellers, namely a six-blade pitched blade turbine (PBT) and a B2 hydrofoil operating in an upward-pumping mode and turbulent regime (300–650 RPM) experimentally. They reported that the power number and axial flow numbers for the EE impeller were higher than two other impellers. Particle image velocimetry (PIV) results showed that the global flow fields were similar for all impellers, and the TKE value produced by all impellers reached a similar value. The authors concluded that there was no conclusive proof that the EE impeller generated a lower shear rate than other studied impellers. Of note, their study was limited to single liquid phase flow. Collingnon et al. [8], however, reported that the EED impeller generated the lowest shear rate among all tested axial impellers when compared at the just-suspended speed. The authors also reported a high suspending capacity of the EE impeller (i.e., microcarrier full suspension was achieved at a very low rotational speed) compared to other studied axial impellers. Bustamante et al. [4] conducted a study in order to evaluate the shear rate generated by the Rushton impeller compared to an EE impeller in an aerated mixing tank for both Newtonian and non-Newtonian fluids. They observed that, for both types of fluids, the oxygen mass transfer coefficient ($k_L a$) ranges for the Rushton and the EEU impeller were similar and higher than those gained from the EED impeller. Furthermore, the EE impeller generated lower shear rate values compared to the systems equipped with the Rushton impeller. The authors also concluded that EEU with good oxygen mass transfer and low shear stress could be considered as an appropriate impeller for bioprocesses including shear-sensitive microorganisms.

The application of double or multiple impellers in bioreactors is prevalent in order to eliminate the dead zones, to improve mixing homogeneity, and to shorten mixing time [6,11]. Using double or multiple impellers in bioreactors, however, introduces more geometrical parameters to be optimized such as impeller type combination/configuration (all radial, all axial, or a combination of axial and radial impellers), identical or different impeller size, spacing between impellers, and number of impellers [12]. Depending on the spacing between impellers and the size of the impellers, the intensity of interaction between impellers can vary and different flow hydrodynamics can be generated. Rutherford et al. [11] classified the flow patterns in a stirred tank equipped with double Rushton impellers as parallel, merging, and diverging flows. The influence of spacing between impellers on the flow pattern of a mixing tank equipped with double-axial impellers was also discussed by Hari-Prajitno et al. [13]. It has been pointed out in literature that, when there is no significant interaction between impellers, the fluid flow generated by each impeller in a multiple-impeller system is similar to the fluid flow generated by that impeller in a single-impeller system [12].

Various experimental techniques such as Electrical Resistance Tomography (ERT) [14], Laser Doppler Velocimetry (LDV) [3], and Particle Image Velocimetry (PIV) [7,10,15] have been used in order to investigate the performance of stirred bioreactors. These experimental methods are extremely time consuming to fully characterize the fluid flow in bioreactors. Although these sophisticated experimental methods can provide a general understanding of the mixing characteristics in the stirred bioreactors, capturing critical fluid-flow data such as the local and temporal heterogeneity of turbulence, shear and normal stress distribution, and hydrodynamic forces acting on cells is challenging or even impossible with the current development of experimental systems. Furthermore, the application of experimental techniques to characterize large-scale bioreactors is not usually practical due to the lack of optical accessibility [16]. In addition, trial and error experiments can be costly, especially when the biological material is expensive and can result in producing a large amount of biological waste. Therefore, simulation methods have recently become increasingly popular in the investigation of stirred bioreactors. Simulation techniques can provide spatially and time-resolved information, which cannot be obtained conveniently through experiments. One of the numerical approaches that has been successfully used in simulation of bioreactors is referred to as Computational Fluid Dynamics (CFD). CFD has shown its capability in predicting the fluid flow and underlying phenomena happening in stirred bioreactors in several studies [16–19]. It should be noted that, despite the popularity of CFD in the simulation of biotechnological apparatus, it needs to be validated by experimental data as there are some assumptions and simplified models in CFD. Depending on the desirable parameters to study, numerical simulations can be performed for a single phase [20], two phases [21], or three phases [22]. Kaiser et al. [20] used single-phase CFD simulations in order to compare the flow pattern of the 2-litre single-use bioreactor and its reusable counterpart cell-culture bioreactors. The bioreactors were equipped with a double impeller (the lower impeller was a Rushton impeller and the upper impeller was a Segment impeller) operating in downward-pumping mode. The impellers had almost identical diameter, and the distance between them was 1.25 times the impeller's diameter. The CFD simulation results showed that the fluid velocity profiles and turbulence distributions were very similar in both bioreactors. In their simulations for both bioreactors, the Segment impeller showed a downward axial flow pattern and the Rushton impeller discharged the flow in the radial direction.

The objective of this study is to demonstrate the application of CFD in investigating the effect of the impeller configuration (a design parameter) and impeller rotational speed (an operational parameter) on the hydrodynamic behaviour and mixing performance of a stirred fermenter equipped with a double impeller. The hydrodynamic behaviour of a bioreactor equipped with the selected impeller configurations has rarely been investigated in literature. In this study, velocity profiles, stress generated by impellers due to the turbulence and velocity gradient, and mixing time were quantified to compare various CFD simulations.

2. Experimental Setup and Measurements

The stirred fermenter under investigation in this study was a 0.02-m^3 Sartorius fermenter available at the Sanofi Pasteur laboratory. This fermenter is used for Tetanus fermentations. Two different double-impeller configurations were used in the current study. One impeller configuration included a Rushton and a Segment impeller (i.e., Ruston was the lower impeller, and Segment was the upper impeller), and another impeller configuration had two Segment impellers. The Segment impellers were manufactured by bbi-biotech GmbH. In all experiments, the total liquid volume was 0.015 m^3, resulting in the initial liquid height to be 1.14 times the fermenter diameter ($L_H = 1.14\ T$). As in Tetanus fermentations, the density and viscosity of the medium is comparable with water. Therefore, room-temperature water was used in all experiments. The dimensions of the fermenter vessel and impellers are summarized in Table 1 and are shown in Figure 1a,b. In all experimental cases, the direction of impeller rotation was selected to generate an upward-pumping flow in the fermenters (i.e., clockwise as shown in Figure 1c,d). The submergence of the upper impeller in all cases was selected to guarantee that the vortex was not formed, and there was no entrainment of air for all operating conditions tested in this study.

Table 1. Dimensions of fermenter vessel and impellers.

Parameter	Value
T (m)	0.263
C (m)	0.088 ($T/3$)
D_R (m)	0.103 (~$T/3$)
D_S (m)	0.132 ($T/2$)
H (m)	$\frac{3}{4}\,D_S$
L_H (m)	$1.14T$
Shaft size (m)	0.018
Baffle width (m)	0.025 (~$T/10$)
Segment blade angle relative to the horizontal	30°

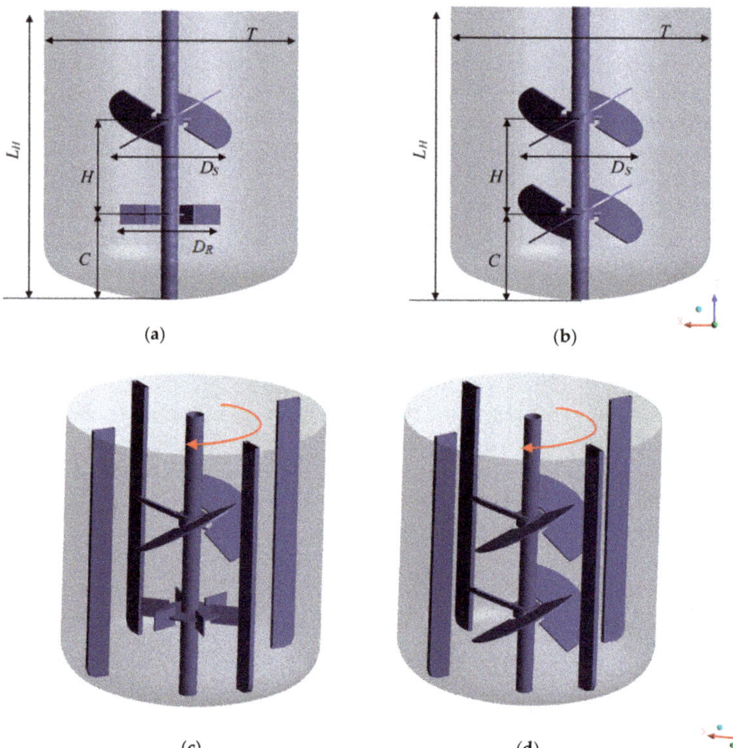

Figure 1. The fermenter geometries (front view) (**a**) with the Segment–Rushton impeller and (**b**) with the Segment–Segment impeller: the baffles were removed for clarity; (**c**,**d**) the fermenter geometries including baffles.

In the current study, six cases were considered to analyze the influence of impeller configuration and RPM on the hydrodynamic behaviour and mixing performance of a laboratory-scale fermenter. The experimental cases are summarized in Table 2.

Table 2. Experimental cases.

	RPM	Impeller Configuration
Case 1	50	
Case 2	100	Segment–Rushton
Case 3	150	
Case 4	50	
Case 5	100	Segment–Segment
Case 6	150	

The torque value for each case was measured experimentally by using a rotary sensor (manufactured by S. Himmelstein) attached to the shaft. The measured experimental torque values were used to calculate power values ($P = 2\pi MN$, where M is torque and N is impeller speed). The obtained power values were used in order to validate the CFD models. The experimental measurements were repeated three times to assure that measured torque values are reproducible.

3. Simulation Methodology and CFD Model Validation

In the current study, the Reynolds–Average Navier–Stokes (RANS) equations were solved to simulate the fluid flow in the stirred fermenter by ANSYS® FLUENT, Release 16.2 (ANSYS, Inc, Canonsburg, PA 15317, USA). The liquid was assumed to be a Newtonian fluid with properties similar to water.

$$\frac{\partial \rho}{\partial t} + \nabla \cdot (\rho \boldsymbol{u}) = 0 \tag{1}$$

$$\frac{\partial (\rho \boldsymbol{u})}{\partial t} + \nabla \cdot (\rho \boldsymbol{u} \otimes \boldsymbol{u}) = -\nabla p + \nabla \cdot \boldsymbol{\tau} + \rho \boldsymbol{g} + \nabla \cdot \boldsymbol{\sigma} \tag{2}$$

where $\rho, \boldsymbol{u}, p, \boldsymbol{g}, \boldsymbol{\tau}$, and $\boldsymbol{\sigma}$ are the fluid density, fluid average velocity, pressure, gravitational acceleration, viscose stress tensor, and Reynolds–Stress tensor, respectively. To simulate the motion of impellers, the Multiple Reference Frame (MRF) method was employed in CFD simulations.

As seen in Table 3, calculating the Reynolds number ($Re = (\rho ND_i^2)/\mu$) showed that the liquid flow inside the tank was in the turbulent regime in all simulation cases. Therefore, the k-ε turbulence model was used to describe the Reynolds stress tensor.

Table 3. *Re* values for different simulation cases.

	Case 1	Case 2	Case 3	Case 4	Case 5	Case 6
$Re(-) \times 10^4$	1.17	2.34	3.50	1.45	2.90	4.3

Other CFD settings and parameters selected in all simulation cases are summarized in Table 4. Six simulation cases similar to the experimental cases were performed in this study. All simulations were performed on High-Performance Computing Virtual Laboratory (HPCVL) Canada.

Initially, the influence of grid numbers on the CFD results (grid independence) was tested. Two rotating zones around the two impellers were defined with the finer mesh sizes. The grid size around the impellers was 1/3 of the grid size in other regions of the simulation domain. Three different simulation cases with 700,000, 1 million, and 1.3 million mesh elements (i.e., unstructured tetrahedral elements) were simulated to find the optimal mesh size (i.e., number of mesh elements). To obtain the desired mesh element sizes and numbers, the sizing function in the ANSYS meshing tool was used. The velocity magnitude on a line crossing the fermenter at the Segment impeller height ($Z = 0.187$ m) was used as a parameter to evaluate the mesh quality. The comparison between cases was performed when a steady-state condition was achieved. Tracking the values of torque obtained from simulations demonstrated that the torque value did not change considerably between 10 to 15 s of simulations. This confirmed that the steady-state condition was reached at around 10 s of simulations. As seen in Figure 2, the instantaneous velocity magnitude values obtained from simulations with 1 and 1.3 million mesh elements showed similar trends and were relatively close, with around a 10% difference in their averaged values. This small discrepancy between results might be tolerated for the sake of shorter computational time obtained from the simulation with 1 million mesh elements. The simulation with 700,000 mesh elements showed a different trend in comparison with other simulation cases. Comparing the velocity magnitude values obtained from simulations with 700,000 with those obtained from simulations with 1 and 1.3 million mesh elements showed relatively high differences in their values. Therefore, 1 million mesh elements were used in all simulations in the current study, as it provided a compromise between computational time and simulation accuracy.

Table 4. Computational fluid dynamics (CFD) simulation settings.

CFD Parameters	
Fluid density (kg/m^3)	1000
Fluid viscosity (Pa s)	0.001
Spatial discretization scheme	QUICK
Pressure–velocity coupling	SIMPLEC
Transient formulation	Second Order Implicit
Time step (s)	0.001

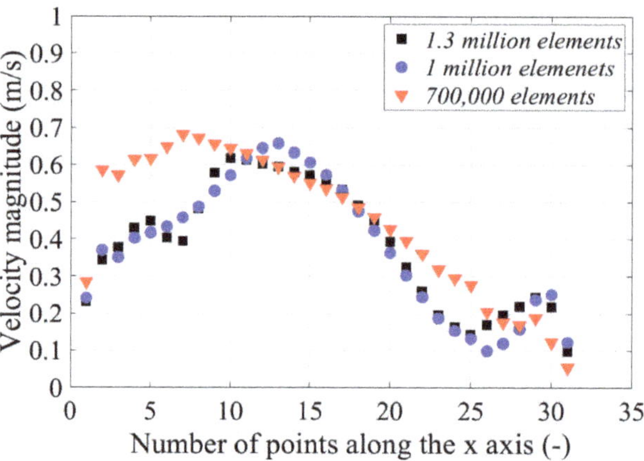

Figure 2. Comparison of instantaneous velocity magnitudes for simulations with different mesh sizes (at 15 s of simulation).

As mentioned previously, in this study, the CFD models were validated by comparing the power values ($P = 2\pi MN$) measured experimentally with power values obtained from the CFD models for various simulation cases as tabulated in Table 5.

As it can be seen, a relatively close agreement between the simulation results and experimental data is observed. This suggests that the CFD model with the selected input parameters and settings could replicate the phenomena happening in the experiments. The observed error in Table 5 can be mainly attributed to the selected mesh size. As previously mentioned, 1 million mesh elements were chosen as an optimal mesh element number as it provided relatively close results to the fine mesh size (1.3 million mesh elements) at a reasonable computational time. Using the finer mesh size in the CFD simulations may decrease the discrepancy between the simulations and experiments at the cost of higher computational time. In addition, in this study, the k-ε turbulence model (the most widely used turbulence model in literature) was used to simulate stirred bioreactors due to its numerical convenience [16,17,20,23,24]. However, as mentioned by Singh et al. [25] in the simulation of a mixing tank with Rushton impeller, the *k-ε* turbulence model may adversely affect the model accuracy. Aubin et al. [26] found that using a more sophisticated turbulence model such as Reynolds stress model in a CFD simulation of a stirred tank had a slight effect on turbulent kinetic energy and mean flow compared to the simulation case using the *k-ε* turbulence model. The authors also mentioned the CFD convergence difficulties when using the Reynolds stress model. Therefore, using more sophisticated turbulence models such as the Reynolds stress model may not help to reduce the error. More accurate modeling approaches such as direct numerical simulation (DNS) or large eddy simulations (LES) at a considerably higher computational time can be considered to reduce the observed error.

Table 5. Comparison between the experimental and CFD results.

	Power (W)		
	Experiment	**CFD**	**Error (%)**
Case 1	0.0436	0.0380	13.1
Case 2	0.279	0.306	9.46
Case 3	1.31	1.14	13.3
Case 4	0.0262	0.0234	10.5
Case 5	0.157	0.166	5.54

4. Results and Discussion

In this section the CFD simulation results are used in order to compare the hydrodynamic behaviour and mixing performance of various simulation cases outlined above.

4.1. Velocity Contours/Vectors and Profiles

The contours and vectors of fluid velocity can be obtained from the simulations. Figures 3 and 4 present the instantaneous liquid velocity contours and vectors on a XZ plane at Y = 0 (i.e., at 15 s of simulations). The following figures help to visualize the flow distribution inside the fermenter tank.

As it can be seen in Figure 3, at low RPM (Case 1), two impellers operated independently. As expected, the Rushton impeller acted as a radial and the Segment impeller acted as an axial impeller (Figure 3a,b). Four liquid circulation loops were observed around the Rushton impeller, and two circulation loops were formed by the Segment impeller. As the RPM was increased to 100 (Case 2), it seemed that the axial flow of the Segment impeller affected the fluid flow around the Rushton impeller and the radial discharge flow of the Rushton flow got distorted. The flow circulation loops around the Rushton impeller were not easily noticeable. As the RPM was further increased to 150 (Case 3), the interaction between the impellers increased considerably, and as seen in Figure 3e,f, the Rushton impeller did not operate as a radial impeller. The high axial velocity created by the Segment impeller impacted the fluid flow around the Rushton impeller and changed its hydrodynamic behaviour. As observed, two impellers interacted and two circulation loops were created in the fermenter tank by the impellers.

Figure 3 shows that not only the spacing between impellers can affect the impellers' interaction as presented previously in literature for a double impeller with two radial impellers [11] but also the impeller's RPM can change the intensity of the interaction between the impellers when radial and axial impellers are used in a double-impeller configuration.

Figure 4 demonstrates that, for simulation cases when two Segment impellers were employed (Cases 4–6), two impellers interacted and two circulation loops were created in the fermenter tank in all cases.

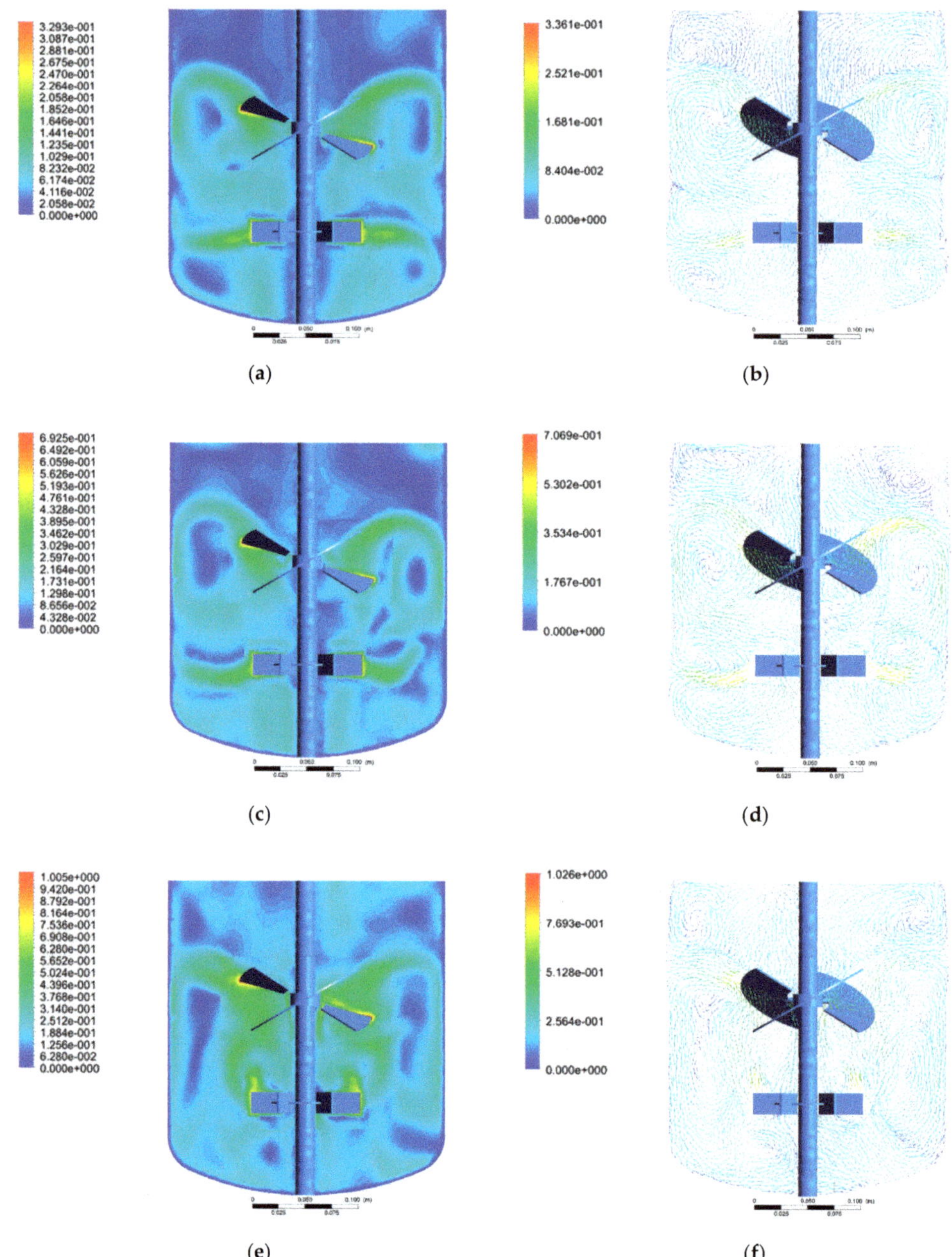

Figure 3. Velocity contour and velocity vector Case 1 (**a**,**b**), Case 2 (**c**,**d**), and Case 3 (**e**,**f**).

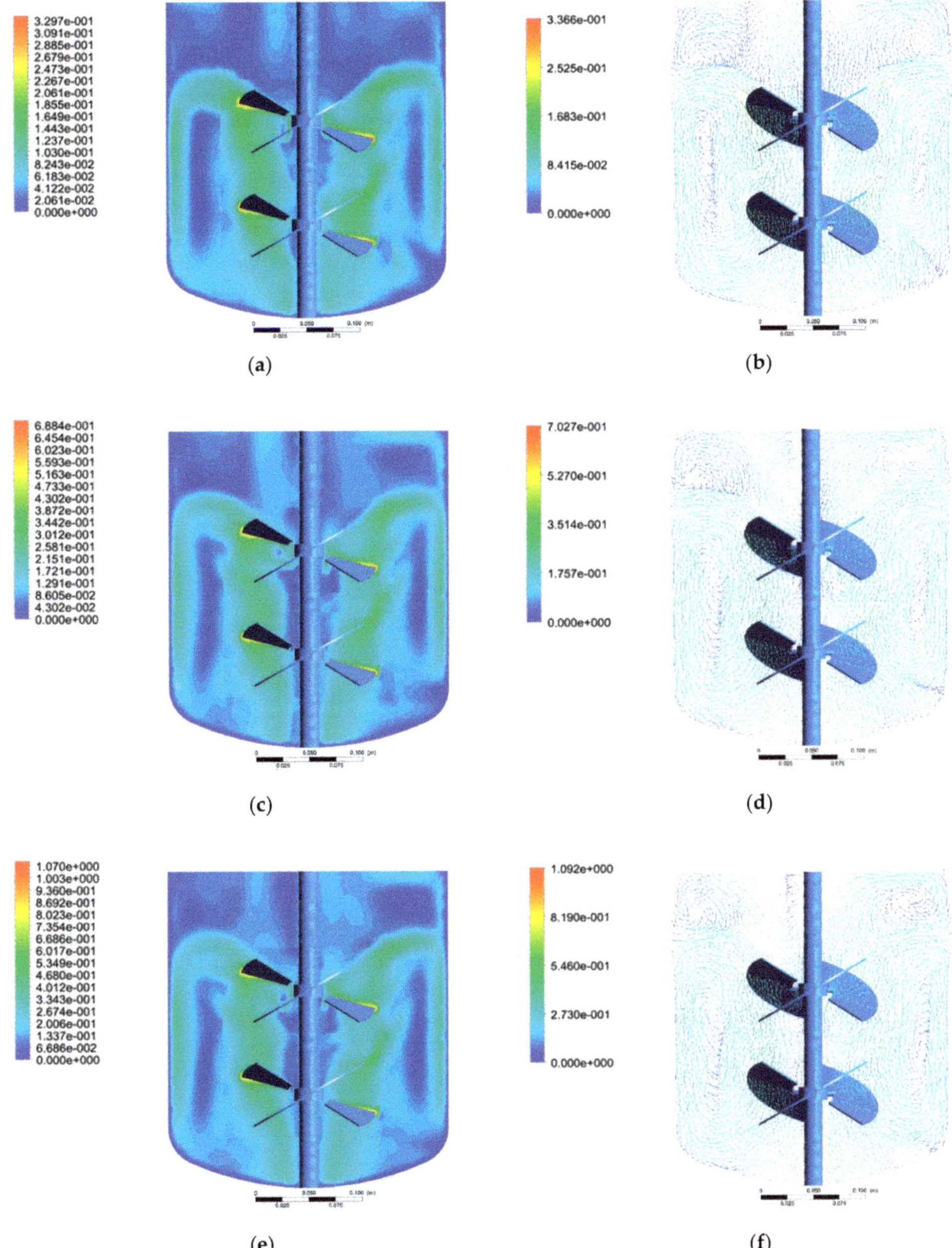

Figure 4. Velocity contour and velocity vector Case 4 (**a,b**), Case 5(**c,d**), and Case 6 (**e,f**).

To further analyze the abovementioned hydrodynamic flow patterns, the axial velocity values on a line between two impellers at $z/L_H \sim 0.45$ were extracted from all simulation cases. As seen in Figure 5a,b, when RPM increased the axial velocity, values increased for both impeller configurations. From Figure 5a, it is also observed that, for the simulation case, with the lowest RPM (50), the axial velocity had extremely small values, demonstrating a minimum interaction between impellers, and that two different fluid-flow compartments were formed by the impellers in the tank. For the simulation case with 150 RPM, however, the axial velocity increased considerably compared to other simulation cases due to the impact of the Segment impeller on the liquid flow around the Rushton impeller.

Comparing the axial velocity values of the two impeller configurations at a constant RPM (Figure 6) showed that, at lower RPM values (Figure 6a,b), the axial velocity values were considerably higher for the Segment–Segment impeller configuration compared to the Segment–Rushton impeller configuration and the axial velocity trends were also different. However, for the highest RPM value studied (150; Figure 6c), the axial velocity values and their trends obtained for both the Segment–Segment and Segment–Rushton impeller configurations became relatively similar. This again showed that the Segment impeller had a pronounced impact on the hydrodynamic behaviour around the Rushton impeller and consequently had a dominant influence on the hydrodynamic behaviour of the stirred bioreactor at a high RPM.

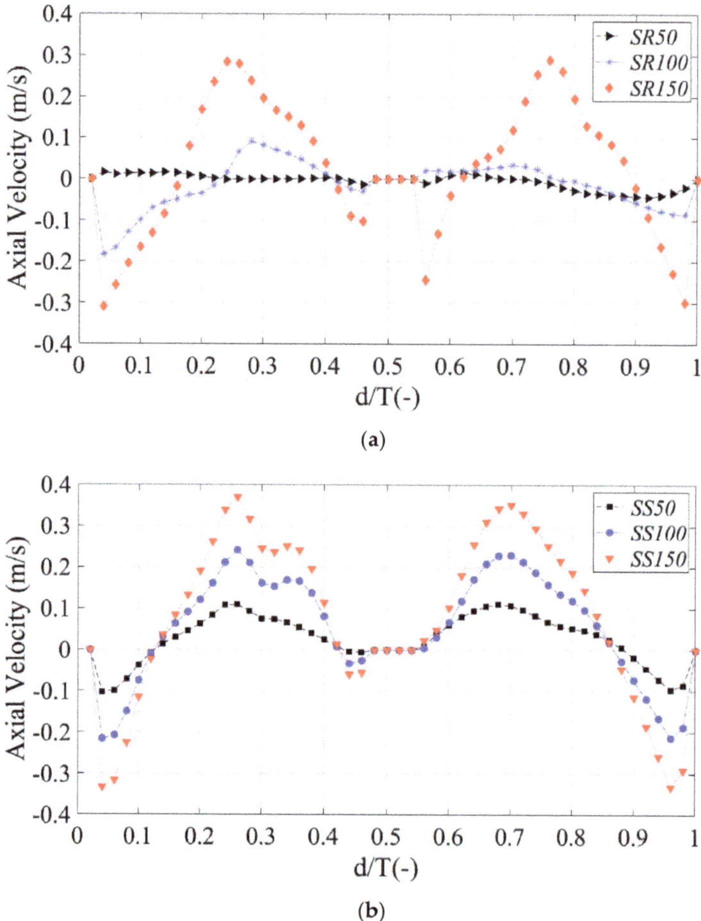

Figure 5. Effect of RPM on axial velocity profiles (**a**) with Segment–Rushton impeller configuration and (**b**) with Segment–Segment impeller configuration (SR and SS stand for Segment–Rushton and Segment–Segment impeller configurations, respectively): The following numbers show the applied RPM).

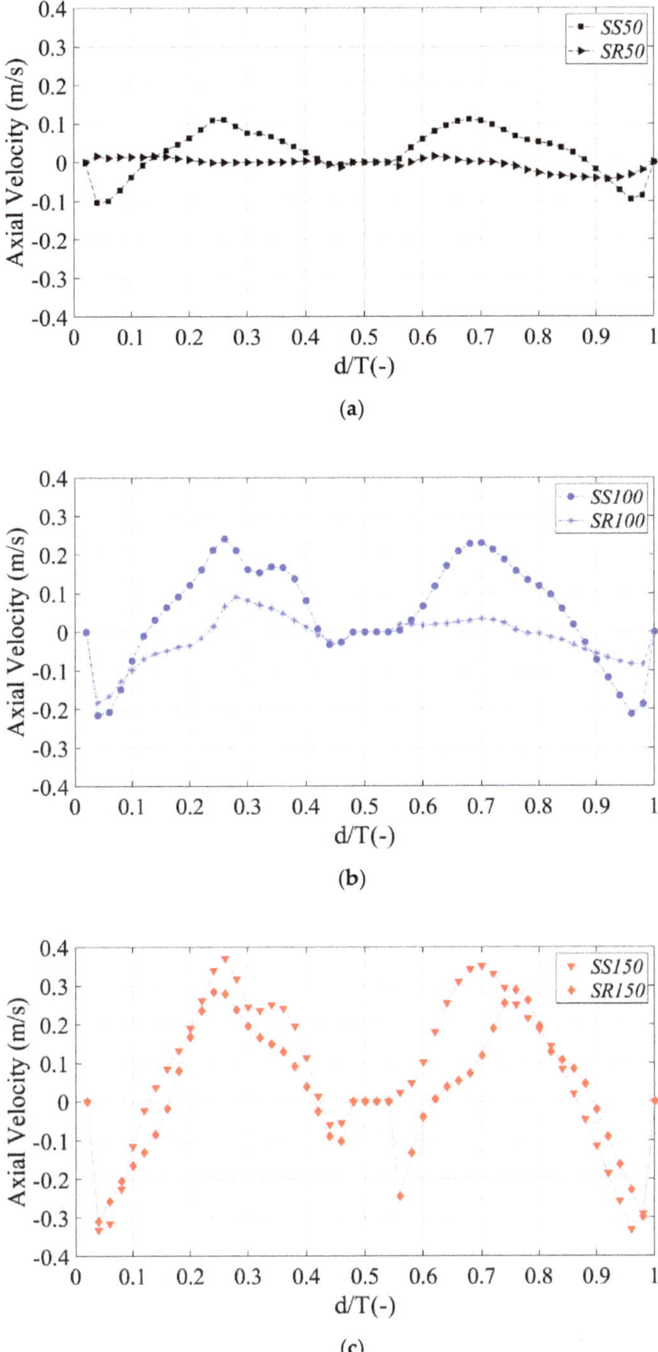

Figure 6. Effect of impeller configuration on axial velocity profiles: (**a**) 50 RPM, (**b**) 100 RPM, and (**c**) 150 RPM (SR and SS stand for Segment–Rushton and Segment–Segment impeller configurations, respectively).

4.2. Power and the Power Number

The torque value of each impeller was obtained from the CFD simulations. The power and the power number for each simulation case were then calculated based on the following equations:

$$P = 2\pi MN \tag{3}$$

$$N_p = \frac{P}{\rho N^3 D_i^5} \tag{4}$$

The power and power number values are presented in Table 6. As it is observed, the Rushton impellers had higher P and N_p values compared to the Segment impellers (i.e., Cases 1–3). It is also seen that, at a constant RPM (Case 1 versus Case 4, Case 2 versus Case 5, and Case 3 versus Case 6), the cases with the Segment–Segment impeller had considerably smaller P_{total} and $N_{p,total}$ compared to the cases with the Segment–Rushton impeller. Moreover, as expected, the higher the RPM, the higher the P_{total} value for both impeller configurations studied. The $N_{p,total}$ values obtained for simulation cases with a specific impeller configuration (when comparing Cases 1– 3 and when comparing Cases 4–6) were relatively equal. It is worth mentioning that the P and N_p values of the lower-Segment impeller were smaller than the P and N_p values of the upper-Segment impeller (i.e., Cases 4–6).

Table 6. Power and power numbers obtained from CFD results for different cases.

		P (W)	N_p	P_{total} (W)	$N_{p,total}$
Case 1	Segment	0.0158	0.680	0.0380	3.98
	Rushton	0.0222	3.30		
Case 2	Segment	0.132	0.712	0.306	3.94
	Rushton	0.174	3.23		
Case 3	Segment	0.527	0.841	1.14	4.20
	Rushton	0.609	3.36		
Case 4	Upper-Segment	0.0161	0.696	0.0234	1.01
	Lower-Segment	0.00729	0.315		
Case 5	Upper-Segment	0.120	0.648	0.166	0.894
	Lower-Segment	0.0456	0.246		
Case 6	Upper-Segment	0.409	0.653	0.547	0.873
	Lower-Segment	0.138	0.220		

4.3. Stress Analysis

The stress acting on cells in a fermenter can be attributed to two parameters, namely the fluid velocity gradient and turbulence [16]. The influence of the fluid-velocity gradient on cells can be related to the strain rate magnitude [16,20]:

$$\text{Strain rate } (D) = \frac{1}{2}\left[(\nabla \otimes u) + (\nabla \otimes u)^T\right] \tag{5}$$

The Kolmogorov length scale, $l_e = \left(\frac{\mu^3}{\rho^3 \varepsilon}\right)^{1/4}$, is commonly calculated as a critical value when the cell damage due to turbulence is studied. According to Odeleye et al. [15] and Nienow [27], if the cell size is smaller than the Kolmogorov length scale, then the cell would not be damaged due to turbulence. On the other hand, Liu et al. [24], Sorg et al. [28], and Tanzeglock et al. [29] stated that the local hydrodynamics within an eddy can impose stress on cells even if they are smaller than the

Kolmogorov length scale. Based on the relation between the size of the cell and the Kolmogorov length scale, the stress acting on cells due to turbulence was formulated as follows [24,28,29]:

$$\tau_t = \frac{5}{2}\mu\sqrt{\frac{\varepsilon}{6v}} \approx \mu\sqrt{\frac{\varepsilon}{v}}, \text{ if } d_{cell} < l_e \tag{6}$$

$$\tau_d = \rho(\varepsilon d_{cell})^{2/3}, \text{ if } d_{cell} > l_e \tag{7}$$

where ε and v represent the turbulence energy dissipation rate and kinematic viscosity, respectively.

In this study, the strain rate magnitudes were calculated for three zones in the fermenter tank: two zones in the vicinity of the impellers and one zone representing the rest of the fermenter tank. As it is seen in Table 7, Case 3 had the highest average strain rate magnitude compared to other cases. Case 4, on the other hand, produced the lowest average strain rate magnitude. At a constant RPM (Case 1 versus Case 4, Case 2 versus Case 5, and Case 3 versus Case 6), the values of average strain rate magnitudes obtained from the CFD simulations with the Segment–Rushton impeller were higher than the average values of strain rate magnitudes obtained from the CFD simulations with the Segment–Segment impeller. It is also observed that the Rushton zone had the highest values of strain rate magnitudes compared to other zones (Cases 1–3). This implies that the probability of a cell being damaged due to the fluid velocity gradient is higher than in the Rushton zone compared to other zones in the tank. Moreover, it can be seen in Table 7 that the higher the RPM selected, the larger the D value produced, regardless of the impeller configuration (when comparing Cases 1–3 and when comparing Cases 4–6). It should also be noted that the lower-Segment impeller generated smaller strain rate values compared to the upper-Segment impeller (Cases 4–6).

Table 7. Values of strain rate magnitude for different simulation cases.

		D (1/s)
Case 1	Rushton zone	18.1
	Segment zone	10.9
	Rest of the tank	6.10
	Average values	11.7
Case 2	Rushton zone	39.8
	Segment zone	21.0
	Rest of the tank	13.0
	Average values	24.6
Case 3	Rushton zone	70.0
	Segment zone	36.2
	Rest of the tank	18.2
	Average values	41.5
Case 4	Upper-Segment zone	10.5
	Lower-Segment zone	9.00
	Rest of the tank	5.10
	Average values	8.20
Case 5	Upper-Segment zone	19.8
	Lower-Segment zone	15.8
	Rest of the tank	9.50
	Average values	15.03
Case 6	Upper-Segment zone	29.2
	Lower-Segment zone	23.1
	Rest of the tank	14.6
	Average values	22.3

Similar to the strain rate calculations, the turbulence energy dissipation rate (ε) values, which are required to calculate both the Kolmogorov length scale and stress due to turbulence, were also calculated for three zones in the system. As reported in Table 8, in all cases, the Kolmogorov length scale values were in the range of 10^{-5} and 10^{-4} m. As the cell size in the Tetanus fermentation is in the range of 10^{-6} m, $d_{cell} < l_e$, and based on Equation (6), the τ_t values were calculated from the CFD results and are presented in Table 8. Generally, the τ_t values in the Segment zones were considerably lower than the τ_t values in the Rushton zones (Cases 1–3). This shows that the chance of a cell being damaged due to turbulence is higher in the Rushton zone compared to the Segment zone. As seen in Table 8, at a constant RPM (Case 1 versus Case 4, Case 2 versus Case 5, and Case 3 versus Case 6), the Segment–Segment impeller configuration generated lower average τ_t values compared to the Segment–Rushton impeller configuration. Based on these results, one can conclude that an increase in RPM led to an increase in the average τ_t value in both impeller configurations studied (when comparing Cases 1–3 and when comparing Cases 4–6). It should also be noted that the τ_t values generated by the lower-Segment impeller were smaller than the τ_t values generated by the upper-Segment impeller (Cases 4–6).

Table 8. Values of ε_{avg}, l_e, and τ_t for different simulation cases.

		ε_{avg} (W/kg) $\times 10^{-4}$	l_e (m) $\times 10^{-5}$	τ_t (Pa) $\times 10^{-2}$
Case 1	Rushton zone	46.6	12.1	6.83
	Segment zone	19.7	15.0	4.44
	Rest of the tank	9.70	17.9	3.11
	Average values	25.3	15.0	4.79
Case 2	Rushton zone	337	7.38	18.3
	Segment zone	124	9.48	11.1
	Rest of the tank	73.7	10.8	8.58
	Average values	178	9.22	12.7
Case 3	Rushton zone	1460	5.11	38.2
	Segment zone	606	6.37	24.6
	Rest of the tank	195	8.47	14.0
	Average values	754	6.65	25.6
Case 4	Upper-Segment zone	17.9	15.4	4.23
	Lower-Segment zone	13.1	16.6	3.62
	Rest of the tank	5.38	20.8	2.32
	Average values	12.1	17.6	3.39
Case 5	Upper-Segment zone	103	9.92	10.2
	Lower-Segment zone	72.4	10.8	8.51
	Rest of the tank	34.5	13.0	5.87
	Average values	70.0	11.3	8.18
Case 6	Upper-Segment zone	294	7.64	17.1
	Lower-Segment zone	212	8.28	14.6
	Rest of the tank	110	9.77	10.5
	Average values	205	8.56	14.1

To visualize the distribution of stress acting on cells due to turbulence, the contours of τ_t values on an XZ plane at Y = 0 (at 15 s of simulations) are presented in Figure 7. As illustrated, the Rushton impeller generated higher τ_t values compared to the Segment impeller. The spatial heterogeneity of stress is clearer for the cases using the Rushton–Segment impeller compared to the cases with the Segment–Segment impeller at a constant RPM (Case 1 versus Case 4, Case 2 versus Case 5, and Case 3 versus Case 6). It is also seen that, as RPM increased, the level of τ_t increased in the stirred fermenter regardless of the impeller configuration.

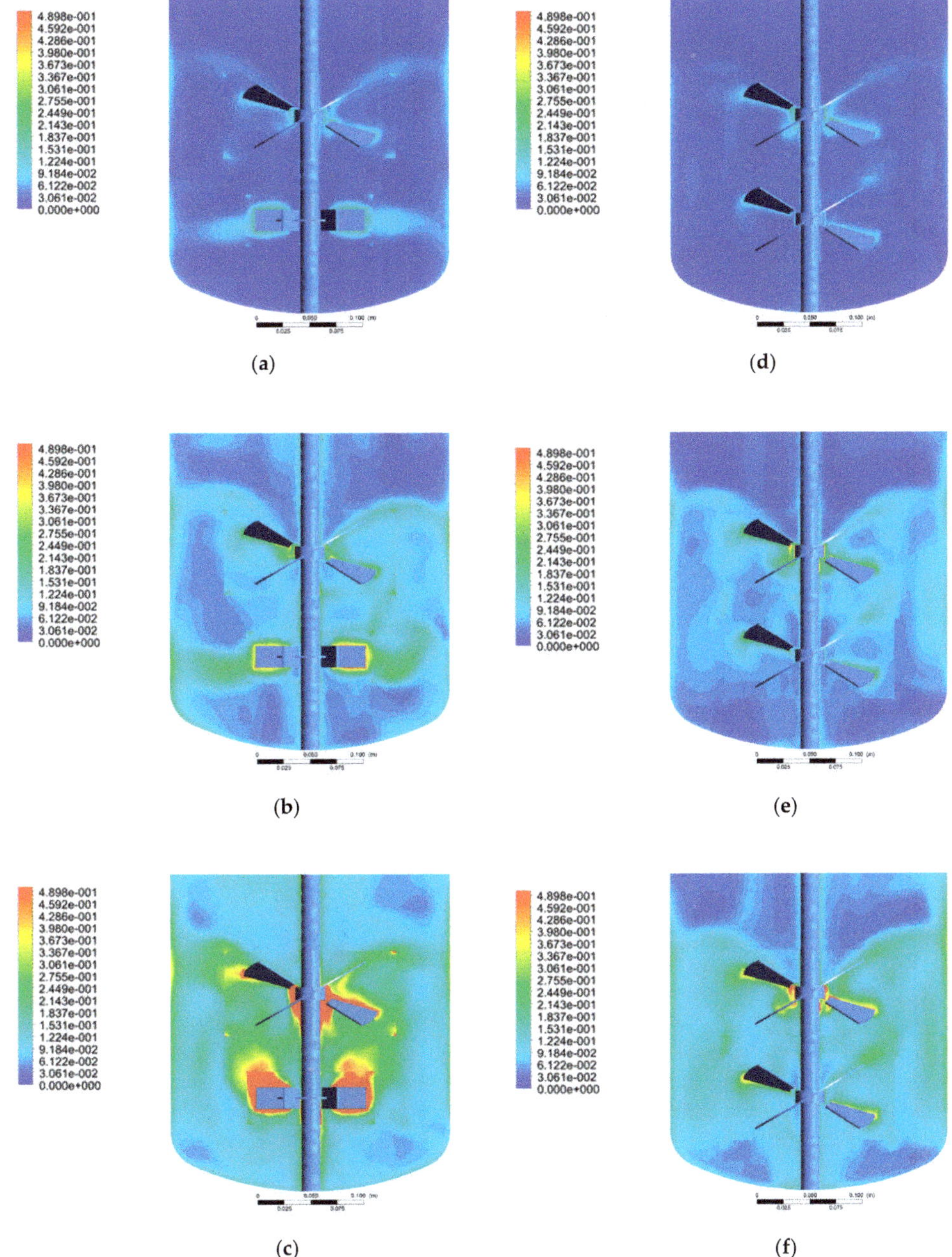

Figure 7. τ_t contours: (**a**) Case 1, (**b**) Case 2, (**c**) Case 3, (**d**) Case 4, (**e**) Case 5, and (**f**) Case 6.

Based on the results presented in this section, it can be concluded that Segment–Segment impeller configuration can be considered as a better candidate for cultivation of stress-sensitive microorganisms compared to the Segment–Rushton impeller.

4.4. Flow Number

The flow numbers have commonly been calculated in research studies to quantify the pumping capacity of impellers [3,12,20]. In this study, to further analyze the CFD simulation results, the radial flow number (Fl_r) and axial flow number (Fl_z) were calculated as follows [20]:

$$Fl_r = \frac{2\pi}{ND_i^3} \int_{z_1}^{z_2} rU_r(z)dz \tag{8}$$

$$Fl_z = \frac{2\pi}{ND_i^3} \int_{r=0}^{r=D_i/2} rU_z(r)dr \tag{9}$$

The Fl_r for each impeller was calculated by integrating the radial velocity on a cylinder around the impeller. The cylinder had a radius equivalent to the radius of the impeller. The length of the cylinder was from the lower edge of the impeller to the upper edge of the impeller. The Fl_z for each impeller was calculated by integrating the axial velocity on a circle located at the upper edge of the impeller. The radius of the circle was equivalent to the radius of the impeller. The flow numbers values are presented in Table 9.

Table 9. Flow numbers for different simulation cases.

		Fl_r	Fl_z	$Fl_s = Fl_r + Fl_z$ (for Each Impeller)	Fl_{total}
Case 1	Rushton	0.408	0.291	0.699	1.05
	Segment	0.0809	0.273	0.354	
Case 2	Rushton	0.425	0.264	0.689	0.964
	Segment	0.0299	0.245	0.275	
Case 3	Rushton	0.00668	0.408	0.415	0.835
	Segment	0.367	0.0516	0.419	
Case 4	Upper-Segment	0.367	0.133	0.501	1.32
	Lower-Segment	0.231	0.586	0.817	
Case 5	Upper-Segment	0.339	0.166	0.506	1.33
	Lower-Segment	0.236	0.548	0.820	
Case 6	Upper-Segment	0.353	0.150	0.503	1.32
	Lower-Segment	0.225	0.589	0.813	

As observed in Table 9, for the simulation cases with the Segment–Rushton impeller and low RPMs (Cases 1 and 2), the Rushton impeller had a higher Fl_r value compared to its Fl_z value and the Segment impeller had a higher Fl_z value compared to its Fl_r value, as expected. For the simulation case with the Segment–Rushton impeller and high RPM (Case 3), however, the interaction between impellers increased significantly and fluid-flow pattern in the vicinity of impellers changed; the Rushton impeller had a higher Fl_z value compared to its Fl_r value and the Segment impeller showed a higher Fl_r value compared to its Fl_z value.

The velocity contours and vectors of these simulation cases (Figure 3) show the abovementioned hydrodynamic shift. In Figure 3a–d, it is seen that both impellers operated relatively independently with a slight interaction. However, as the RPM increased to 150 (Figure 3e,f), the high axial velocity created by the Segment impeller drastically affected the fluid flow around the Rushton impeller and a high axial velocity was also seen close to the Rushton impeller. Therefore, in this case, the Rushton impeller did not act as a radial impeller. The Fl_{total} total value of Case 3 was quite different compared to the Fl_{total} total values of Cases 1 and 2. This difference can be attributed to the different hydrodynamic behaviour of Case 3 compared to Cases 1 and 2.

For the simulation cases with the Segment–Segment impeller (Cases 4–6), the lower-Segment impeller had a considerably higher Fl_z value compared to its Fl_r value and the upper-Segment impeller had a noticeably higher Fl_r value than its Fl_z value. The Fl_{total} values were almost equal for all these simulation cases regardless of the RPM value. It was also observed that, in these cases, the value of Fl_s obtained for the lower-Segment impeller was higher than the value of Fl_s obtained for the upper-Segment impeller.

Table 9 also shows that, at a constant RPM (Case 1 versus Case 4, Case 2 versus Case 5, and Case 3 versus Case 6), the simulation cases with the Segment–Segment impeller had higher Fl_{total} compared to the cases with the Segment–Rushton impeller.

4.5. Mixing Time

The capability of a stirred fermenter to efficiently blend the vessel content and to achieve the uniform/homogeneous mixing environment is commonly evaluated by mixing time [16]. In the current study, the mixing time (θ_m) was calculated by the injection of a tracer at the top of the fermenter vessel when the fully developed flow domain was reached. The distribution of the tracer inside the vessel was estimated by solving the time-dependent species transport equation [14,30]:

$$\frac{\partial}{\partial t}(\rho Y_i) + \nabla \cdot (\rho U Y_i) = -\nabla \cdot J_i \tag{10}$$

where Y_i represents the mass fraction of species i, J_i is the diffusion flux of species i in turbulent flows, and it is calculated based on the following equation:

$$J_i = -\left(\rho D_{i,m} + \frac{\mu_t}{Sc_t}\right)\nabla Y_i \tag{11}$$

where $D_{i,m}$ is the mass-diffusion coefficient, μ_t is the turbulent viscosity, and Sc_t is the turbulent Schmidt number. In all simulations, the value of $D_{i,m}$ was taken as 10^{-9} (m^2/s) [14] and Sc_t equals 0.7 (which is the default value in the CFD software) [20]. Five monitoring points (as shown by "+" in Figure 8) were defined in the vessel in order to track the changes of the tracer mass fraction during mixing. These monitoring points were selected in order to have a point above the upper impeller, a point below the lower impeller, two points at the impellers level, and a point between the impellers. This selection of monitoring points enabled to collect data regarding the tracer mass fraction in the entire tank height.

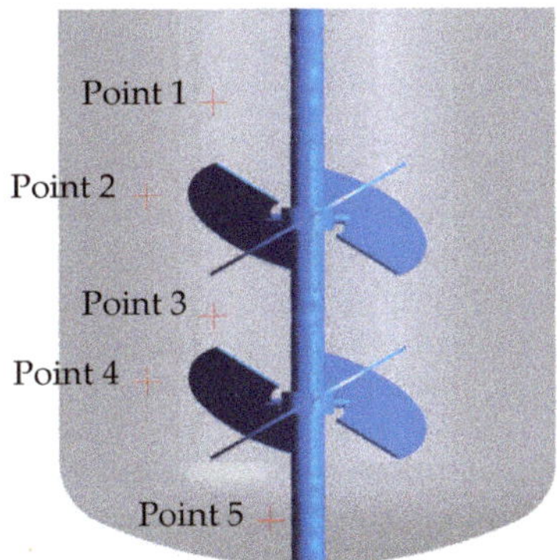

Figure 8. Location of monitoring points for the mixing-time calculation.

A typical tracer response curve for all monitoring points is seen in Figure 9. For the closest monitoring points to the injection point (Point 1), the tracer mass fraction increased rapidly at the beginning of the simulation and then decreased toward 1 as the mixing time progressed. The tracer mass fraction at other points increased gradually toward 1. The mixing time was calculated as the time period between the start of the injection of the tracer and the time when the mass fraction of the tracer at all monitoring points reached 100 ± 5% of the steady-state value [14].

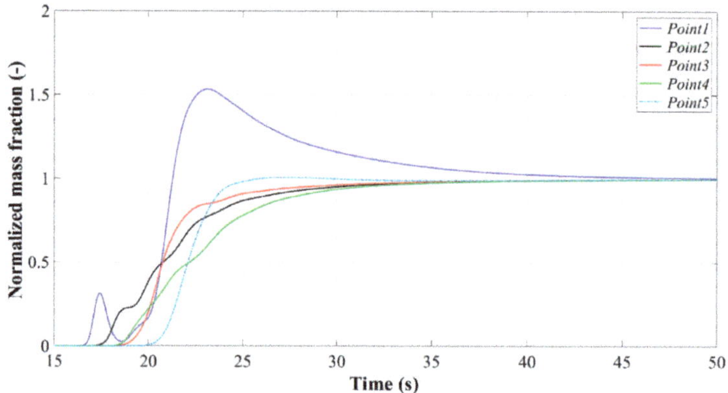

Figure 9. Tracer response curve for different points (Case 5 simulation).

The mixing time values are presented in Table 10. As expected, the mixing time decreased as RPM was increased, regardless of the impeller configuration (when comparing Cases 1–3 and when comparing Cases 4–6). It is also seen that, for simulation cases with low RPMs (Case 1 versus Case 4 and Case 2 versus Case 5), the mixing time obtained for the Segment–Segment impeller was considerably smaller than the mixing time obtained for the Segment–Rushton impeller. However, when a high RPM value was set in the simulations (i.e., when comparing Case 3 with Case 6), the simulation case with the Segment–Rushton impeller (Case 3) had a remarkably smaller mixing time than the simulation case with the Segment–Segment impeller (Case 6). Despite a low Fl_{total} obtained from Case 3, this simulation case showed the smallest mixing time, which can be attributed to the extremely high level of turbulence obtained in this case compared to other cases.

Table 10. Mixing time values for different simulation cases.

	Case 1	Case 2	Case 3	Case 4	Case 5	Case 6
$\theta_m(s)$	61.4	37.2	7.81	34.3	21.5	17.2

Calculating the energy consumption, which is the product of mixing time and total power consumption [12] (Table 11), also showed that, at low RPMs (Case 1 versus Case 4 and Case 2 versus Case 5), the simulation cases with the Segment–Segment impeller were more efficient (lower energy consumption) compared to the cases with the Segment–Rushton impeller. On the other hand, for a high RPM value (150), the simulation case with the Segment–Rushton impeller (Case 3) was slightly more efficient than the simulation case with the Segment–Segment impeller (Case 6).

Table 11. Energy consumption values for different simulation cases.

	Case 1	Case 2	Case 3	Case 4	Case 5	Case 6
$P_{total} \times \theta_m$ (J)	2.33	11.4	8.87	0.788	3.56	9.39

5. Conclusions

In this study, CFD simulations were applied to comprehensively investigate the influence of impeller configuration and RPM on the mixing performance and hydrodynamic behaviour of a laboratory-scale bioreactor operating in the turbulent regime. Two different impeller configurations (Segment–Segment and Segment–Rushton) and three different RPM values (50, 100, and 150) were used in the CFD simulations. To validate the CFD model, the power values obtained from the CFD models were compared with the experimentally measured power values. The power values were in close agreement. This indicated that the CFD models could represent the experimentations accurately. For simulations with the Segment–Rushton impeller configuration, it was observed that, at the lowest RPM value (50), two impellers operated independently with minimum interaction. The Rushton and Segment impellers discharged the flow in radial and axial directions, respectively. However, as RPM was increased, the intensity of the interaction between the impellers increased. At the highest RPM value, the high axial flow generated by the Segment impeller altered the fluid hydrodynamics around the Rushton impeller considerably and the Rushton impeller did not operate as a radial impeller. For simulations with the Segment–Segment impeller configuration, two impellers interacted and two circulation loops were created in the fermenter tank in all cases. CFD simulations demonstrate that, at a constant RPM, the Segment–Rushton impeller configuration had higher P_{total} and $N_{p,total}$ values, average strain rate magnitude, average τ_t values, and lower Fl_{total} values compared to the Segment–Segment impeller configuration. It then could be concluded that the Segment–Rushton impeller configuration may not be suitable for cultivation of shear-sensitive microorganisms. Calculating the mixing time and energy consumption showed that, at low RPM values (50 and 100), the Segment–Segment impeller configuration had a better performance. However, at the highest RPM, the Segment–Rushton impeller had a shorter mixing time and energy consumption, which could be attributed to the high turbulent flow generated by this type of impeller configuration. The CFD simulation results also showed that increasing the RPM increased the P_{total} values, average strain rate magnitude, and average τ_t values and shortened the mixing time regardless of the impeller configuration employed.

This study shows that CFD can be used as a valuable tool to obtain detailed information regarding the stirred bioreactors which might otherwise be challenging or impossible to obtain experimentally. With the rapid advancement in computational facilities, more accurate but computationally demanding simulation techniques such as LES and DNS can be used in future studies to simulate stirred bioreactors. These approaches can provide a more comprehensive understanding of the system.

Author Contributions: Conceptualization, M.E., M.T., and F.E.-M.; methodology, M.E., M.T., and F.E.-M.; software, M.E.; validation, M.E., R.M.V., and A.C.; formal analysis, M.E., M.T., R.M.V., A.C., and F.E.-M.; resources, F.E.-M., M.T., R.M.V., and A.C.; writing—original draft preparation, M.E.; writing—review and editing, F.E.-M. and M.T.; supervision, F.E.-M. and M.T.; funding acquisition, F.E.-M. and M.T.

Funding: The financial support provided through the Mitacs accelerate program is gratefully acknowledged.

Conflicts of Interest: The authors declare no conflict of interest.

Nomenclature

C (m)	*Clearance*	N_p (-)	*Power number*
D (s^{-1})	*Strain rate*	P (W)	*Power*
D_i (m)	*Impeller diameter*	Sc_t (-)	*Turbulent Schmidt number*
D_R (m)	*Rushton impeller diameter*	T (m)	*Tank internal diameter*
D_S (m)	*Segment impeller diameter*	U (m/s)	*Fluid velocity*
$D_{i,m}$ (m^2/s)	*Mass-diffusion coefficient*	U_r (m/s)	*Radial velocity*
d_{cell} (m)	*Cell diameter*	U_z (m/s)	*Axial velocity*
Fl_r (-)	*Radial flow number*	Y_i (-)	*Mass fraction of species i*
Fl_z (-)	*Axial flow number*	ε (W/kg)	*Turbulence dissipation rate*
H (m)	*Spacing between impellers*	ρ (kg/m^3)	*Fluid density*
J_i (kg/m^2 s)	*Diffusion flux of species i*	μ (Pa s)	*Dynamic viscosity*
L_H (m)	*Liquid height*	μ_t (Pa s)	*Turbulent viscosity*
l_e (m)	*Kolmogorov length scale*	ν (m^2/s)	*Kinematic viscosity*
M (N m)	*Torque*	$\tau_t,\ \tau_d$ (Pa)	*Stress acting on cell due to turbulence*
N (s^{-1})	*Impeller speed*	θ_m (s)	*Mixing time*

References

1. Amanullah, A.; Buckland, B.C.; Nienow, A.W. Mixing in the Fermentation and Cell Culture Industries. In *Handbook of Industrial Mixing*; John Wiley & Sons: Hoboken, NJ, USA, 2004; pp. 1071–1170.
2. Zhou, T.; Zhou, W.; Hu, W.; Zhong, J. Bioreactors, cell culture, commercial production. *Encycl. Ind. Biotechnol. Bioprocess Biosep. Cell Technol.* **2009**, 1–18. [CrossRef]
3. Patwardhan, A.W.; Joshi, J.B. Relation between Flow Pattern and Blending in Stirred Tanks. *Ind. Eng. Chem. Res.* **1999**, *38*, 3131–3143. [CrossRef]
4. Bustamante, M.; Cerri, M.; Badino, A.C. Comparison between average shear rates in conventional bioreactor with Rushton and Elephant ear impellers. *Chem. Eng. Sci.* **2013**, *90*, 92–100. [CrossRef]
5. Jüsten, P.; Paul, G.C.; Nienow, A.W.; Thomas, C.R. Dependence of mycelial morphology on impeller type and agitation intensity. *Biotechnol. Bioeng.* **1996**, *52*, 672–684. [CrossRef]
6. Vrábel, P.; Van Der Lans, R.G.; Luyben, K.C.; Boon, L.; Nienow, A.W. Mixing in large-scale vessels stirred with multiple radial or radial and axial up-pumping impellers: Modelling and measurements. *Chem. Eng. Sci.* **2000**, *55*, 5881–5896. [CrossRef]
7. Zhu, H.; Nienow, A.W.; Bujalski, W.; Simmons, M.J. Mixing studies in a model aerated bioreactor equipped with an up- or a down-pumping 'Elephant Ear' agitator: Power, hold-up and aerated flow field measurements. *Chem. Eng. Res. Des.* **2009**, *87*, 307–317. [CrossRef]
8. Collignon, M.-L.; Delafosse, A.; Crine, M.; Toye, D. Axial impeller selection for anchorage dependent animal cell culture in stirred bioreactors: Methodology based on the impeller comparison at just-suspended speed of rotation. *Chem. Eng. Sci.* **2010**, *65*, 5929–5941. [CrossRef]
9. Venkat, R.V.; Chalmers, J.J. Characterization of agitation environments in 250 mL spinner vessel, 3 L, and 20 L reactor vessels used for animal cell microcarrier culture. *Cytotechnology* **1996**, *22*, 95–102. [CrossRef]
10. Simmons, M.; Zhu, H.; Bujalski, W.; Hewitt, C.; Nienow, A. Mixing in a Model Bioreactor Using Agitators with a High Solidity Ratio and Deep Blades. *Chem. Eng. Res. Des.* **2007**, *85*, 551–559. [CrossRef]
11. Rutherford, K.; Lee, K.C.; Mahmoudi, S.M.S.; Yianneskis, M. Hydrodynamic characteristics of dual Rushton impeller stirred vessels. *AIChE J.* **1996**, *42*, 332–346. [CrossRef]
12. Kazemzadeh, A.; Ein-Mozaffari, F.; Lohi, A.; Pakzad, L. Effect of Impeller Spacing on the Flow Field of Yield-Pseudoplastic Fluids Generated by a Coaxial Mixing System Composed of Two Central Impellers and an Anchor. *Chem. Eng. Commun.* **2017**, *204*, 453–466. [CrossRef]
13. Hari-Prajitno, D.; Mishra, V.P.; Takenaka, K.; Bujalski, W.; Nienow, A.W.; Mckemmie, J. Gas-liquid mixing studies with multiple up- and down-pumping hydrofoil impellers: Power characteristics and mixing time. *Can. J. Chem. Eng.* **1998**, *76*, 1056–1068. [CrossRef]
14. Kazemzadeh, A.; Ein-Mozaffari, F.; Lohi, A.; Pakzad, L. Investigation of hydrodynamic performances of coaxial mixers in agitation of yield-pseudoplasitc fluids: Single and double central impellers in combination with the anchor. *Chem. Eng. J.* **2016**, *294*, 417–430. [CrossRef]
15. Odeleye, A.; Marsh, D.; Osborne, M.; Lye, G.; Micheletti, M. On the fluid dynamics of a laboratory scale single-use stirred bioreactor. *Chem. Eng. Sci.* **2014**, *111*, 299–312. [CrossRef] [PubMed]
16. Werner, S.; Kaiser, S.C.; Kraume, M.; Eibl, D. Computational fluid dynamics as a modern tool for engineering characterization of bioreactors. *Pharm. Bioprocess.* **2014**, *2*, 85–99. [CrossRef]
17. Kelly, W.J. Using computational fluid dynamics to characterize and improve bioreactor performance. *Biotechnol. Appl. Biochem.* **2008**, *49*, 225. [CrossRef] [PubMed]
18. Delafosse, A.; Liné, A.; Morchain, J.; Guiraud, P. LES and URANS simulations of hydrodynamics in mixing tank: Comparison to PIV experiments. *Chem. Eng. Res. Des.* **2008**, *86*, 1322–1330. [CrossRef]
19. Haringa, C.; Mudde, R.F.; Noorman, H.J. From industrial fermentor to CFD-guided downscaling: What have we learned? *Biochem. Eng. J.* **2018**, *140*, 57–71. [CrossRef]
20. Kaiser, S.C.; Löffelholz, C.; Werner, S.; Eibl, D. CFD for Characterizing Standard and Single-Use Stirred Cell Culture Bioreactors. In *Computational Fluid Dynamics Technologies and Applications*; IntechOpen: London, UK, 2011.
21. Hashemi, N.; Ein-Mozaffari, F.; Upreti, S.R.; Hwang, D.K. Hydrodynamic characteristics of an aerated coaxial mixing vessel equipped with a pitched blade turbine and an anchor. *J. Chem. Technol. Biotechnol.* **2018**, *93*, 392–405. [CrossRef]

22. Azargoshasb, H.; Mousavi, S.M.; Jamialahmadi, O.; Shojaosadati, S.A.; Mousavi, S.B. Experiments and a three-phase computational fluid dynamics (CFD) simulation coupled with population balance equations of a stirred tank bioreactor for high cell density cultivation. *Can. J. Chem. Eng.* **2016**, *94*, 20–32. [CrossRef]
23. Panneerselvam, R.; Savithri, S.; Surender, G.D. CFD modeling of gas–liquid–solid mechanically agitated contactor. *Chem. Eng. Res. Des.* **2008**, *86*, 1331–1344. [CrossRef]
24. Liu, Y.; Wang, Z.-J.; Xia, J.-Y.; Haringa, C.; Liu, Y.-P.; Chu, J.; Zhuang, Y.-P.; Zhang, S.-L. Application of Euler Lagrange CFD for quantitative evaluating the effect of shear force on *Carthamus tinctorius* L. cell in a stirred tank bioreactor. *Biochem. Eng. J.* **2016**, *114*, 209–217. [CrossRef]
25. Singh, H.; Fletcher, D.F.; Nijdam, J.J. An assessment of different turbulence models for predicting flow in a baffled tank stirred with a Rushton turbine. *Chem. Eng. Sci.* **2011**, *66*, 5976–5988. [CrossRef]
26. Aubin, J.; Fletcher, D.F.; Xuereb, C. Modeling turbulent flow in stirred tanks with CFD: The influence of the modeling approach, turbulence model and numerical scheme. *Exp. Therm. Fluid Sci.* **2004**, *28*, 431–445. [CrossRef]
27. Nienow, A.W. Reactor Engineering in Large Scale Animal Cell Culture. *Cytotechnology* **2006**, *50*, 9–33. [CrossRef]
28. Sorg, R.; Tanzeglock, T.; Périlleux, A.; Solacroup, T.; Broly, H.; Soos, M.; Morbidelli, M. Minimizing hydrodynamic stress in mammalian cell culture through the lobed Taylor-Couette bioreactor. *Biotechnol. J.* **2011**, *6*, 1504–1515. [CrossRef]
29. Tanzeglock, T.; Šoóš, M.; Stephanopoulos, G.; Morbidelli, M. Induction of mammalian cell death by simple shear and extensional flows. *Biotechnol. Bioeng.* **2009**, *104*, 360–370. [CrossRef]
30. *Fluent, 12.0 Theory Guide*; Ansys Inc.: Canonsburg, PA, USA, 2009.

 *processes*

Article

Numerical Simulation of Combustion in 35 t/h Industrial Pulverized Coal Furnace with Burners Arranged on Front Wall

Jiade Han [1], Lingbo Zhu [1], Yiping Lu [1,*], Yu Mu [1], Azeem Mustafa [2] and Yajun Ge [1]

[1] Department of Mechanical and Power Engineering, Harbin University of Science and Technology, Harbin 150080, China; jdhan@hrbust.edu.cn (J.H.); z1543536222@163.com (L.Z.); muyu1239@163.com (Y.M.); ge_yajun@163.com (Y.G.)

[2] School of Energy Science and Engineering, Harbin Institute of Technology, Harbin 150001, China; azeem.mustafa6@gmail.com

* Correspondence: luyiping@hrbust.edu.cn

Received: 1 August 2020; Accepted: 2 October 2020; Published: 10 October 2020

Abstract: Coal-fired industrial boilers should operate across a wide range of loads and with a higher reduction of pollutant emission in China. In order to achieve these tasks, a physical model including two swirling burners on the front wall and boiler furnace was established for a 35 *t/h* pulverized coal-fired boiler. Based on Computational Fluid Dynamics (CFD) theory and the commercial software ANSYS Fluent, mathematical modeling was used to simulate the flow and combustion processes under 75% and 60% load operating conditions. The combustion characteristics in the furnace were obtained. The flue gas temperature simulation results were in good agreement with experimental data. The simulation results showed that there was a critical distance L along the direction of the furnace depth (x) and Hc along the direction of the furnace height (y) on the burner axis. When $x < L$, the concentration of NO decreased sharply as the height increased. When $y < Hc$, the NO concentration decreased sharply with an increase in the y coordinate, while increasing dramatically with an area-weighted average gas temperature increase in the swirl combustion zone. This study provides a basis for optimizing the operation of nitrogen-reducing combustion and the improvement of burner structures.

Keywords: swirling burner; combustion characteristics; CFD; industrial pulverized coal furnace

---

1. Introduction

At present, with the increasing awareness of environmental protection, the pollutant emissions from coal-fired power stations, industrial boilers and industrial furnaces are widely concerning. When this fuel-burning equipment is used for industrial activities, it is crucial to better understand the combustion characteristics inside the furnace. The distribution characteristics of pollutant emissions (e.g., NO emissions), mechanism of NO formation, and emission reduction technology have become some of the hottest issues worldwide [1,2]. There has been increasing cooperation among countries to improve the performance of coal-fired boilers and reduce emissions [3,4]. Combustion is a complex phenomenon with flow–heat–mass transfer and chemical reactions. With the development of powerful computer hardware and advanced numerical techniques, simulation methods have been adopted to solve the problem of actual boiler combustion. More accurate and reliable results can be obtained [5,6]. Therefore, this kind of numerical simulation can provide theoretical guidance for the modification of burner structure and boiler operation.

In recent years, various forms of burners and graded air supplying methods have been developed in various countries to decrease nitrogen emissions from combustion flames using numerical techniques together with experimental methods. For example, Hwang et al. [7] performed a full-scale Computational Fluid Dynamics (CFD) simulation of the combustion characteristics of their designed burner and an air staging system for an 870 MW pulverized boiler for low rank coal, and the results of the simulation agreed well with measurements obtained under normal operating conditions, e.g., temperature. Noor et al. [8] studied the influence of primary airflow on coal particle size and coal flow distribution in coal-fired boilers. Silva-Daindrusiak et al. [9], on the other hand, used numerical methods to simulate the three-dimensional flow field in the combustion chamber and heat exchanger of a 160 MW pulverized coal furnace with the objective of identifying factors of inefficiency. The code was built and combined together with the commercial software ANSYS Fluent. Constenla et al. [10] predicted the flow of a reactive gas mixture with pulverized coal combustion occurring in a tangentially fired furnace under actual operating conditions, and shared the experience of ensuring the stable convergence of the equations. Similar to the work mentioned above [10], Srdjan Beloševi et al. [11] simulated the emission of nitrogen oxides and sulfides in a lignite boiler, performed with an in-house developed numerical code. Wang et al. [12] developed a low-volatile coal combustion system with four corners for the burner and two tangential air distribution systems on the four walls, and with the main combustion nozzle arranged on the sidewall near the center of the flame. On this basis, the factors that influence the NO_X generation and the flow field during multi-scale fractional combustion were studied. Sung et al. [13] investigated the combustion performance and NO_X emissions of nontraditional ring-fired furnaces, including models with an additional inner water wall, based on a traditional 500 MW tangentially fired furnace, using commercial CFD code.

Many researchers have contributed to the study of mechanisms to reduce pollutant production. For example, Lisandy et al. [14] established drop-tube furnace (DTF) modeling using one-dimensional numerical simulation for the rapid, good-accuracy prediction of the output gas and unburned carbon (UBC) concentrations. The results showed that the existence of a high CO gas concentration would assist in suppressing the NO formation during the combustion of coal char particles of pulverized coal. The models were crucial for accurate predicted simulation results. Zhang et al. [15] proposed a char combustion order reduction model, successfully implemented into the CFD software, Fluent, and applied to a parallel CFD simulation of a 600 MW boiler. For the efficient utilization of variable quality fuels, a three-dimensional numerical model and experiments were used by Shen et al. [16] to simulate the flow and combustion of binary coal blends under simplified blast furnace conditions, and they found that the chemical interactions between two components in terms of particle temperature and volatile content are responsible for the synergistic effect.

In summary, it can be seen that most of the published studies have carried out experiments using air depth classification technology that changes the burner structure and fractional rate of air supply. Furthermore, numerical simulations of the characteristics of the flow field and combustion processes, as well as reaction mechanism, model and pollutant emission prediction research both have been carried out. However, there have been few studies on the combustion situation of industrial pulverized coal boilers at different loads.

The main objective of the present paper is to present a CFD numerical method for studying the combustion performance in 35 t/h industrial pulverized coal boilers with two swirl burners arranged on the wall. The fluid flow and temperature field of the flue gas near the burner and the distribution characteristics of the combustion products (i.e., NO, O_2, CO, CO_2 and unburned carbon concentrations) were investigated at two loads to provide basic data for optimizing the combustion operation for nitrogen reduction in industry boilers.

2. Physical Model and Meshing

2.1. Physical Model

In the present work, we studied the combustion characteristics of a 35 t/h pulverized coal boiler with two swirl burners on the front wall. The combustion characteristics were closely related to the boiler structure of the chamber, burner and air staging system designed, the operating conditions, and other factors. The burner used by the boiler was mainly composed of an ignition device, central tube, primary air (PA) duct, internal secondary air (SA) duct and external SA duct. Figure 1 shows that the central air (CA) duct was arranged in the PA duct, which was equipped with a conical nozzle. As Figure 1b shows, there were 12 fixed guide vanes uniformly arranged along the circumferential direction in the inner and outer SA ducts, respectively, to achieve the air staging supply of SA of the inner and outer swirls. The CA can effectively regulate the center location of the combustion flame; moreover, it is also a key factor affecting flashback characteristics. An air staging system is also helpful for adjusting the radial velocity profile of the burner exit.

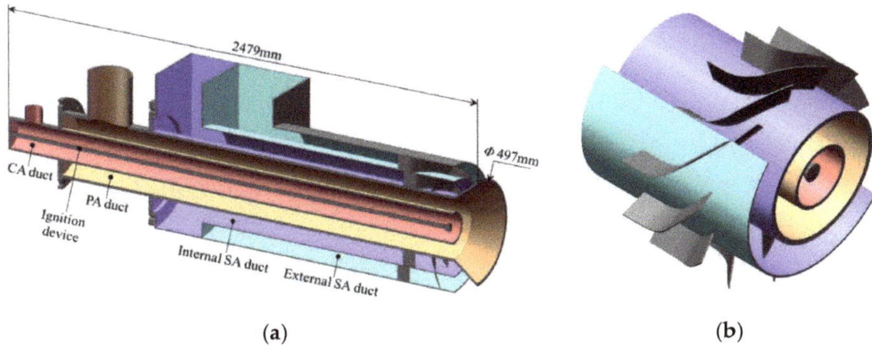

Figure 1. Schematic diagram of swirl burner. (**a**) Structure of swirl burner; (**b**) Vane geometry.

The furnace height, width and depth were 13.115, 4.24 and 4.37 m, respectively. The nozzles of the two burners were arranged on the water-cooling wall surface of the front wall, the horizontal spacing of the two burner central axes was 1.7 m, and the distance from the burner central axis to the furnace bottom was 4.35 m, as shown in Figure 2a. Figure 2b shows a local view of the part connecting the burners with the furnace front wall, and we can see how the SA was discharged into the furnace through the fixed vanes. To undertake the simulation, the commercial software SOLIDWORKS was used to establish the overall computational domain of the two swirling burners and furnace.

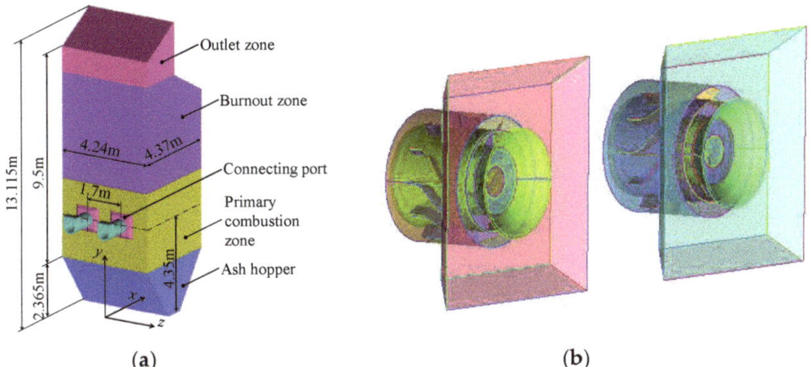

Figure 2. Computational domain. (**a**) Computational domain; (**b**) Local view of the part connecting the burners with the furnace front wall.

2.2. Meshing

A structured mesh was generated by using the commercial software ICEM, and after several trial calculations, the corresponding mesh was refined for the burner and the primary combustion zone with a large magnitude of the gradient of physical quantities. An initial mesh with about 1.40 million cells was first created in the computational domain. The number of mesh cells was then increased to 2.3, 4.0 million, respectively. The Grid Convergence Index (GCI) was used to quantify the grid independence [17]. The GCI_{12} for fine and medium grids was 1.21%. The GCI_{23} for medium and coarse grids was 3.09%. The value of $GCI_{23}/(r^p GCI_{12})$ was 1.015, which was approximately 1 and indicates that the solutions were well within the asymptotic range of convergence. The mesh-independence test was verified by comparing the temperature of the monitoring point (y = 10.5 m, x = 0.21 m) simulated on three grids and calculated via Richardson extrapolation; see Figure 3. Finally, the numerical results showed that the number of mesh cells was approximately 2.3 million, as shown in Figure 4.

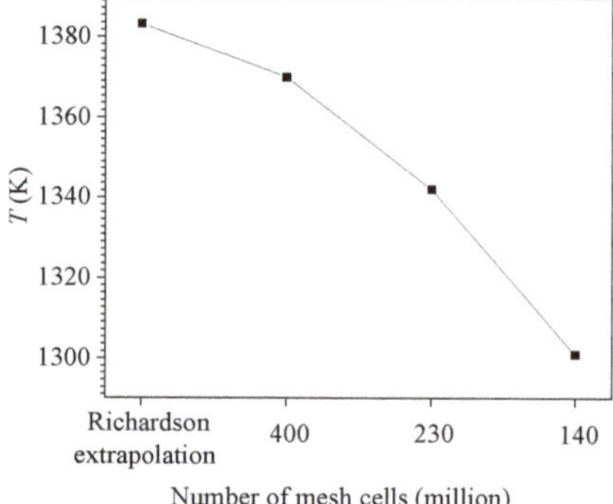

Figure 3. The temperature of a monitoring point simulated on three grids and calculated via Richardson extrapolation.

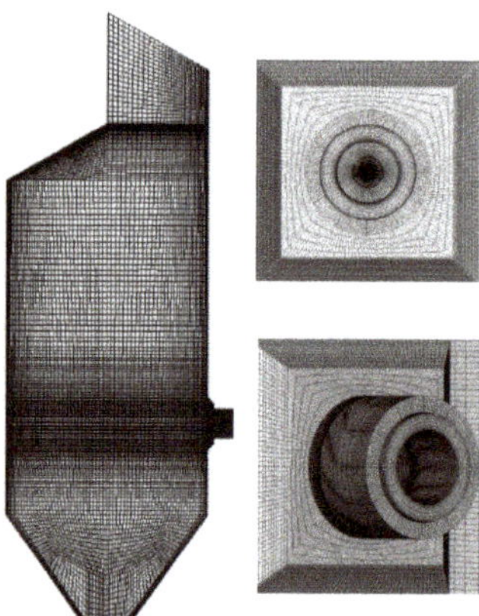

Figure 4. Meshing of computational domain.

3. Mathematical Model and Solution Conditions

In this study, the commercial software Fluent 16.0 was used to predict the behavior of reacting gases and the coal combustion in the furnace; the CFD parameters had been widely used and were referenced in many coal combustion simulation studies. Due to the inside of the furnace having a strongly swirling flow, the $k - \varepsilon$ realizable model seemed to be a suitable model for calculating the turbulent viscosity, μ_t and other parameters of gas turbulence [6,10]. During the combustion of coal particles, several processes occur, which was difficult to model; a non-premixed combustion model was used, and a stochastic tracking model was applied to analyze pulverized-coal flows, while calculations of gas/particle two-phase coupling employed the particle-source-in-cell method [18]. The release rate for the volatiles was expressed with the two competing rates model; Arrhenius equations are used for predicting this chemical reaction rate. The gas-phase turbulent combustion of volatiles was modeled by employing probability density function (PDF) theory; char combustion was expressed with a diffusion/kinetics model. The P-1 model was used to account for the radiative heat transfer effect. Detailed equations for mass, momentum, energy and species transport were provided by [6,19]. These models provided good approximate solutions for a full-scale boiler simulation.

Here, the specific models and equations in Fluent that were used, such as for the mass, momentum, energy, chemical species etc., are listed as follows:

$$\frac{\partial \rho}{\partial t} + \frac{\partial (\rho u_i)}{\partial x_i} = 0, \tag{1}$$

$$\frac{\partial \rho u_i}{\partial t} + \frac{\partial \rho u_i u_j}{\partial x_j} = \frac{\partial \tau_{ij}}{\partial x_j} - \frac{\partial p}{\partial x_i} + \rho f_i, \tag{2}$$

$$\frac{\partial \rho h}{\partial t} + \frac{\partial \rho u_i h}{\partial x_i} = \frac{\partial q_i^{res}}{\partial x_j} + \frac{\partial p}{\partial t} + u_i \frac{\partial p}{\partial x_i} + \tau_{ij} \frac{\partial p}{\partial t} + S_h, \tag{3}$$

$$\frac{\partial \rho m_l}{\partial t} + \frac{\partial (\rho m_l u_i + j_i)}{\partial x_i} = S_l, \tag{4}$$

$$\frac{\partial}{\partial t}(\rho k) + \frac{\partial}{\partial x_i}(\rho k u_i) = \frac{\partial}{\partial x_j}\left[\left(\mu + \frac{\mu_t}{\sigma_k}\right)\frac{\partial k}{\partial x_j}\right] + G_k + G_b - \rho \varepsilon - Y_M + S_k, \tag{5}$$

$$\frac{\partial}{\partial t}(\rho \varepsilon) + \frac{\partial}{\partial x_i}(\rho \varepsilon u_i) = \frac{\partial}{\partial x_j}\left[\left(\mu + \frac{\mu_t}{\sigma_\varepsilon}\right)\frac{\partial \varepsilon}{\partial x_j}\right] + \rho C_1 S_\varepsilon - \rho C_2 \frac{\varepsilon^2}{k + \sqrt{v\varepsilon}} + C_{1\varepsilon}\frac{\varepsilon}{k}C_{3\varepsilon}G_b + S_\varepsilon, \tag{6}$$

$$\frac{du_p}{dt} = F_D(u - u_p) + \frac{g_x(\rho_p - \rho)}{\rho_p} + F_x,$$

$$F_D = \frac{18\mu}{\rho_p d_p^2}\frac{C_D R_e}{24}, C_D = a_1 + \frac{a_2}{R_e} + \frac{a_3}{R_e}, \tag{7}$$

$$q_r = -\frac{1}{3(\alpha + \sigma_s) - C\sigma_s}\nabla G = \Gamma \nabla G,$$

$$\nabla(\Gamma \nabla G) - \alpha G + 4\pi\sigma T^4 = S_G, \Gamma = -\frac{1}{3(\alpha + \sigma_s) - C\sigma_s}, \tag{8}$$

$$k_n = A_n exp(-E/RT_p), n = 1, 2,$$

$$dV/dt = dV_1/d\tau + dV_2/d\tau = (\alpha_1 A_1 + \alpha_2 A_2)W, \tag{9}$$

$$\frac{dm_p}{dt} = -\pi d_p^2 p_{ox}\frac{D_0 R}{D_0 + R},$$

$$D_0 = C_1 \frac{(T_P - T_\infty)}{d_p}, R = C_2 e^{-(E/RT_p)}, \tag{10}$$

where $x_{i,j}$ = Cartesian coordinates; t = time; $u_{i,j}$ = the velocity in directions i and j, respectively; P = the pressure; ρ/ρ_p = the gas/partial density; $\tau_{i,j}$ = the viscous stress tensor; and f_i = the external force. In the law of conservation of energy formulation 3, h = the specific enthalpy, q_i^{res} is associated with energy transfer by the conduction and diffusion flux of matter, and S_h = a source of energy due to chemical reactions and radiative heat transfer [6]. m_l = the mass concentration of the components l, j_i = a weight average flow in the $_i$-th direction, S_l = components of source term l; k, ε = the turbulence kinetic energy and its rate of dissipation,; more details ae given in [19]. Only steady state was modeled and analyzed in this study.

The coal combustion simulation required that values be assigned to many input parameters. The coal properties are listed in Table 1. The particle size of the coal at each nozzle follows the Rosin–Rammler distribution; the particle size here was 75–100 μm. The test results showed that the end of the flame in the furnace was closer to the back wall at rated load, and the boiler mainly worked when the peak load was adjusted to under 75% or 60% for low-load operating conditions in the context of safety. Under different load conditions, the boundary conditions for each burner are listed in Table 2. The flame length of the flue gas in the furnace was relatively within the safe range, where the PA volume accounted for 20% of the total air volume, and the SA volume of both the internal and external air volume accounted for 40% of the total air volume under the current operating load, respectively.

Table 1. Coal properties.

Proximate Analysis (As Received) (wt.%)	
Moisture	3.40
Volatile matter	31.58
Fixed carbon	56.64
Ash	8.38
Ultimate Analysis (Dry Basis) (wt.%)	
C	71.71
H	4.30
N	1.01
O	10.90
S	0.30
Low heating value	27.95MJ/kg

Table 2. Settings of the burner inlet parameters.

Load (%)	Boiler Capacity (t/h)	Mass Flow Rates of Coal in a Single Burner (kg/s)	Temperature of PA Stream and Pulverized Coal (°C)	Temperature of Internal and External SA Streams (°C)
75	26.25	3.96	140	250
60	21	3.17	140	250

The water tubes in the furnace were simplified as walls; a "no-slip", "no-turbulence" and "no-mass flow" boundary condition was employed along the wall for the gas phase. Because the temperature difference of the wall was not large, the wall temperature could be assumed as uniform and was set as 254 °C, higher than the saturation temperature (204 °C) of the water and steam at the working pressure (16 MPa) of the boiler, because the temperature difference of the wall was small. The thermal radiation coefficient set for the wall regions was 0.6. The outlet boundary was the flue gas passage at the lateral wall near the top of the boiler, where the mean static pressure was −50 Pa.

The mass, pressure, momentum, chemical species and energy equations were each discretized using a second-order upwind scheme; a separation and implicit solution scheme based on a pressure solver was adopted. In solving algebraic equations, the SIMPLE algorithm for the pressure correction was applied to couple the velocity and pressure fields [20,21]. Many important scientific and engineering applications involving turbulent flows require the direct integration of the modeled turbulent equations to a surface boundary. Turbulent flows involve boundary layer separation or complex alterations of the surface transport properties [22]. The near-wall modeling significantly impacts the fidelity of numerical solutions, inasmuch as walls are the main source of mean velocity and turbulence. Here, the standard wall function approach was used for near-wall treatment; also, the local mesh refinement of the boundary layer near the wall surface was performed by adaption to make the dimensionless distance y^+ meet the standard wall function requirements ($y^+ > 30$) in the globe range. At section $y = 4.35$, the y^+ value is 50.9, for example.

Before solving the combustion process, the simulation of the cold flow field was conducted, and the convergence criterion adopted was the RMS (Root Mean Square of the residual values). After the convergence of the flow field iteration, energy, radiation, species and discrete-phase model (DPM) equations were then activated. The number of iterations was set until the solution with successive under-relaxation satisfied the pre-specified tolerance. Some settings and parameters of the Non-Premixed combustion and DPM models are listed in Table 3.

Table 3. Settings and parameters of Non-Premixed combustion and discrete-phase model (DPM) models.

Non-Premixed Combustion Model	Chemistry State Relation	Chemistry Energy Treatment	Chemistry Stream Option	Operating Pressure	PDF Opinion
Settings	Chemistry equilibrium	Non-adiabatic	Empirical stream	101,325Pa	Inlet Diffusion
Discrete Phase Model	**DPM Iteration Interval**	**Max. Number of Steps**	**Specify Length Scale Length Scale**	**Turbulent Dispersion Discrete**	**Number of Tries**
Settings	20	5000	0.005	Random walk model	10

Finally, the NOx was computed using the solution obtained with the previous steps. NOx formation includes thermal NOx and fuel NOx but hardly any prompt NOx. In this study, the formation of prompt NOx was neglected in calculations, and only NO production was taken into account because the NOx emitted into the atmosphere from combusting fuels consists mostly of NO. The concentration of thermal NOx was calculated using the extended Zeldovich mechanism (specifically, $N_2 + O \rightarrow NO + N$, $N + O_2 \rightarrow NO + O$, $N + OH \rightarrow NO + H$) [23]. The fuel NO$x$ concentration was calculated using De Soete's model [23,24].

4. Results and Discussion

4.1. Analysis of the Fluid Flow and Temperature Field of the Burners

In order to not only reflect the temperature value of the high-temperature flue gas generated by the combustion reaction near the furnace burner region, but also show the characteristics of the vortex flow field more clearly, Figure 5 shows the velocity vector and gas temperature profile of the swirl burner in the longitudinal section of the primary zone where $z = 0.85$ m (a), center axes of horizontal section where $y = 4.35$ m (b), and the gas temperature contour at different y planes of the furnace under two kinds of loads (c).

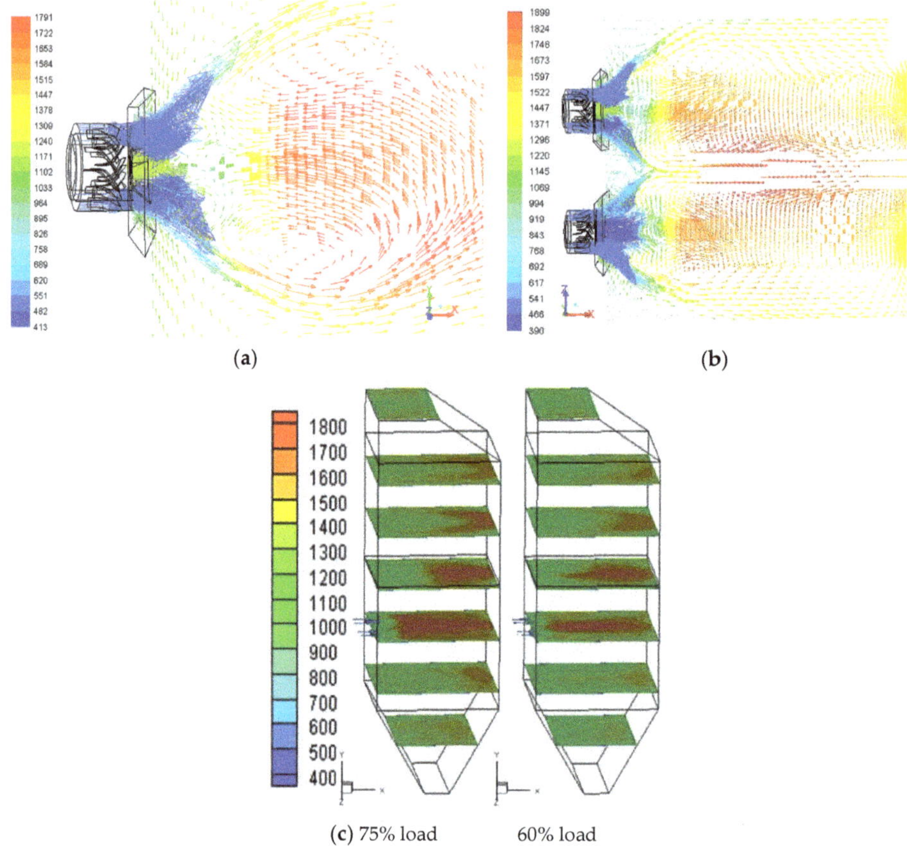

(c) 75% load   60% load

Figure 5. Velocity vector colored by temperature in a longitudinal section of the primary zone at $z = 0.85$ m (**a**) and in the horizontal section $y = 4.35$ m (**b**), and gas temperature contour on different level sections of the furnace (**c**) with the variation of load (K).

As can be seen from Figure 5, the temperature of the inner and outer SA and CA of the front wall burners were the lowest, with an order of magnitude above 390 K. The geometry of the outer side of the duct end of each burner was a conical structure, which generated reverse pressure action along the jet direction, forming intense eddy currents caused by the entrainment and jetting. The lower temperature mixture of SA near the side walls, and the unburned fuel and combustion products flowed to the exit region of the burner. The eddy currents were conducive to the preheating and stable ignition of pulverized coal particles carried by the PA, forming an ideal flow characteristic of a swirl burner. Then, the gas temperature near the nozzle center increased dramatically. In the middle position near the exit of the burner, two vortexes were small. It can be seen from Figure 5a,b that the fluid of the fuel gas zones of two burners were mixed with each other by jet flow, their flames intermingled and connected together, forming a high temperature zone of intense combustion, with the temperature partially reaching 1800 K. These showed that the heat and mass transfer in the jets were strong, and combustion reactions occur mainly in this area. The flue gas temperature was basically symmetrical on both sides of the wall in the horizontal direction. Figure 5c depicts the temperature characteristics along the level of the furnace. It can be seen that with the variation of the load, the area and length of the high-temperature area of the flame in the furnace were increased, and the temperature was gradually decreased from the primary combustion zone to the exit.

4.2. Analysis of Main Combustion Species in the Center Axis of the Burner

Figure 6 shows the variation curves of the molar fractions of the combustion products O_2, CO_2 and CO and concentration of NO per unit volume in the standard state under two kinds of loads on the sampling line of the central axis of the burner. According to the distribution of the main species, the intensity and characteristics of the reaction between the PA and pulverized coal combustion on the burner axis can be determined.

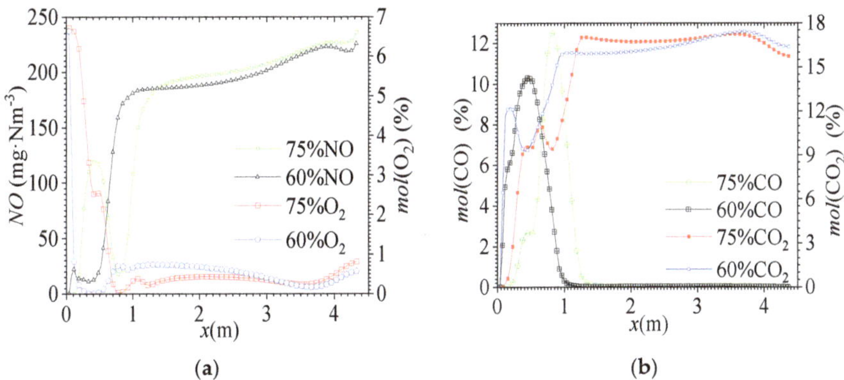

Figure 6. Species variation along the center axis of burner. (**a**) NO concentration and O_2 molar fractions; (**b**) CO and CO_2 molar fractions.

It can be seen from Figure 6 that the curves can be divided into two regions with obvious characteristics. The critical distance L was taken as the demarcation point of two regions, and L grows longer with an increase in load. In this paper, at 60% load, $L = 1$ m; at 75% load, $L = 1.3$ m. The gas area where $x < L$ was in the diffusion combustion zone and burned under two loads. The molar fraction of O_2 decreased sharply with the x-coordinate, reaching a minimum value (Figure 6a). Meanwhile, the molar fraction of CO changed according to an approximate parabola law in this area, and the peak value was higher than 60% of the load at 75% load. At the same time, the molar fraction of CO_2 at the corresponding coordinate increased sharply, reaching a first peak, and the CO_2 molar fraction generated by combustion at the 60% operating condition was higher than that at 75% load. Then, as the x-coordinate increased, the CO_2 concentration firstly decreased and then increased dramatically, reaching a second peak, and the CO_2 molar fraction produced by combustion under 60% load was lower than that under 75% load until the furnace outlet front area (Figure 6b).

It was found that at 75% load, the NO concentration had the same characteristics as that of CO_2, showing a sharp increase first and then a sharp decrease; the first peak reached 120 mg/Nm^3 when x was between 0 and 0.5 m. This is because the temperature of the flue gas increased with the load. Then, the magnitude of the NO concentration increased sharply after a minimum value appeared near 0.8 m. At distance L, a sharp increase of 180 g/Nm^3 was observed at two loads, with values fluctuating wildly. The NO concentration curve at 60% load was different from in the above conditions, increasing with the depth of the furnace.

It can be seen from Figure 6b that the reducing gas CO's concentration was higher in the species at the corresponding position at 75% load. Because the O_2 concentration at the corresponding position was lower, the volatiles from pulverized coal were ignited and burned here. Due to the incomplete combustion, the CO concentration increased. At the same time, a high concentration of CO gas helps to inhibit the NO formation during the combustion of coal char particles [14], resulting in the concentration of NO per unit volume in standard state decreasing slightly with an increase in load. Despite the fact that different combustion reactions and components in the primary combustion zone existed, when $x > L$, the NO concentration gradually increased with an increase in furnace depth.

Meanwhile, the CO molar fraction was very small. The change in the CO_2 molar fraction with horizontal distance from the burner and load was not obvious, and the value decreased after $x > 4$ m until the furnace outlet.

4.3. Analysis of Flue Gas Temperature and Species NO along the Furnace Height

In order to quantitatively analyze the variation characteristics of the gas temperature and NO concentration along the height direction of the furnace, several horizontal sections were taken in the height direction. The area-weighted average gas temperature and NO concentration under standard state of the flue gas on each horizontal section were calculated, as shown in Figure 7.

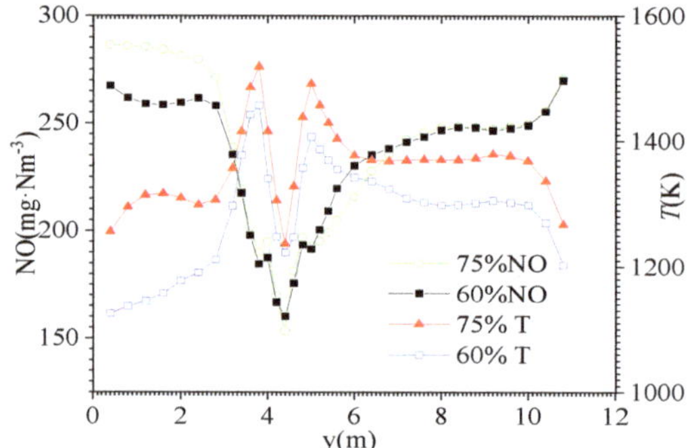

Figure 7. Average gas temperature and NO concentration with boiler height.

From Figure 7, it can be seen that the area-weighted average gas temperature near the center cross section of the swirl burner, $y = 4.35$ m, is lower, which is due to the lower temperature of the incoming air and pulverized coal. The heating process and the incomplete combustion process all were carried out around the central axis and height (y) direction. With the increase in the swirl radius, the SA was gradually replenished; the mixing of the pulverized coal and air was more and more sufficient. The exothermic reaction of combustion was intense, and the combustion was strongest at the heights of about 3.8 and 5 m, respectively. The flame temperature increased sharply and reached the peak temperature value at two positions, while the temperature gradient near the peak position was very large. This phenomenon was matched with the flow field shown in Figure 5. The lowest gas temperature at 60% load appeared at the bottom of the cold ash bucket; it was different from that at 75% load. In the burn-out zone, the temperature change along the height direction was relatively gentle, while the temperature gradient was large near the 0.8 m area of the furnace top. Most flue gas with high temperature flowed out through the nose region of the furnace.

By contrast, it can be found that the distribution of the gas temperature and NO concentration under the two loads showed opposite change laws partially along the height direction, and the change was the lowest near the level section of the burner axis. This was because the reductive gas CO generated by incomplete combustion reduced most of the NOx that had been generated, and the high concentration of CO was the main factor that inhibited the generation of NOx. Taking the height (y) at the center axis of the burner as the Hc ($y = 4.35$ m), when $y < Hc$, the concentration of NO decreased sharply with an increase in the y coordinate, and the concentration of NO in the ash bucket was higher at 75% than that at 60% load; when $y > Hc$, the concentration of NO in the swirling combustion zone increased sharply with the rise in temperature but insignificantly with the load; the concentration of NO in the outlet of the furnace was basically the same. In general, the NO concentration was relatively high in the burnout area near the furnace outlet and the lower ash bucket area. In the later stage of the combustion reaction near the furnace outlet, the temperature was higher, the pulverized coal fuel was almost exhausted, and excess air reacted with NO_2 to form a large amount of NOx. With the increase in the load, the combustion time and the stroke of pulverized coal particles in the furnace increased obviously. The above analysis mainly studied the NO generation process. Considering that the furnace outlet is the main position for environmental parameter monitoring, the NO, oxygen and carbon burnout rates according to the simulation are listed in Table 4.

Table 4. Important parameters of furnace outlet at standard state.

Load (%)	NO (mg/Nm^3)	O$_2$ (%)	Carbon Burn-off Rate (%)
75	226.4	6.60	99.19
60	219.2	6.45	99.04
Rate of change (%)	+3.3	−2.3	−0.15

By comparison, it can be found that at the outlet of the furnace under standard state conditions, the concentration of NO (the emission concentration was slightly higher than 200 mg/Nm^3), the volume fraction of O$_2$ and the burnout rate of carbon all decreased with the load reduction. In general, the percentage of change with load did not exceed 5%. In the actual operation process, the air–fuel ratio [6] can be adjusted to meet the requirements of environmental protection. The above study provided basic data for the follow-up energy saving and environmental protection technologies of industrial boilers with similar typical energy consumption equipment.

4.4. Comparison of Simulation Results and Experimental Data

When the 35 t/h pulverized coal industrial test boiler was operating under 60% and 75% load conditions, a gas pump thermocouple embedded in the sidewall of the furnace exit region was used to detect the steady-state temperature at y = 10.5 m of the furnace exit temperature region. A total of 12 points were measured in each working condition. The steady-state temperature data of the flue gas were obtained from the position of the measuring point. The simulation results obtained under the same conditions were compared with the measured values. Figure 8 shows a comparison of the predicted temperature field within the boiler and the real-case results for the same position. The simulated and measured temperature values agreed well where x > 0.1 m; the maximum errors of the numerical calculation and measurement values were 5.2% and 2.4%, respectively. An exception was in the boundary zone, where x < 0.08 m, and the results were relatively accurate within the permissible limit. This error was partly due to the assumption of constant particle emissivity [25].

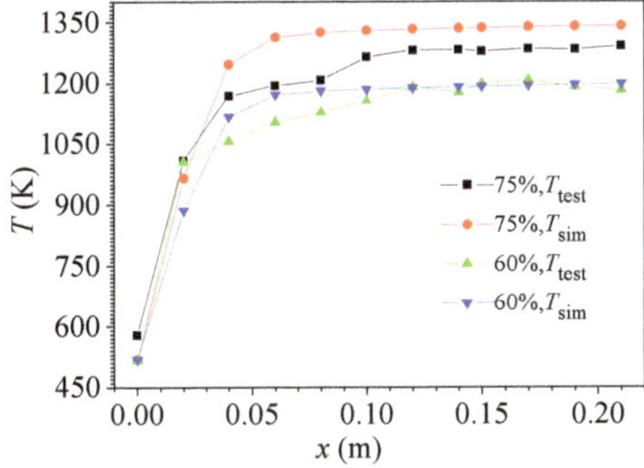

Figure 8. Comparison of flue gas temperature between test data and numerical results.

5. Conclusions

In this paper, a CFD numerical method was presented to study the combustion performance in 35 t/h industrial pulverized coal boilers with two swirl burners arranged on the wall. The fluid flow and temperature field of the flue gas near the burner and the distribution characteristics of the combustion products (i.e., NO, O_2, CO, CO_2 and unburned carbon concentrations) were investigated at two loads to provide basic data for optimizing the combustion operation for nitrogen reduction in industrial boilers. The conclusions were as follows:

1. The swirl burner of this work can produce an obvious vortex flow region caused by the entrainment, which is conducive to the preheating and stable ignition of pulverized coal particles.
2. Along the direction of the furnace depth (x) in the burner axis, there was a critical distance L; when $x < L$, the O_2 molar fraction decreased sharply with the x coordinate. When $x > L$, the NO concentration increased gradually with an increase in furnace depth; the increase value was very small with load. L increased with an increase in load.
3. In the furnace along the height direction (y), take the height of the center axis of the burner as the H_C ($y = 4.35$ m). When $y < H_C$, the NO concentration decreased sharply with an increase in the y coordinate, and the NO concentration at 75% load was greater than that at 60% load in the ash bucket area; when $y > H_C$, the NO concentration increased sharply with the temperature increasing in the swirl combustion zone with no obvious change, and the NO concentration was basically the same at the furnace outlet. The average temperature of the flue gas in the furnace became lower at the Hc position with an increase in the swirl radius of the SA diffuse flow; there were two peak temperatures at two locations.

Author Contributions: Y.L., J.H. and Y.G. are the main authors of this manuscript. All the authors contributed to this manuscript. Y.L. and Y.G. conceived the novel idea, and J.H. performed the analysis; L.Z. and Y.M. analyzed the data and contributed analysis tools; L.Z. and Y.L. wrote the entire paper. A.M. and L.Z. checked, reviewed and revised the paper; Y.L. performed final proofreading, and she supervised this research. All authors have read and agreed to the published version of the manuscript.

Funding: The authors would like to acknowledge the funding received from the Hi-Tech Research and Development Program of China (2018YFF0216001). The authors also wish to thank Wang L. and Zhang S.S. for their continuous support and useful discussions during the development of the numerical models used in this study.

Conflicts of Interest: The authors declare no conflict of interest.

References

1. Hu, F.; Li, P.F.; Wang, K. Evaluation, development, and application of a new skeletal mechanism for fuel-NO formation under air and oxy-fuel combustion. *Fuel Process. Technol.* **2020**, *199*, 1–17. [CrossRef]
2. Wen, X.; Luo, K.; Yujuan, L. Large eddy simulation of a semi-industrial scale coal furnace using non-adiabatic three-stream flamelet/progress variable model. *Appl. Energy* **2016**, *183*, 1086–1097. [CrossRef]
3. Bermúdez, A.; Ferrín, J.L.; Liñán, A.; Saavedra, L. Numerical simulation of group combustion of pulverized coal. *Combust. Flame* **2011**, *158*, 1852–1865. [CrossRef]
4. Mikulcic, H.; Von-Berg, E.; Wang, X. Using an Advanced Numerical Technique for Improving Pulverized Coal Combustion inside an Industrial Furnace. *CET J.* **2017**, *61*, 235–240.
5. Li, S.; Zhong, G.F.; Xue, N.D. Influence of combustion system retrofit on NOx formation characteristics in a 300 MW tangentially fired furnace. *Appl. Therm. Eng.* **2016**, *98*, 766–777.
6. Askarova, A.; Bolegenova, S.; Maximov, V. Numerical modeling of pulverized coal combustion at thermal power plant boilers. *J. Therm. Sci.* **2015**, *24*, 275–282. [CrossRef]
7. Hwang, M.Y.; Ahn, S.G.; Jang, H.C. Numerical study of an 870MW wall-fired boiler using De-NOx burners and an air staging system for low rank coal. *J. Mech. Sci. Technol.* **2016**, *30*, 5715–5725. [CrossRef]
8. Mohd Noor, N.A.W.; Hassan, H.; Hashim, M.F. Preliminary investigation on the effects of primary airflow to coal particle distribution in coal-fired boilers. *J. Phys. Conf. Series* **2017**, *822*, 21–23.
9. Silva-Daindrusiak, C.V.; Sbeskow, M.L. CFD analysis of the pulverized coal combustion processes in a 160 MW tangentially-fired-boiler of a thermal power plant. *J. Braz. Soc. Mech. Sci.* **2010**, *32*, 427–436. [CrossRef]
10. Constenla, I.; Ferrín, J.L.; Saavedra, L. Numerical study of a 350 MW tangentially fired pulverized coal furnace of the As Pontes Power Plant. *Fuel Process. Technol.* **2013**, *116*, 189–200. [CrossRef]

11. Belošević, S.; Tomanović, I.; Crnomarković, N. Numerical study of pulverized coal-fired utility boiler over a wide range of operating conditions for in-furnace SO_2/NOx reduction. *Appl. Therm. Eng.* **2016**, *9*, 657–669. [CrossRef]
12. Wang, J.; Zheng, K.; Singh, R. Numerical simulation and cold experimental research of a low-NOx, combustion technology for pulverized low-volatile coal. *Appl. Therm. Eng.* **2017**, *11*, 498–510. [CrossRef]
13. Sung, Y.; Choi, C.; Moon, C. Nitric oxide emissions and combustion performance of nontraditional ring-fired boilers with furnace geometries. *J. Mech. Sci. Technol.* **2015**, *29*, 3489–3499. [CrossRef]
14. Lisandy, K.Y.; Kim, J.W.; Lim, H. Prediction of unburned carbon and NO formation from low-rank coal during pulverized coal combustion: Experiments and numerical simulation. *Fuel* **2016**, *185*, 478–490. [CrossRef]
15. Zhang, Z.; Li, Z.S.; Cai, N.S. Reduced-order model of char burning for CFD modeling. *Combust. Flame* **2016**, *16*, 83–96. [CrossRef]
16. Shen, Y.S.; Guo, B.Y.; Yu, A.B.; Zulli, P. A three-dimensional numerical study of the combustion of coal blends in blast furnace. *Fuel* **2009**, *88*, 255–263. [CrossRef]
17. Roache, P.J. Perspective: A method for uniform reporting of grid refinement studies. *J. Fluids Eng.* **1994**, *116*, 405–413. [CrossRef]
18. Feng, R.; Li, Z.; Liu, G. Numerical simulation of flow and combustion characteristics in a 300 MW down-Fired boiler with different overfire air angles. *Energy Fuels* **2011**, *25*, 1457–1464.
19. ANSYS Inc. *ANSYS FLUENT Theory Guide*; ANSYS Inc.: Canonsburg, PA, USA, 2010; pp. 200–396.
20. Patankar, S.V. *Numerical Heat Transfer and Fluid Flow*; Hemisphere Publishing Corporation: New York, NY, USA, 1980; pp. 104–156.
21. Wang, Q.; Guan, J.; Sun, D.H. A simulation of flow field on furnace for horizontal coal smokeless combustion boiler. *Proc. CSEE* **2003**, *23*, 164–167.
22. Sarkar, A.; So, R.M.C. A critical evaluation of near-wall two-equation models against direct numerical simulation data. *Int. J. Heat Fluid Flow* **1997**, *18*, 197–208. [CrossRef]
23. Hill, S.C.; Smoot, L.D. Modeling of nitrogen oxides formation and destruction in combustion systems. *Prog. Energy Combust. Sci.* **2000**, *26*, 417–458. [CrossRef]
24. De Soete, G.G. Overall reaction rates of NO and N_2 formation from fuel nitrogen. In Proceedings of the 15th International Symposium on Combustion, Tokyo, Japan, 25–31 August 1974; pp. 1093–1102.
25. Laubscher, R.; Rousseau, P. Numerical investigation on the impact of variable particle radiation properties on the heat transfer in high ash pulverized coal boiler through co-simulation. *Energy* **2020**, *195*, 117006. [CrossRef]

 *processes*

Article

Computational Fluid Dynamics Simulation of Gas–Solid Hydrodynamics in a Bubbling Fluidized-Bed Reactor: Effects of Air Distributor, Viscous and Drag Models

Ramin Khezri [1], Wan Azlina Wan Ab Karim Ghani [1,2,*], Salman Masoudi Soltani [3,*], Dayang Radiah Awang Biak [1], Robiah Yunus [1], Kiman Silas [1], Muhammad Shahbaz [4] and Shiva Rezaei Motlagh [1]

[1] Department of Chemical and Environmental Engineering, Faculty of Engineering, Universiti Putra Malaysia, Serdang 43400, Malaysia

[2] Sustainable Process Engineering Research Center, Faculty of Engineering, Universiti Putra Malaysia, Serdang 43400, Malaysia

[3] Department of Chemical Engineering, Brunel University London, Uxbridge UB8 3PH, UK

[4] Division of Sustainable Development, College of Science and Engineering, Hamad Bin Khalifa University (HBKU), Qatar Foundation, Doha 34110, Qatar

* Correspondence: wanazlina@upm.edu.my (W.A.W.A.K.G.); salman.masoudisoltani@brunel.ac.uk (S.M.S.); Tel.: +603-97696287 (W.A.W.A.K.G.); +44(0)1895265884 (S.M.S.)

Received: 8 July 2019; Accepted: 7 August 2019; Published: 8 August 2019

Abstract: In this work, we employed a computational fluid dynamics (CFD)-based model with a Eulerian multiphase approach to simulate the fluidization hydrodynamics in biomass gasification processes. Air was used as the gasifying/fluidizing agent and entered the gasifier at the bottom which subsequently fluidized the solid particles inside the reactor column. The momentum exchange related to the gas-phase was simulated by considering various viscous models (i.e., laminar and turbulence models of the re-normalisation group (RNG), k-ε and k-ω). The pressure drop gradient obtained by employing each viscous model was plotted for different superficial velocities and compared with the experimental data for validation. The turbulent model of RNG k-ε was found to best represent the actual process. We also studied the effect of air distributor plates with different pore diameters (2, 3 and 5 mm) on the momentum of the fluidizing fluid. The plate with 3-mm pores showed larger turbulent viscosities above the surface. The effects of drag models (Syamlal–O'Brien, Gidaspow and energy minimum multi-scale method (EMMS)) on the bed's pressure drop as well as on the volume fractions of the solid particles were investigated. The Syamlal–O'Brien model was found to forecast bed pressure drops most consistently, with the pressure drops recorded throughout the experimental process. The formation of bubbles and their motion along the gasifier height in the presence of the turbulent flow was seen to follow a different pattern from with the laminar flow.

Keywords: gasification; fluidized bed; CFD; hydrodynamics; multiphase flow

---

1. Introduction

Owing to the renewable, sustainable and environmentally-friendly nature of biomass-based energy generation, over the past few years, it has attracted a great deal of interest within both the scientific community and the industrial sector. Biomass fuel can dampen the dependency on the energy production from fossil fuels, and hence, mitigate the associated adverse environmental emissions [1–4]. Biomass gasification is still facing various technical challenges, repelling potential commercialization.

Such challenges comprise the complex chemical reactions propagation inside the fluidizing zone and the restrictions imposed through process control and difficulty in the prediction of momentum patterns caused by the turbulent behaviour of the gas flow and moving particles [5]. The complicated processes—inherently associated with embedded limitations—would make it almost economically and environmentally unviable to experimentally investigate all the factors in play, and therefore, necessitate the development of reliable mathematical models.

Reliable models can facilitate the investigation of thermochemical processes associated with biomass gasification; likewise, they help to study and evaluate the effects of the operational parameters on the gasifier performance and product quality [6]. Numerical models and simulation studies for biomass gasification using either Eulerian [7–11] or Lagrangian (discrete element) [12–14] models have been the focus of many researchers in recent years. A large number of numerical modelling and simulation studies have been reported in literature on the complex behaviour of gas–solid mixing, as well as on the fluidization hydrodynamics, in order to formulate the problems and to devise solutions. Furthermore, using computational fluid dynamics (CFD), model-based turbulent flow calculations can be done more efficiently and in a quicker manner. Standard (STD) k-ε, re-normalisation group (RNG) k-ε and k-ω are among the most commonly-used models used in systems involving fluid flows with large Reynolds numbers; e.g., flows in many fluidized beds. The RNG k-ε model is a modified STD k-ε model and in many cases, produces improved results.

In a previous study, the hydrodynamics of gas–solid mixing in fluidized bed reactors was evaluated through numerical simulation and experimental techniques in order to investigate the effect of drag models on the hydrodynamics of fluidization [15]. In another study, a CFD model was developed to determine the fluid-particle interaction during fast pyrolysis inside a fluidized bed reactor with a 150 gh^{-1} capacity [16]. The model was capable of calculating the heat, mass and momentum transport between the fluidization agent and the bed material (i.e., silica sand). However, the characterization of momentum in both studies was merely limited by a laminar model. In another study a three dimensional CFD model was developed to simulate a high-flux circulating, fluidized bed riser [17]. The effects of different drag models on the hydrodynamics of the gas–solid fluidization were investigated. The results showed that the drag models of Gidaspow and Syamlal–O'Brien were able to accurately predict the solid volume fraction values, at almost every region inside the fluidized bed except for the zone near the bottom surface with the high concentration of the solids. In another study, a CFD simulation of flow through the gas distributor in a multi-stage biomass gasifier was conducted by adopting the Eulerian-multiphase model approach, where the effects of different bed materials, and a distributor perforated area, on the hydrodynamics of a solid–gas fluidized bed were examined [18]. Although the most critical aspects affecting the fluidized bed hydrodynamics were carefully addressed in the study, it lacked a study to reveal the impacts of different drag models.

This study aimed to simulate the hydrodynamics of gas–solid fluidization. The simulation results were experimentally validated on a bubbling fluidized bed gasifier. The effects of different momentum characteristics (i.e., laminar and turbulent models) on the hydrodynamics of gas–solid mixing, as well as on the fluidization of the dispersed solid phase, were studied. In addition, the pressure drop profile obtained from different drag models, including laminar, RNG k-ε and k-ω turbulent were compared with the experimentally-derived data. Furthermore, the effect of air distributors with different pore sizes was studied numerically to determine the turbulent viscosity of the fluid passing through the pores of the distributor plate. Ultimately, the motion patterns of the particles and bubble formation linked to the hydrodynamics of granular solid fluidization were investigated. The developed hydrodynamic model, upon successful experimental validation, is able to effectively pinpoint the optimum operating conditions for the investigated parameters, including the superficial velocity, bed expansion, pressure drop gradient throughout the reactor interior, and the momentum exchanged between the particles and the media. The above parameters are not easy to measure experimentally, yet are significant enough to be considered for accurate reactor optimization.

2. Modelling and Experimental Methods

2.1. Governing Equations

The basic equations used to in the development of the numerical models will be introduced and explained in this section. The numerical formulae are the governing equations used by Ansys Fluent Solver to calculate the corresponding parameters.

2.1.1. Gas–Particle Flow Equations

An unsteady-state Eulerian–Eulerian multiphase flow model was used to compute the transient nature of gas–solid in the fluidization of granular solids. The different phases in the Eulerian model were treated mathematically as interpenetrating continua. The phasic volume fraction was considered to differentiate the volume occupied by each phase as a function of time and space, implying that although the involved phases were interacting with each other, the volume of each phase could not be occupied by another phase [19]. Conservation equations were derived for all phases and closed through constitutive relations defined via the application of kinetic theory. The gas phase was modelled for laminar, RNG k-ε turbulent and k-ω turbulent characteristics, while the particle phase was modelled using the kinetic theory of granular flow. The governing conservation equations are presented in Table 1.

Table 1. Governing equations of momentum and continuity for gas–solid phases [20–25].

Gas Phase Conservation Equations	
Continuity equation	$\frac{\partial}{\partial t}(\delta_g \rho_g) + \frac{\partial}{\partial y}(\delta_g \rho_g \overline{u}) = 0$
Momentum equation	$\frac{\partial}{\partial t}(\rho \vec{V}) + \nabla.(\rho \vec{V}\vec{V}) = -\nabla P + \nabla.(\overline{\overline{\tau}}) + \rho \vec{g} + \vec{F}$
Solid Phase Conservation Equations	
Continuity equation	$\frac{\partial}{\partial t}(\delta_s \rho_s) + \frac{\partial}{\partial y}(\delta_s \rho_s \overline{u}) = 0$
Momentum equation	$\frac{\partial}{\partial t}(\rho_s \delta_s v) + \frac{\partial}{\partial y}(\rho_s \delta_s vv) =$ $-\frac{\partial P_s}{\partial y} - \frac{\partial(\delta_s s)}{\partial y} + \delta_s \rho_s g + \beta_g(\overline{u} - v)$

The volume fraction, density and instantaneous velocity of gas/solid phase in the governing equations in Table 1 are represented by δ, ρ and u, respectively. Subscripts g and s denote gas and solid phases, respectively.

2.1.2. Gas–Solid Interaction

The momentum exchange coefficient of solid–gas, K_{sg}, is defined as [18]:

$$K_{sg} = \frac{\varepsilon_s \rho_s f}{\tau_s} \tag{1}$$

where the particulate relaxation time, τ_s, is defined as:

$$\tau_s = \frac{\rho_s d_s^2}{18\mu_g} \quad (d_s \text{ is the particle size in the solid phase.}) \tag{2}$$

2.1.3. Drag Models

One of the significant terms in describing the hydrodynamics of fluidized bed in a Eulerian–Eulerian approach is the drag force. The significance of drag force is due to its effect on the characteristics of the interaction between the gas and solid phases. The drag models employed in this study to calculate the gas–solid momentum exchange coefficient are the Syamlal–O'brien, Gidaspow and energy minimum multi-scale method (EMMS). The Gidaspow model is a combination of the Wen and Yu model, and the Ergun equation [13,24], and is well-suited for densely packed fluidized bed applications:

$$
K_{sg} = \begin{cases} \frac{3}{4}C_D\left(\frac{\delta_s\delta_g\rho_g\left|\vec{v}_s-\vec{v}_g\right|}{d_s}\right)\delta_g^{-2.65}, & \delta_g > 0.8 \\[2em] 150\left(\frac{\delta_s(1-\delta_g)\mu_g}{\delta_g d_s^2}\right) + 1.75\left(\frac{\delta_s\rho_g\left|\vec{v}_s-\vec{v}_g\right|}{d_s}\right), & \delta_g < 0.8 \end{cases}
\tag{3}
$$

where, C_D, is the drag coefficient and defined for a smooth particle as [26]:

$$
C_D = \begin{cases} \frac{24}{\delta_g Re_s}\left[1 + 0.15\left(\delta_g Re_s\right)^{0.687}\right], & Re < 1000 \\[1.5em] 0.44, & Re \geq 1000 \end{cases}
\tag{4}
$$

where, Re_s, is a function of slip velocity and is described as:

$$
Re_s = \frac{\rho_g d_s\left|\vec{v}_s - \vec{v}_g\right|}{\mu_g}
\tag{5}
$$

The Syamlal–O'Brien model is described as [15]:

$$
K_{gs} = \frac{3\alpha_s\alpha_g\rho_g}{4v_{r,s}^2 d_s}C_d\left(\frac{Re_s}{v_{r,s}}\right)\left|\vec{V}_s - \vec{V}_g\right|
\tag{6}
$$

where, $v_{r,s}$, is the terminal velocity correlation for the solid phase as described below:

$$
v_{r,s} = 0.5\left(A - 0.06Re_s + \sqrt{\left(0.06Re_s{}^2\right) + 0.12Re_s(2B - A) + A^2}\right)
\tag{7}
$$

$$
A = \alpha_g^{4.14}, B = 0.8\alpha_g^{1.28} \text{ for } \alpha_g \leq 0.85 \text{ and } B = 0.8\alpha_g^{2.65} \text{ for } \alpha_g > 0.85
\tag{8}
$$

The EMMS model is described as [27]:

$$
K_{gs} = \frac{3}{4}c_D\frac{\alpha_s\alpha_g\rho_g\left|\vec{V}_s - \vec{V}_g\right|}{d_s}\alpha_g^{-2.65}H_d \text{ for } \alpha_g \geq 0.75
\tag{9}
$$

$$
K_{gs} = 150\frac{\alpha_s^2\mu_g}{\alpha_g d_s^2} + 1.75\frac{\alpha_s\rho_g\left|\vec{V}_s - \vec{V}_g\right|}{d_s} \text{ for } \alpha_g < 0.75
\tag{10}
$$

The heterogeneity index, H_d, is calculated based on a proposed method elsewhere [28].

2.1.4. Turbulent Models

The transport equations for the RNG k-ε and k-ω models are written as below, where the buoyancy effect is neglected [29]:

$$
\frac{\partial}{\partial t}(\rho k) + \frac{\partial}{\partial x_i}(\rho k u_i) = \frac{\partial}{\partial x_j}\left[\left(\mu + \frac{\mu_t}{\sigma_k}\right)\frac{\partial k}{\partial x_j}\right] + P_k + \rho\varepsilon
\tag{11}
$$

$$
\frac{\partial}{\partial t}(\rho\varepsilon) + \frac{\partial}{\partial x_i}(\rho\varepsilon u_i) = \frac{\partial}{\partial x_j}\left[\left(\mu + \frac{\mu_t}{\sigma_\varepsilon}\right)\frac{\partial\varepsilon}{\partial x_j}\right] + C_{\varepsilon 1}\frac{\varepsilon}{k}P_k - C_{\varepsilon 2}^*\rho\frac{\varepsilon^2}{k}
\tag{12}
$$

where $C^*_{\varepsilon 2} = C_{\varepsilon 2} + \frac{C_\mu \eta^3 \left(1 - \frac{\eta}{\eta_0}\right)}{1 + \beta \eta^3}$, $\eta = \frac{Sk}{\varepsilon}$, $S = \left(2 S_{ij} S_{ij}\right)^{1/2}$ and P_k is the production rate of turbulence. Transport equations for the k-ω model suggested by Wilcox (1991) [30] are described as:

$$\frac{\partial(\rho k)}{\partial t} + \frac{\partial}{\partial x_j}(\rho U_j k) = \frac{\partial}{\partial x_j}\left[\left(\mu + \frac{\mu_t}{\sigma_k}\right)\frac{\partial k}{\partial x_j}\right] + P_k - \beta \rho k \omega + P_{kb} \quad \text{"K-equation"} \tag{13}$$

$$\frac{\partial(\rho \omega)}{\partial t} + \frac{\partial}{\partial x_j}(\rho U_j \omega) = \frac{\partial}{\partial x_j}\left[\left(\mu + \frac{\mu_t}{\sigma_\omega}\right)\frac{\partial \omega}{\partial x_j}\right] + \alpha \frac{\omega}{k} P_k - \beta \rho \omega^2 + P_{\omega b} \quad \text{"ω-equation"} \tag{14}$$

The numerical values of the coefficients for the RNG k-ε and k-ω models are available in the literature [31].

2.2. Models' Descriptions and Simulation Methods

2.2.1. Initial Conditions, Boundary Conditions, Geometry and Grids

The geometry, boundary conditions and grids of the reactor at the initial state are defined and shown in Figure 1a,b, in order to study the fluidization of the granular biomass particles.

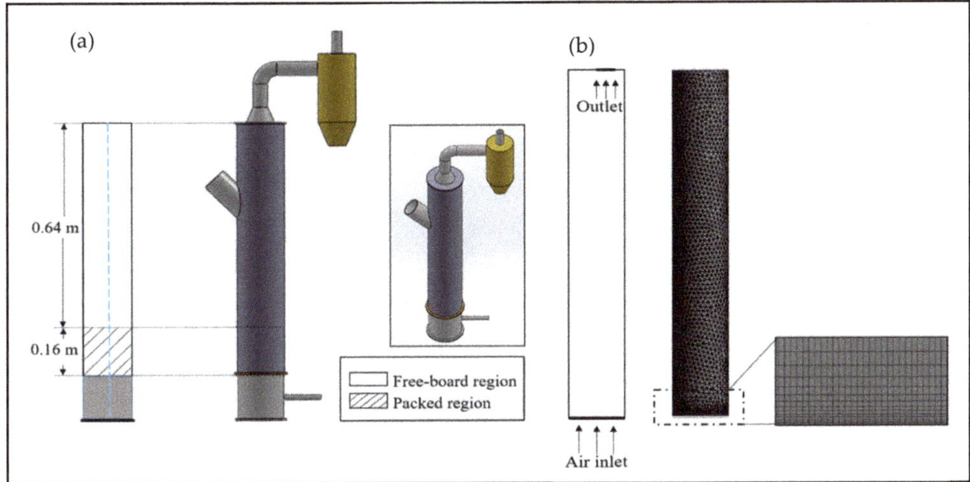

Figure 1. (**a**) Reactor geometry, (**b**) boundary zones and grid design at initial state (static bed).

Air (i.e., the fluidizing agent) enters the reactor from the bottom and passes through a distributor plate prior to its injection into the chamber. The producer gas exits from the top of the reactor where the pressure is set to atmospheric with no backflow pressure. The walls are thermally insulated and defined with a specularity coefficient of $\phi = 0.5$ for the solid phase, as recommended for multiphase granular flow on bubbling fluidized beds and a no-slip condition for air [32]. Up to 20% of the reactor volume was initially packed with bed material (i.e., silica sand). The volume fraction of solid particles within the patched zone was adjusted to 60%. The ambient air temperature and pressure were set at the reference values.

In order to avoid instability and poor convergence in the simulation of the multiphase flows, a small time step-size of 0.001 s with 100 iterations per time-step was used until the relative convergence criterion of 1×10^{-3} was satisfied. The physical specifications of the bed material are summarised in Table 2.

Table 2. Physical specifications of bed materials (silica sand particle) [33].

Physical Properties	Unit	Quantity
Density (Dry basis)	(kg m^{-3})	2636
Particle diameter	(μm)	550
Molecular weight	(kg kmol^{-1})	60

A fine grid was selected to capture the steep gradients between the neighbouring areas. The non-uniform grid spacing was generated using Ansys Desing Modeler (Figure 1). The second order upwind scheme was used to discretise the convective terms in the momentum and energy equations. The values and parameters used in the grid generation and numerical model development are shown in Table 3.

Table 3. Parameters used in grid generation and model development of fluidization hydrodynamics.

Model Statistics	
Model basis	transient
No. of elements	58,000
Grid structure	Unstructured
Turbulent multiphase model	Dispersed
Turbulent dispersion model	Adopted from Burns et al. [34]

2.2.2. The Effect of Viscous Models on Pressure Drop Profile

The pressure drop profiles against different superficial velocities were derived for each viscous model (i.e., laminar, RNG k-ε and k-ω) and were then compared against the experimental data. It must be noted that the occurrence of pressure drop was due to continual fission and fusion of the bubbles in the fluidization zone [35]. The pressure drop values were recorded over the hypothetical vertical line aligned in the centre of the reactor's geometry (Figure 1a).

The best viscous model was selected based on how accurately it could describe the pressure drop profile when compared with the experimentally-obtained data. That model was subsequently used for further analysis. In order to identify such a model, a statistical method was employed to evaluate a two-sample-T-test and to calculate the mean squared error (MSE) for each curve. In order to ensure that the bed has been effectively fluidized, the velocities <minimum fluidization velocity (Umf) were discarded in the statistical comparison.

2.2.3. Experimental Validation of the Numerical Analyses

Controlled experiments were conducted using a fluidized bed reactor in order to validate the simulation results. A reactor with similar dimensions and physical characteristics assumed in the modelling work, was used in experimental data collection. Pressure drop profiles were generated from the cold test evaluation when the reactor was initially packed with bed material with similar particle diameter (i.e., 550 μm) as assumed in the modelling, to the same level (i.e., 20% of total reactor volume) as described earlier in this work. Further details on the reactor specifications and experimental setup can be found in our previous study [36]. Each incremental drop in pressure throughout the reactor was caused by the gradual increase in the superficial velocity of the air between the air distributor and the reactor outlet.

2.2.4. The Effect of Air Distributor on Fluid Properties

The nature of the turbulence caused by the passage of the fluid through the air distributor was investigated. The area below and above the air distributor plate located at the bottom of the reactor was modelled. The dynamics of the air passing through the distributor plates (with pore diameters of 2, 3 and 5 mm) was selected for our simulation. The properties of the fluid with a superficial velocity of 0.2 ms^{-1} was described by the RNG K-ε turbulent model in this section. The selected superficial velocity was about five times greater than the minimum fluidization velocity to ensure complete fluidization [9,37].

A mesh-structured air distributor plate was placed at the bottom of the reactor (Figure 2a). It was necessary to provide a regulated momentum of air inside the reactor chamber to avoid the falling back of the solid particles into the distributor zone, which could consequently have blocked the air nozzles. Distributor plates with different pore sizes (Figure 2b) were used in this simulation to investigate the effect of each perforated plate on the turbulent viscosity of the air. Although the diameter of the plate's air holes was variable, the total area of the pores was intended to be constant by adjusting the number of the pores accordingly.

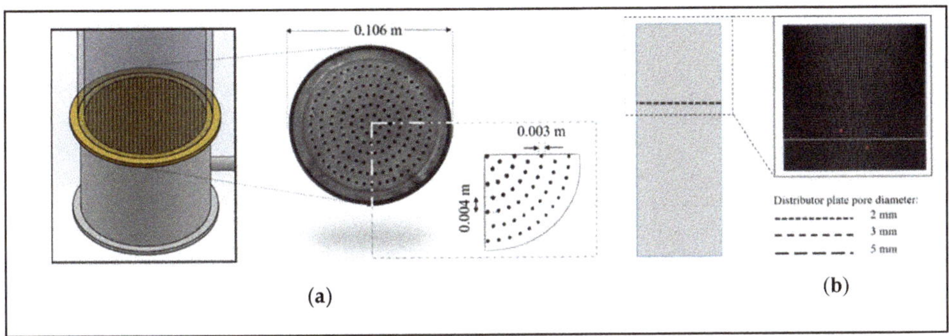

Figure 2. Geometry, boundary condition and grid design for the air distributor plate used in (**a**) experimentation and (**b**) numerical studies.

2.2.5. The Effect of Drag Models on Fluidization Hydrodynamics

Drag force is an key dominant force between the solid particles and the gas phase in a fluidized bed reactor [38]. Typically, the drag law describing the inter-phase momentum exchange is described by experimentally-developed models. However, for a fluidized bed with particular characteristics, it was necessary to investigate whether such models are still applicable. Therefore, three drag models (i.e., Syamlal–O'Brien, Gidaspow and EMMS) were selected in this study to numerically investigate the effect of drag functions on the fluidization hydrodynamics. Initially, the pressure drop throughout the bed was recorded over time for each drag model. That helped to understand when the impact of pressure drop began to diminish and a steady-state was reached. That indicates complete fluidization where the bed has been expanded to its maximum target capacity. At that point, the hydrodynamics of gas–solid fluidization can be defined on a time-averaged basis. Subsequently, the effect of drag models on the volume fraction of the particles was investigated.

2.2.6. Integrated Turbulent Model and Model Validation

A comprehensive model of the fluidization of the granular particles was be developed via the integration of the optimum parameters obtained for the most suitable viscosity, drag and pore size of the air distributor pore models for each stage. The contours of particles' volume fraction obtained from the turbulent model were compared with the standard laminar model in a similar setup. It must be noted that the term "solid" in this study refers to the particle materials in the bed (i.e., granular silica sands). It is of key importance to achieve a well-developed fluidized bed in terms of bed expansion and pressure gradient prior to injecting the biomass into the reactor. Moreover, in the actual process, the biomass particles degrade immediately as injected into the hot reactor and undergo a drastic loss in density while converting into the solid biochar within a very short period of time. The density of the produced char is about 40 times less than the bed material (i.e., 2636 kg m^{-1}) (e.g., Napier grass biochar has the density of 71 kg m^{-1} [36]). Therefore, it was not possible for the fluidizing agent to overcome the drag force, and as a consequence, the particles would fly out of the reactor. Therefore, the characteristics of fluidization, such as the minimum fluidization condition, maximum bed expansion and the bubble formation are mostly governed by the motion pattern of the sand particles and are less effected by the biomass particles.

Once the final integrated model has been set up, it is capable of monitoring the solid volume fraction profile inside the reactor throughout the gasification process. It enabled us to also predict the parameters associated with the particles' motion and the physical aspects of the fluidization, including the kinetic parameters, pressure-related characteristics, formation and transportation of bubbles and the parameters associated with the fluidized bed expansion.

2.2.7. Grid Resolution Study

Grid resolution study was conducted in order to check the accuracy of the results. This was done by pursuing the balance between coarseness of the mesh elements and the computation time required for a solution. Indeed, coarser grids require less computation time and vice versa. The procedure starts with selecting a coarse grid with the least number of elements and performing the computation until the results have been converged with lower relative errors than the pre-defined limit (i.e., 5×10^{-4}) for all models involved in this study. Similar grid sizes and numbers of elements were used in both studies of fluidization hydrodynamics and evaluation of temperature gradient. Therefore, a single grid study was valid for both models. The results of grid resolution tests are shown in Figures 3 and 4, respectively.

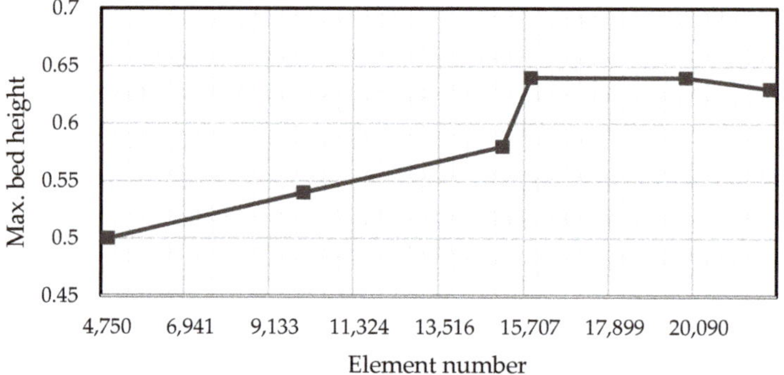

Figure 3. Grid resolution test for laminar flow.

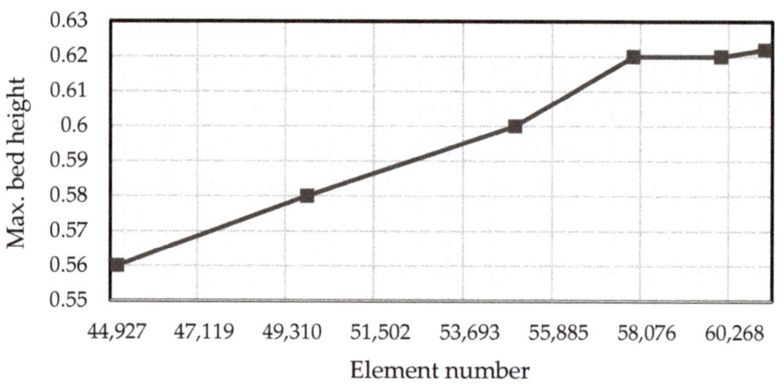

Figure 4. Grid resolution test for re-normalisation group (RNG) K-ε turbulent flow.

The output factor selected in this grid study was the maximum bed height and is related to the particles' superficial velocities along the y-axis. Maximum bed expansion occurs when a particle with the highest y value in all time steps has the velocity of zero. Zero y-velocity with highest possible elevation means that the particle is not travelling to any higher zones, and thus, merely bounces within the fluidization region. An accurate grid quality for a reasonable computation time ensures the highest bed expansion and better fluidization characteristics. For the laminar model, the computations were completed for six runs while each run was committed to a gradual increase in element numbers and eventually converged within a specific time window. Figure 3 shows that the bed height does not significantly increase with finer grid structure above 16,000 elements. Any further increase in element quantities above this point leads to an unnecessarily prolonged computing time. The same trend was observed in the turbulent model but with 58,000 elements (Figure 4). Those optimum grid qualities were employed in the evaluation of the models in this work.

3. Result and Discussion

3.1. The Effect of Viscous Models on Pressure Drop Profiles

The modelled pressure drop profiles against the superficial velocities corresponding to each viscous model are shown in Figure 5.

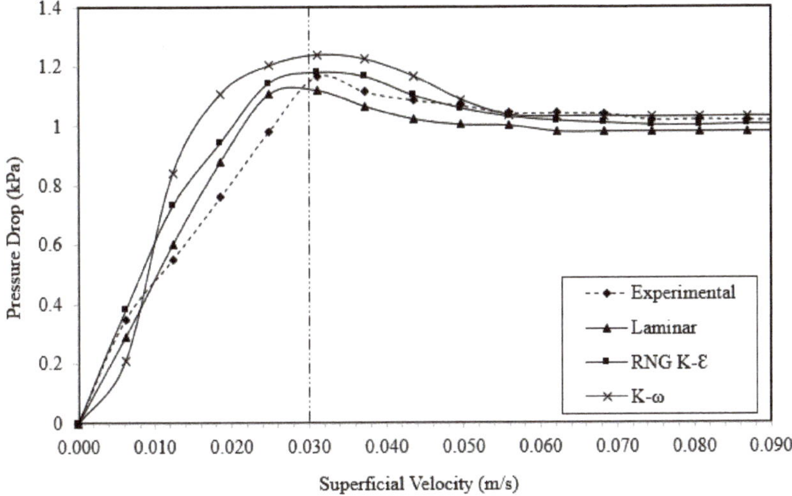

Figure 5. Pressure drop versus superficial velocity for different fluid regimes.

In the laminar regime, the uniform flow of the inlet air stream shows insignificant acceleration, and hence, appears to follow the Hagen–Poiseuille equation order [39]. According to Figure 5, although different pressure drop values are indicated by each model, the minimum fluidization velocities (Umf) are just slightly different. The values of Umf are 0.030, 0.029, 0.031 and 0.037 (ms^{-1}) for experimental, laminar, RNG K-ε and K-ω models, respectively.

At lower superficial velocities, an increase in the pressure drop in the laminar model is seen with a constant slope until the maximum Δp is reached. This can be due to the uniform and vibrating motion of the static bed as well as the regular pressure drop gradient [40]. For the velocities below Umf, an increase in the superficial velocity leads to an increase in the pressure drop; however, the slope was observed to be steeper for the RNG K-ε and K-ω models compared to the laminar model. That can be explained by the different functions of each model in describing the particle interactions and parameters, such as friction coefficient, shear stress, formation of eddies and the associated transferred energy [41]. All models tend to converge and surpass at the onset of fluidization when the superficial velocity is over 0.03 ms^{-1}. The Δp is then recorded at constant pressure of 1.016 kPa during the experimental trial: 0.980 kPa for laminar, 1.001 kPa for RNG K-ε and 1.032 kPa for K-ω models.

According to the statistical analyses, only the two laminar and RNG K-ε models with *p*-values of <0.05 were capable of adequately representing the experimental data. The RNG K-ε with a smaller MSE of 539 was selected as the most desirable model for this study to further describe the turbulent characteristics of the fluidization hydrodynamics.

3.2. The Effect of Air Distributor on the Fluid Properties

In this study, the effect of air distributor on the fluid properties was investigated and an effective turbulent viscosity was calculated to analyse the transfer of momentum caused by turbulent eddies. The corresponding contours of the turbulent viscosity across the distributor area are presented in Figure 6.

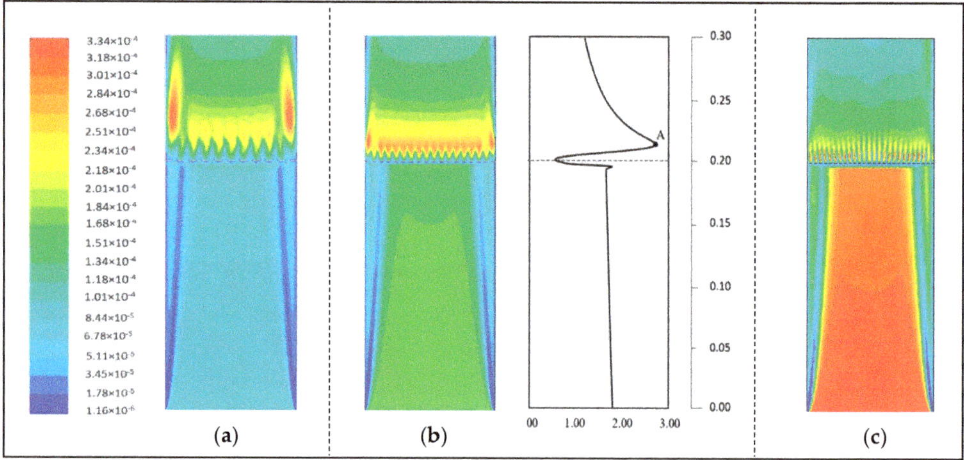

Figure 6. Effect of distributor plate with pore diameters of (a) 5 mm, (b) 3 mm and (c) 2 mm on the turbulent viscosity of the fluid.

The air distributor with larger pores (Figure 6a) was observed to be less effective in increasing the magnitude of the kinetic energy of the fluid. Moreover, the pressure gradient is not uniformly distributed above the surface of the plate. The energy that is introduced by eddies are maximized around the corners which may result in pushing the solid particles to travel along the wall zones and less towards the centre of the reactor. This may cause a sudden pressure drop across the centre which simultaneously reduces the bed expansion. Plates with pore diameters of <5 mm are, therefore, desirable. However, according to Figure 6c, very small pore diameters also tend to be non-ideal as they increase the turbulent viscosity, as well as the gas pressure below the distributor plate. Therefore, they create the red zone—an indication of the air failing to pass through the nozzles. This is due to the fact that the sudden pressure drop imposed by the very small pores leads to the overall pressure of the fluid below the plate to significantly increase [42]. Larger eddies are created across the plate and the chance of a back pressure developing increases. The velocity of the fluid above the distributor plate, on the other hand, is too small to vibrate the solid particles, hence is not able to expand the bed. A plate with a pore diameter of 3 mm (Figure 6b) was observed to have a better turbulent viscosity profile than the other two cases discussed. The distribution of energy above the plate is uniform and the viscosities are maximal (i.e., a 1.8 fold increase). Although the intensity of turbulent viscosity is higher around the wall area, it is relatively constant towards the centre of the plate.

A plot of the turbulent viscosity against the centre line of Y-coordinates is shown in Figure 6b. The turbulent viscosity of the air was slightly reduced from 1.79×10^{-4} to 1.65×10^{-4} kg(ms)$^{-1}$ from the heights of 0 to 0.18 m. However, as the fluid touches the distributor plate, the turbulent viscosity, encountered by an intensive pulse, passes through the grate pores. At the beginning, it rapidly increases to 1.7×10^{-4} kg(ms)$^{-1}$ at a height of 0.196 m and then dramatically drops to 0.58×10^{-4} kg(ms)$^{-1}$ at a height of 0.2 m. It then increases to the maximum value of 2.69×10^{-4} kg(ms)$^{-1}$ at point "A" at a height of 0.217 m and then falls again as it exits from the pores of the mesh structured plate, and flows along the height of reactor until it reaches 1.21×10^{-4} kg(ms)$^{-1}$ at the height of 0.3 m. The resultant pulse was similar to the performance of an air nozzle, indicating a rapid rise in the kinetic energy by applying a sudden pressure drop to the fluid that is passing through the narrow channels of the nozzle [43]. For a better visualization of the phenomena, a closer view of the fluid momentum when passing through the nozzles is shown in Figure 7 with the assistance of velocity vectors.

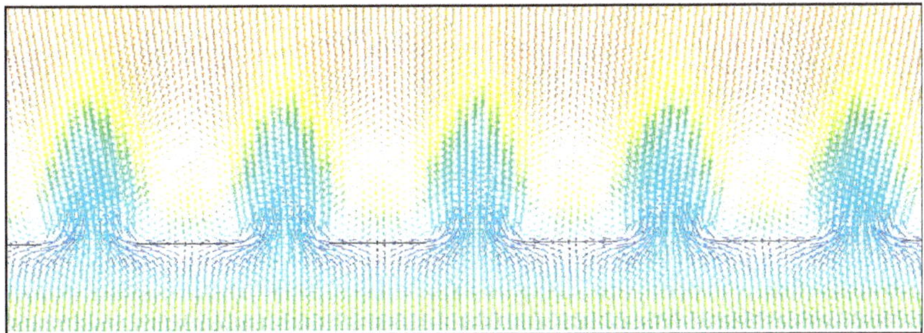

Figure 7. Vector velocity coloured by turbulent viscosity on the pores of the distributor plate.

A sudden pressure drop is observed when the upwards-moving fluid reaches the small air nozzles. The air molecules are contracted as a result and their kinetic energy reduces accordingly while passing through the nozzles. Molecules with a high potential energy eject from the pores to above the distributor plate as their energy converts to kinetic energy in a very short time span, creating the eddies with the highest motions [42]. Since the fluid velocity is adequate and the pore diameters are defined properly, the turbulent viscosity profile becomes consistent from across the top of the plate and creates a steady pressure profile for fluidization. An air distributor plate with a pore diameter of 3 mm was, therefore, considered desirable in this study and will be used for further hydrodynamics investigations. Also, all the calculated quantities corresponding to the turbulent fluidizing fluid have been employed to derive the properties at point "A" in Figure 6b.

3.3. The Effect of Drag Models on Fluidisation Hydrodynamics

The results of pressure drop versus time for the three drag models used in this study are shown in Figure 8. It is observed that the pressure drop corresponding to all the three drag models have considerably reduced and hit a minimum at the beginning of the fluidization in approximately 1.5 s.

However, the pressure drop began to slightly increase until it reached a relatively steady-state condition with least fluctuation after almost 3 s.

Figure 9 indicates the variation of time-averaged solid volume fraction as the result of using the selected drag models along the height of the reactor at the central axis. It is seen that the models predicted by Syamlal–O'Brien and Gidaspow showed relatively lower solid volume fractions at any heights below 0.4 m as compared to the EMMS. However, the overall bed expansion for the model predicted by the EMMS was smaller (i.e., 0.54 m) than those forecast by Syamlal–O'Brien (i.e., 0.58) and Gidaspow (i.e., 0.6). Furthermore, the solid volume fraction at the surface of the fluidized bed gradually dropped to zero for the models predicted by Syamlal–O'Brien and Gidaspow, which seemed to be more physically realistic than those predicted by the EMMS.

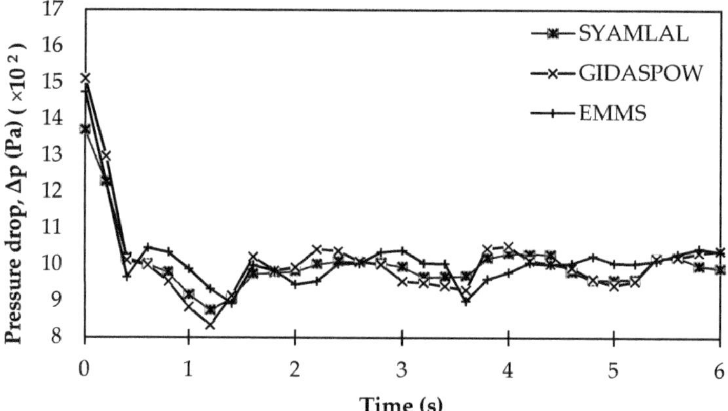

Figure 8. Comparison of simulated bed pressure drop using Syamlal–O'Brien, Gidaspow and energy minimum multi-scale (EMMS) drag models (U = 0.2 ms^{-1}; i.e., ~5 Umf).

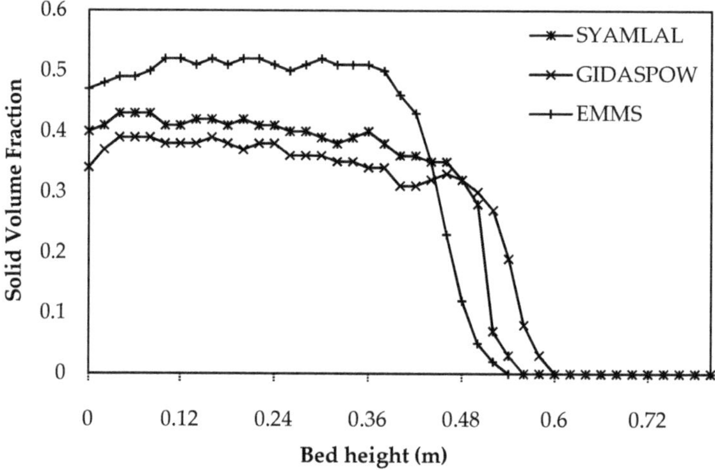

Figure 9. Effect of drag models on the variation of time-averaged solid volume fraction along the reactor height at the central axis.

The EMMS model unrealistically predicts too sharp of a change in the solid volume fraction at similar zones. According to the pressure drop values corresponding to the three drag models (Figure 8), the formation of bubbles and the motion of particles are expected to follow a similar trend; however, the effects of each on the bed expansion and solid volume fraction are significant. The pressure drop corresponding to the minimum fluidization condition obtained experimentally (i.e., 1.166 kPa) was compared to that from the simulation of the drag models. The value obtained from the Syamlal–O'Brien model (i.e., 1.123 kPa) is observed to be more consistent with the experimental data, and therefore, it was considered as the reference drag model in the development of the hydrodynamic model.

3.4. Development of the Fluidisation Hydrodynamics Turbulent Model

The hydrodynamics of fluidization was simulated with the RNG K-ε turbulent model and the Syamlal–O'Brien drag model. An air distributor with a pore diameter of 3 mm was assumed to define the momentum exchange between the gas and solid phases, inter-particle interactions, bubble formation and motion pattern of the bed material in the state of fluidization. The initial air velocity and the turbulent viscosity were 0.2 ms^{-1} and 2.69 × 10^{-4} kg(ms)$^{-1}$, respectively. The process was simulated for the initial 3 s. The contours for the solid volume fractions corresponding to different time steps-starting from the initial condition (i.e., t = 0) and continuing for three seconds of complete fluidization, are illustrated in Figure 10.

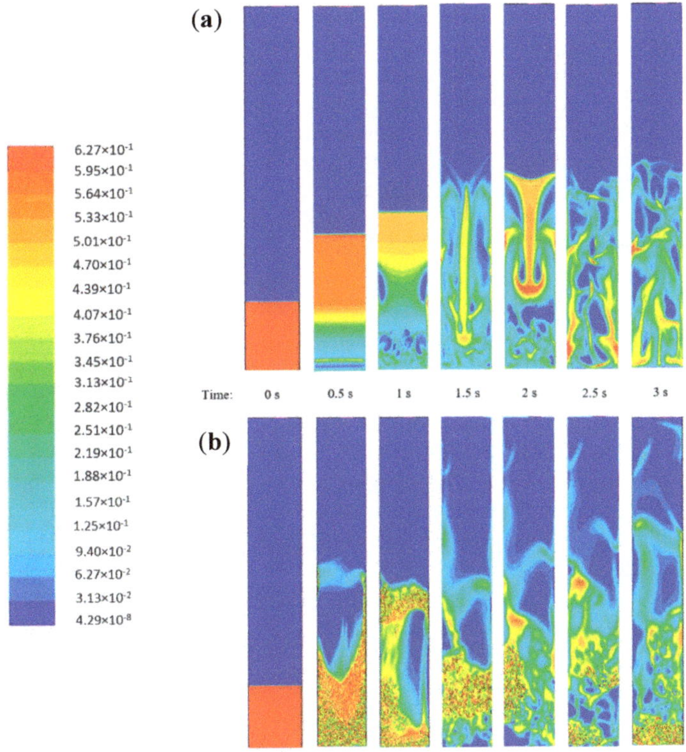

Figure 10. Solid volume fraction contours in a (**a**) laminar model and a (**b**) RNG k-ε turbulent model.

The formations of bubbles as well as the maximum bed expansion were noticeable at each time step. Both systems gradually reached a steady state and became fully-developed within three seconds once the fluidization was initiated. The particles in the laminar model (Figure 10a) are seen to have an axisymmetric distribution. The bubbles began to form at the surface of the air distributor and were enlarged as they moved upwards due to the surface tension gradient. Similar hydrodynamic behaviour has been reported previously [10,13]. However, as depicted in Figure 10b, the formation and development of bubbles as described by the RNG k-ε turbulent model seems to be non-uniform when compared to that described by the laminar model. Large quantities of small bubbles are firstly formed at the surface of the air distributor as opposed to the laminar model. These bubbles then gradually distributed towards the bed medium—more likely closer to the wall. The enlargement of the bubbles close to the wall was due to the friction caused by the wall roughness. The momentum imparted to the walls was introduced by the specularity coefficient for the solid phase at wall conditions. However, the bubbles in the middle section of the chamber were developed in size and quantity due to the interaction with the smaller nearby bubbles. All bubbles formed eventually tend to migrate to and collapse at the bed surfaces. Similar behaviour of bubbling fluidized beds is described in a previous study [44] as the bubbles started to form and expand in diameters prior to upwards movement through the bed medium towards the free board region. The movements of the bubbles, as they collide with the stationary particles, transfers the kinetic energy and causes them to travel in random directions along the height of the gasifier and hence, establishing the fluidization phenomenon.

4. Conclusions

A numerical model with a Eulerian—Eulerian multiphase approach was employed to simulate biomass gasification in a bubbling fluidized bed reactor. Different turbulent models were investigated to study the hydrodynamics of the gasification process. The pressure drop was modelled along the gasifier's height for various superficial velocities, and the results were compared and validated against the experimentally-obtained data. The values corresponding to the RNG k-ε turbulent model were seen to have a better description of the experimental results. Moreover, the impact of the air distributor, with different pore diameters on the momentum characteristics of the air as it enters the gasifier, was investigated. The distributor plate with a pore diameter of 3 mm was selected for generating the largest turbulent viscosities above the distributor's plate (i.e., 2.69×10^{-4} kg(ms)$^{-1}$) with a uniform distribution profile. The drag models of Syamlal–O'brien, Gidaspow and EMMS were investigated in solid fluidization in terms of pressure drop profiles and the volume fraction. In comparison, the Syamlal–O'Brien model corresponding to a pressure drop of 1123 Pa was observed to be the most consistent with the experimental data (i.e., 1166 Pa). Upon the integration of the models' outputs, the contours for the solid volume fractions were generated and compared with the laminar model. The formation of the bubbles and their movement along the gasifier's height as well as the pressure drop profiles in the presence of the turbulent flow followed a different pattern compared to the laminar flow. The developed hydrodynamic model was able to predict the numerical terms associated with the particles' motions and the physical aspects of fluidization, including the kinetics, pressure-related characteristics, formation, and transportation of bubbles and fluidized bed expansion. The derived hydrodynamic model can also help to reduce the number of empirical trials required for process optimization for a fluidized bed reactor—providing accurate details on the physical aspects of fluidization which are challenging to experimentally measure.

Author Contributions: Conceptualization, R.K. and W.A.W.A.K.G.; Funding Acquisition, W.A.W.A.K.G.; Methodology, R.K.; Supervision, W.A.W.A.K.G., D.R.A.B., and R.Y.; Writing—Original Draft, R.K.; Writing—Review & Editing, W.A.W.A.K.G., S.M.S., M.S., S.R.M., and K.S.

Funding: This work has been funded and supported under the Ministry of Higher Education (MOHE) Malaysia via LRGS grant (Grant No. LRGS/2013/UKM/PT). Also, this research has been partially funded by the Engineering and Physical Sciences Research Council through the BEFEW project (Grant No. EP/P018165/1).

Acknowledgments: The authors would like to acknowledge the Chemical and Environmental Engineering Departments at University Putra Malaysia to support conducting this research.

Conflicts of Interest: The authors declare no conflict of interest.

References

1. Khezri, R.; Azlinaa, W.; Tana, H.B. An Experimental Investigation of Syngas Composition from Small-Scale Biomass Gasification. *Int. J. Biomass Renew.* **2016**, *5*, 6–13.
2. Ruiz, J.A.; Juárez, M.C.; Morales, M.P.; Muñoz, P.; Mendívil, M.A. Biomass gasification for electricity generation: Review of current technology barriers. *Renew. Sustain. Energy Rev.* **2013**, *18*, 174–183. [CrossRef]
3. Santos, J.; Ouadi, M.; Jahangiri, H.; Hornung, A. Integrated intermediate catalytic pyrolysis of wheat husk. *Food Bioprod. Process.* **2019**, *114*, 23–30. [CrossRef]
4. Maiti, S.; Sarma, S.J.; Brar, S.K.; Le Bihan, Y.; Drogui, P.; Buelna, G.; Verma, M. Agro-industrial wastes as feedstock for sustainable bio-production of butanol by Clostridium beijerinckii. *Food Bioprod. Process.* **2016**, *98*, 217–226. [CrossRef]
5. Sansaniwal, S.K.; Rosen, M.A.; Tyagi, S.K. Global challenges in the sustainable development of biomass gasification: An overview. *Renew. Sustain. Energy Rev.* **2017**, *80*, 23–43. [CrossRef]
6. Puig-Arnavat, M.; Bruno, J.C.; Coronas, A. Review and analysis of biomass gasification models. *Renew. Sustain. Energy Rev.* **2010**, *14*, 2841–2851. [CrossRef]

7. Couto, N.; Silva, V.; Monteiro, E.; Brito, P.S.D.; Rouboa, A. Experimental and Numerical Analysis of Coffee Husks Biomass Gasification in a Fluidized Bed Reactor. *Energy Procedia* **2013**, *36*, 591–595. [CrossRef]

8. Armstrong, L.M.; Gu, S.; Luo, K.H. Effects of limestone calcination on the gasification processes in a BFB coal gasifier. *Chem. Eng. J.* **2011**, *168*, 848–860. [CrossRef]

9. Sau, D.C.; Biswal, K.C. Computational fluid dynamics and experimental study of the hydrodynamics of a gas–solid tapered fluidized bed. *Appl. Math. Model.* **2011**, *35*, 2265–2278. [CrossRef]

10. Liu, H.; Yoon, S.; Li, M. Three-dimensional computational fluid dynamics (CFD) study of the gas–particle circulation pattern within a fluidized bed granulator: By full factorial design of fluidization velocity and particle size. *Dry. Technol.* **2017**, *35*, 1043–1058. [CrossRef]

11. Hosseini, S.H.; Zivdar, M.; Rahimi, R. CFD simulation of gas–solid flow in a spouted bed with a non-porous draft tube. *Chem. Eng. Process. Process Intensif.* **2009**, *48*, 1539–1548. [CrossRef]

12. Silaen, A.; Wang, T. Effect of turbulence and devolatilization models on coal gasification simulation in an entrained-flow gasifier. *Int. J. Heat Mass Transf.* **2010**, *53*, 2074–2091. [CrossRef]

13. Xie, J.; Zhong, W.; Jin, B.; Shao, Y.; Liu, H. Simulation on gasification of forestry residues in fluidized beds by Eulerian–Lagrangian approach. *Bioresour. Technol.* **2012**, *121*, 36–46. [CrossRef] [PubMed]

14. Abani, N.; Ghoniem, A.F. Large eddy simulations of coal gasification in an entrained flow gasifier. *Fuel* **2013**, *104*, 664–680. [CrossRef]

15. Taghipour, F.; Ellis, N.; Wong, C. Experimental and computational study of gas-solid fluidized bed hydrodynamics. *Chem. Eng. Sci.* **2005**, *60*, 6857–6867. [CrossRef]

16. Papadikis, K.; Gu, S.; Bridgwater, A.V. CFD modelling of the fast pyrolysis of biomass in fluidised bed reactors. Part B: Heat, momentum and mass transport in bubbling fluidised beds. *Chem. Eng. Sci.* **2009**, *64*, 1036–1045. [CrossRef]

17. Armellini, V.A.D.; Di Costanzob, M.; Castroa, H.C.A.; Vergel, J.L.G.; Mori, M.; Martignonic, W.P. Effect of different gas-solid drag models in a high-flux circulating fluidized bed riser. *Chem. Eng.* **2015**, *43*, 1627–1632.

18. Dhrioua, M.; Hassen, W.; Kolsi, L.; Anbumalar, V.; Alsagri, A.S.; Borjini, M.N. Gas distributor and bed material effects in a cold flow model of a novel multi-stage biomass gasifier. *Biomass Bioenergy* **2019**, *126*, 14–25. [CrossRef]

19. Mehrabadi, M. *Fluid-Phase Velocity Fluctuations in Gas-Solid Flows*; Iowa State University Digital Repository: Ames, IA, USA, 2012.

20. Liu, H.; Cattolica, R.J.; Seiser, R.; Liao, C. Three-dimensional full-loop simulation of a dual fluidized-bed biomass gasifier. *Appl. Energy* **2015**, *160*, 489–501. [CrossRef]

21. Liu, H.; Elkamel, A.; Lohi, A.; Biglari, M. Computational Fluid Dynamics Modeling of Biomass Gasification in Circulating Fluidized-Bed Reactor Using the Eulerian–Eulerian Approach. *Ind. Eng. Chem. Res.* **2013**, *52*, 18162–18174. [CrossRef]

22. Cai, J.; Wang, Y.; Zhou, L.; Huang, Q. Thermogravimetric analysis and kinetics of coal/plastic blends during co-pyrolysis in nitrogen atmosphere. *Fuel Process. Technol.* **2008**, *89*, 21–27. [CrossRef]

23. Huang, J.; Lu, Y.; Wang, H. Fluidization of Particles in Supercritical Water: A Comprehensive Study on Bubble Hydrodynamics. *Ind. Eng. Chem. Res.* **2019**, *58*, 2036–2051. [CrossRef]

24. Yancheshme, A.A.; Zarkesh, J.; Rashtchian, D.; Anvari, A. CFD Simulation of Hydrodynamic of a Bubble Column Reactor Operating in Churn-Turbulent Regime and Effect of Gas Inlet Distribution on System Characteristics. *Int. J. Chem. React. Eng.* **2016**, *14*, 213–224. [CrossRef]

25. Yu, G.; Ni, J.; Liang, Q.; Guo, Q.; Zhou, Z. Modeling of Multiphase Flow and Heat Transfer in Radiant Syngas Cooler of an Entrained-Flow Coal Gasification. *Ind. Eng. Chem. Res.* **2009**, *48*, 10094–10103. [CrossRef]

26. Wang, S.; Luo, K.; Hu, C.; Fan, J. CFD-DEM study of the effect of cyclone arrangements on the gas-solid flow dynamics in the full-loop circulating fluidized bed. *Chem. Eng. Sci.* **2017**, *172*, 199–215. [CrossRef]

27. Wu, Y.; Liu, D.; Ma, J.; Chen, X. Effects of gas-solid drag model on Eulerian-Eulerian CFD simulation of coal combustion in a circulating fluidized bed. *Powder Technol.* **2018**, *324*, 48–61. [CrossRef]

28. Wang, J.; Ge, W.; Li, J. Eulerian simulation of heterogeneous gas–solid flows in CFB risers: EMMS-based sub-grid scale model with a revised cluster description. *Chem. Eng. Sci.* **2008**, *63*, 1553–1571. [CrossRef]

29. Ku, X.; Li, T.; Løvås, T. Eulerian–Lagrangian Simulation of Biomass Gasification Behavior in a High-Temperature Entrained-Flow Reactor. *Energy Fuels* **2014**, *28*, 5184–5196. [CrossRef]

30. Wilcox, D. A half century historical review of the k-omega model. In Proceedings of the 29th Aerospace Sciences Meeting, Reno, NV, USA, 7–10 January 1991; p. 615.

31. Perini, F.; Zha, K.; Busch, S.; Reitz, R. *Comparison of Linear, Non-Linear and Generalized RNG-Based K-Epsilon Models for Turbulent Diesel Engine Flows*; SAE: Warrendale, PA, USA, 2017.

32. Lan, X.; Xu, C.; Gao, J.; Al-Dahhan, M. Influence of solid-phase wall boundary condition on CFD simulation of spouted beds. *Chem. Eng. Sci.* **2012**, *69*, 419–430. [CrossRef]

33. Khezri, R.; Azlina, W.; Ab, W.; Ghani, K. Computational Modelling of Gas-Solid Hydrodynamics and Thermal Conduction in Gasification of Biomass in Fluidized-Bed Reactor. *Chem. Eng. Trans.* **2017**, *56*, 1879–1884.

34. Burns, A.D.; Frank, T.; Hamill, I.; Shi, J.-M. The Favre averaged drag model for turbulent dispersion in Eulerian multi-phase flows. In Proceedings of the 5th International Conference on Multiphase Flow (ICMF), Yokohama, Japan, 30 May–4 June 2004; Volume 4, pp. 1–17.

35. Jiang, F.; Zhao, P.; Qi, G.; Li, N.; Bian, Y.; Li, H.; Jiang, T.; Li, X.; Yu, C. Pressure drop in horizontal multi-tube liquid–solid circulating fluidized bed. *Powder Technol.* **2018**, *333*, 60–70. [CrossRef]

36. Khezri, R.; Wan Ab Karim Ghani, W.A.; Awang Biak, D.R.; Yunus, R.; Silas, K. Experimental Evaluation of Napier Grass Gasification in an Autothermal Bubbling Fluidized Bed Reactor. *Energies* **2019**, *12*, 1517. [CrossRef]

37. Pérez, N.P.; Pedroso, D.T.; Machin, E.B.; Antunes, J.S.; Verdú Ramos, R.A.; Silveira, J.L. Fluid dynamic study of mixtures of sugarcane bagasse and sand particles: Minimum fluidization velocity. *Biomass Bioenergy* **2017**, *107*, 135–149. [CrossRef]

38. Hua, L.; Zhao, H.; Li, J.; Wang, J.; Zhu, Q. Eulerian–Eulerian simulation of irregular particles in dense gas–solid fluidized beds. *Powder Technol.* **2015**, *284*, 299–311. [CrossRef]

39. Gooch, J.W. Hagen-Poiseuille equation. In *Encyclopedic Dictionary of Polymers*; Springer: Berlin/Heidelberg, Germany, 2011; p. 355.

40. Ismail, T.M.; Abd El-Salam, M.; Monteiro, E.; Rouboa, A. Eulerian—Eulerian CFD model on fluidized bed gasifier using coffee husks as fuel. *Appl. Therm. Eng.* **2016**, *106*, 1391–1402. [CrossRef]

41. Bakshi, A.; Altantzis, C.; Bates, R.B.; Ghoniem, A.F. Study of the effect of reactor scale on fluidization hydrodynamics using fine-grid CFD simulations based on the two-fluid model. *Powder Technol.* **2016**, *299*, 185–198. [CrossRef]

42. Fu, W.; Li, Y.; Liu, Z.; Wu, H.; Wu, T. Numerical study for the influences of primary nozzle on steam ejector performance. *Appl. Therm. Eng.* **2016**, *106*, 1148–1156. [CrossRef]

43. Bear, J. *Dynamics of Fluids in Porous Media*; Courier Corporation: Chemsford, MA, USA, 2013; ISBN 0486131807.

44. Seo, M.W.; Nguyen, T.D.B.; Lim, Y., II; Kim, S.D.; Park, S.; Song, B.H.; Kim, Y.J. Solid circulation and loop-seal characteristics of a dual circulating fluidized bed: Experiments and CFD simulation. *Chem. Eng. J.* **2011**, *168*, 803–811. [CrossRef]

Article

Simulation Study on Gas Holdup of Large and Small Bubbles in a High Pressure Gas–Liquid Bubble Column

Fangfang Tao, Shanglei Ning, Bo Zhang, Haibo Jin * and Guangxiang He *

Beijing Key Laboratory of Fuels Cleaning and Advanced Catalytic Emission Reduction Technology, School of Chemical Engineering, Beijing Institute of Petrochemical Technology, Beijing 102617, China

* Correspondence: jinhaibo@bipt.edu.cn (H.J.); hgx@bipt.edu.cn (G.H.); Tel.: +86-010-8129-2208 (H.J.)

Received: 1 August 2019; Accepted: 29 August 2019; Published: 4 September 2019

Abstract: The computational fluid dynamics-population balance model (CFD-PBM) has been presented and used to evaluate the bubble behavior in a large-scale high pressure bubble column with an inner diameter of 300 mm and a height of 6600 mm. In the heterogeneous flow regime, bubbles can be divided into "large bubbles" and "small bubbles" by a critical bubble diameter dc. In this study, large and small bubbles were classified according to different slopes in the experiment only by the method of dynamic gas disengagement, the critical bubble diameter was determined to be 7 mm by the experimental results and the simulation values. In addition, the effects of superficial gas velocity, operating pressure, surface tension and viscosity on gas holdup of large and small bubbles in gas–liquid two-phase flow were investigated using a CFD-PBM coupling model. The results show that the gas holdup of small and large bubbles increases rapidly with the increase of superficial gas velocity. With the increase of pressure, the gas holdup of small bubbles increases significantly, and the gas holdup of large bubbles increase slightly. Under the same superficial gas velocity, the gas holdup of large bubbles increases with the decrease of viscosity and the decrease of surface tension, but the gas holdup of small bubbles increases significantly. The simulated values of the coupled model have a good agreement with the experimental values, which can be applied to the parameter estimation of the high pressure bubble column system.

Keywords: high pressure bubble column; the critical bubble diameter; the gas holdup; the large bubbles; the small bubbles

---

1. Introduction

As a common multi-phase reactor, the bubble column reactor has been widely used in petrochemical, fermentation, waste water treatment, mineral processing and metallurgy industries due to its lack of mechanically operated parts, large phase-contacting area, easy operation, high mass transfer and heat transfer efficiency [1–3]. The flow structure in the bubble column is greatly influenced by gas–liquid properties, gas flow rate, bubble size and distributor design [4]. Gas holdup is the volume fraction of gas in the total volume of gas–liquid phase in the bubble column, which is one of the most important parameters to characterize the hydrodynamic characteristics of the bubble column. It is closely related to the bubble size and the superficial gas velocity [5]. Moreover, these parameters will be more or less affected by the hydrodynamic characteristics, such as the style of the gas distributor [6], the column diameter, the liquid height [7] and the liquid properties [8]. It is a focus issue for the design, optimization and scale-up of the bubble column reactor to optimize these operation parameters to improve the gas holdup and phase interface area in the column.

From the mesoscopic scale, there are multi-scales of bubbles in the bubble column. In the homogeneous flow regime, the small sized and uniform bubbles are generated by the gas distributor, and the bubbles rise slowly. However, in the heterogeneous flow regime, because of coalescence and breakup, bubbles can be divided into "large bubbles" and "small bubbles" by a critical bubble diameter dc with a two-bubble-class hydrodynamic model [9–11]. Generally, the small bubbles are in the range of 3–6 mm [10], and the large bubbles are typically in the range of 10–30 mm [10,12]. Therefore, it plays an important role in studying the gas holdup of large and small bubbles, as well as in accurately predicting flow patterns and gas dynamics in the mesoscopic scale [13]. In this study, large and small bubbles were classified only by the different slopes measured by the dynamic gas disengagement method in the experiment [9]. At present, the effects of different factors on gas holdup have been mainly investigated through experimental techniques (dynamic gas disengagement, differential pressure [14], electrical resistance tomography, conductivity probe [15], etc.) and computational fluid simulation. Among them, Besagni et al. [16] used experimental measurements and image analysis to investigate flow regime transition, bubble column hydrodynamics, bubble shapes, and size distributions. Gemello et al. [17] used two optical probes to investigate the effects of contaminants and spargers on the bubble size, and they found that adding contaminants and alcohol in the bubble column inhibited bubble coalescence and caused the decrease of bubble sizes. The data obtained by experimental techniques can provide reliable tools for the validation of the CFD-PBM model. Zhang et al. [18] used dynamic gas disengagement (DGD) to investigate the effects of surface tension, viscosity, pressure and superficial gas velocity on the gas holdup of large and small bubbles. Yang et al. [19] showed that the critical bubble sizes in the acetic acid-air experimental system are 5 and 6 mm by using DGD. The effects of operating pressure (0–2.0 MPa) and superficial gas velocity (0.08–0.32 m/s) on the gas holdup of large and small bubbles were also investigated. The results showed that with the increase of pressure, the gas holdup of small bubbles increased obviously, and with the increase of superficial gas velocity, the gas holdup of large bubbles changed greatly while the gas holdup of small bubbles did not change significantly. Xing [20] also used the DGD method and computational fluid dynamics (CFD) simulations to investigate the influence of superficial gas velocity on the characteristics of bubbles in a deionized water–air system, and they showed a powerful function in different fields for estimating the hydrodynamic behavior in the bubble column, providing a reasonable basis for the design and amplification of the bubble column [21]. CFD is helpful to understand the characteristics of fluid flow by combining simulation results with the experimental results [22], the wake acceleration effect of large and small bubbles, as well as the effect of operating pressure on bubble collapse were considered by Yang et al. [23]. They also used the modified CFD-PBM coupling model to investigate the effect of pressure variation on gas holdup under different superficial gas velocities (0.04–0.16 m/s). The results showed that the gas holdup of small bubbles increased rapidly with the increase of pressure, while the gas holdup of large bubbles basically remained unchanged or decreased slightly.

Most of the previous simulations have focused on changes in total gas holdup under different conditions. However, the distribution characteristics of gas holdup of large and small bubbles are rarely simulated.

The aim of this study was to divide the bubbles into large and small bubbles from the mesoscopic scale, and the coupling CFD-PBM model was used to simulate the results of the literature and present the effect of superficial gas velocity, operating pressure, surface tension, viscosity and other conditions on the gas holdup of large and small bubbles. Through numerical simulation, it was found that the simulation results were basically consistent with the experimental results, and the effect of superficial gas velocity on the gas holdup of large and small bubbles can be well predicted.

2. Experimental Setup

The experimental setup is shown in Figure 1. The material of the bubble column was stainless steel to meet the high pressure experimental conditions. The inner diameter of the reactor was 300 mm, and the height was 6600 mm. The conductivity probe, the differential pressure method and the ERT (electrical resistance tomography) were installed in the height range of 2500–3100 mm on both sides of the bubble column to get the gas holdup. The three measuring methods in different plane had a good reliability in measuring gas holdup, as shown in Figure 2a,b. The experimental system was an air–water system under operating pressure (0.5–2.0 MPa) and superficial velocity (0.16–0.32 m/s). The hydrodynamic parameters in the high pressure bubble column were measured and analyzed by different testing techniques. Among them, two ERT electrode matrices were mounted on two cross-sections of the bubble column height of 3000 (Plane 1) and 2600 mm (Plane 2), and each electrode matrix was composed of 16 electrodes installed in the bubble column.

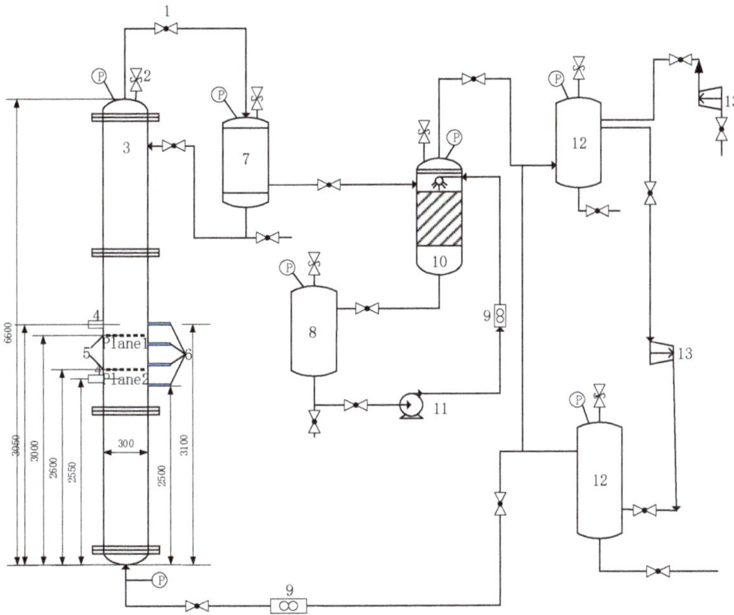

Figure 1. Schematic diagram of the experiment. **1**: Valve; **2**: Safe vale; **3**: Bubble column; **4**: Conductivity probe; **5**: Electrodes of electrical resistance tomography (ERT); **6**: Differential pressure measuring pin; **7**: Gas–liquid separation tank; **8**: Liquid storage tank; **9**: Vortex flowmeter; **10**: Absorption column; **11**: Pump; **12**: Gas storage tank; **13**: Air compressor.

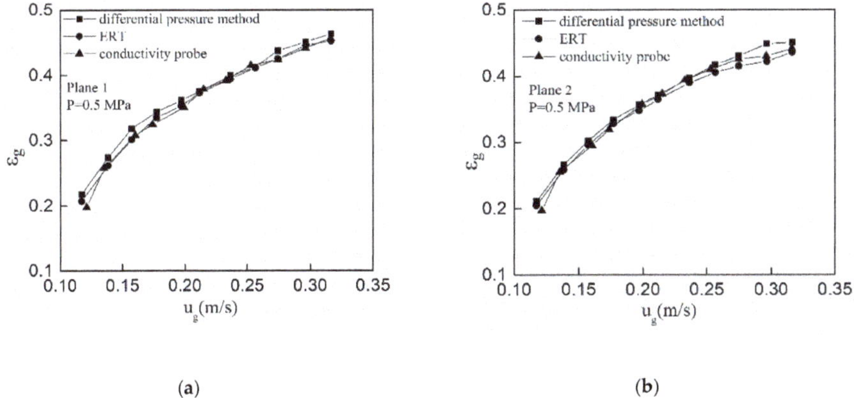

(**a**)   (**b**)

Figure 2. (**a**) The gas holdup ε_g measured by three methods at plane 1 under different superficial gas velocities; (**b**) The gas holdup ε_g measured by three methods at plane 2 under different superficial gas velocities.

3. Mathematical Model

3.1. Two-Fluid Model

In this study, using the FLUENT 15.0 as the platform. In the Euler-Euler model, both water and bubbles were considered as the continuous phase of the calculation zone. The model equation mainly included the continuum equation and the momentum conservation equation. The specific expression is as follows:

Continuum equation: (i = liquid or gas phase)

$$\frac{\partial \alpha_i}{\partial t} + \nabla \cdot (\alpha_i U_i) = 0 \tag{1}$$

Momentum conservation equation: (i = liquid or gas phase)

$$\frac{\partial \alpha_i \rho_i U_i}{\partial t} + \nabla \cdot (\alpha_i \rho_i U_i U_i) = -\alpha \cdot \nabla p + \nabla \cdot (\alpha_i \tau_i) + (-1)^i F_i + \alpha_i \rho_i g \tag{2}$$

3.2. Interphase Force

3.2.1. Drag Force

This study was based on the bubble swarm drag model of Roghair et al. [24]. In the heterogeneous flow regime bubbles in the column can coalescence and break up, it can be seen that the bubble size distribution in the column was wide, which was different from the drag force of single bubble size. The drag force of the bubble swarm is not only related to the liquid phase but is also affected by the interaction between the bubbles. The experimental values of Qin [25] under different superficial gas velocities (0.088–0.317 m/s) and pressures (0.1–2.0 MPa) were used to correct a semi-empirical bubble swarm drag force, as shown in Table 1. The density correction term ρ_g/ρ_0 was introduced into the bubble swarm model, and the modified bubble group drag force model is shown in Equation (3).

Table 1. Bubble group drag coefficient for numerical simulation.

P(MPa)	ρ_g/ρ_0	u_g	$d_{B,exp}$	$\varepsilon_{g,exp}$	C_D	$\varepsilon_{g,sim}$	Error
0.1	1	0.088	9.64	0.17	0.70	0.18	0.17%
		0.132	9.91	0.22	0.40	0.21	−1.94%
		0.154	10.05	0.23	0.28	0.22	−2.08%
		0.199	10.32	0.25	0.21	0.25	0.51%
0.5	5	0.199	9.67	0.35	0.51	0.36	0.42%
		0.233	9.84	0.39	0.46	0.39	−0.83%
		0.275	10.08	0.42	0.41	0.42	−0.81%
		0.317	10.30	0.45	0.36	0.44	−0.97%
1.0	10	0.199	9.39	0.35	0.51	0.36	0.42%
		0.233	9.83	0.49	0.68	0.49	−0.93%
		0.275	10.13	0.51	0.55	0.50	−1.09%
		0.317	10.22	0.52	0.46	0.53	−0.78%
1.5	15	0.199	9.15	0.50	0.88	0.51	0.50%
		0.233	9.69	0.53	0.75	0.53	−0.96%
		0.275	9.93	0.55	0.63	0.55	0.11%
		0.317	10.06	0.56	0.49	0.56	0.82%
2.0	20	0.199	8.98	0.54	0.97	0.54	0.37%
		0.233	9.14	0.56	0.78	0.56	−0.27%
		0.275	9.36	0.57	0.63	0.57	−0.64%
		0.317	9.50	0.58	0.52	0.58	−0.41%

$$\frac{C_D}{C_{D,\infty}(1-\varepsilon_G)} = 3.94\left(\frac{\rho}{\rho_0}\right)^{-0.70} + \left[\frac{97.20}{E_0} - 19.84\left(\frac{\rho}{\rho_0}\right)^{-0.73}\right]\varepsilon_G \tag{3}$$

3.2.2. Turbulent Dispersion Force

Generally, the fluid flow in the bubble column is in a turbulent state. In order to describe the turbulent action in the liquid phase, it is necessary to introduce a turbulent dispersion force, because the introduction of a turbulent diffusion force contributes to uniform calculation results of the gas holdup in the column, which makes it more consistent with the experimental values. In this section, we used the turbulent dispersion force proposed by Lopez de Bertodano [26]. The specific expressions are listed as follows:

$$F_{TD,L} = -\Gamma_{TD,G} = C_{TD}\rho_L k_L \nabla \varepsilon_G \tag{4}$$

where C_{TD} is the turbulent diffusion force coefficient, and its default value is 1.

It is difficult to converge when the numerical simulation is carried out directly using Equation (4). In the FLUENT 15.0 platform, the limit function $f_{TD,limiting}$ is added to the model. Therefore, the modified model of the turbulent dispersion force is given as follows:

$$F_{TD,L} = -F_{TD,G} = f_{TD,limiting}C_{TD}\rho_L k_L \nabla \varepsilon_G \tag{5}$$

$$f_{TD,limiting}(\varepsilon_G) = \max\left(0, \min\left(1, \frac{\varepsilon_{G,2} - \varepsilon_G}{\varepsilon_{G,2} - \varepsilon_{G,1}}\right)\right) \tag{6}$$

where $\varepsilon_{G,1}$ is 0.3 and $\varepsilon_{G,2}$ is 0.7.

3.2.3. Horizontal Lift Force

When the bubble moves upward in the column, the pressure distribution around the bubble is unbalanced due to the asymmetry of the liquid in the direction of movement of the bubble [27]. This produces a horizontal lift force perpendicular to the direction of motion of the bubble. Drew [28] proposed that the lift force experienced in the dispersed phase of the continuous liquid phase is listed as follows:

$$F_L = -C_L \varepsilon_G \rho_L (u_G - u_L) \times (\nabla \cdot u_L) \tag{7}$$

where C_L is the horizontal lift coefficient.

Zhang [29] believed that in the fully developed area, the forces acting on the bubble mainly include the turbulent diffusion force and horizontal lift, and the horizontal lift coefficient C_L and the turbulent diffusion force coefficient C_{TD} are closely related according to the conservation of momentum. The specific expression is as follows:

$$\frac{C_L}{C_{TD}} = -0.2\frac{\varepsilon_L^2}{\varepsilon_L} \tag{8}$$

3.2.4. Wall Lubrication Force

Due to the effect of the wall, the liquid around the bubble is asymmetrical, so the bubble is subjected to a force away from the wall. This force is called wall lubrication force. Nguyen et al. [30] verified that liquid velocity relies on wall lubrication force. Therefore, the model of the wall lubrication force used in this part of the simulation was Tomiyama's equation [31]

$$F_{WL} = C_{WL}\rho_L \varepsilon_G |(u_L - u_G)|^2 n_W \tag{9}$$

The specific expression of C_{WL} is:

$$C_{WL} = C_w \frac{d_b}{2}\left(\frac{1}{y_w^2} + \frac{1}{(D-y_w)^2}\right)$$ (10)

The definition of C_W is:

$$C_w = \begin{cases} 0.47 & Eo < 1 \\ e^{-0.933Eo+0.179} & 1 \le Eo \le 5 \\ 0.00599Eo - 0.0187 & 5 < Eo \le 33 \\ 0.179 & 33 \le Eo \end{cases}$$ (11)

The expression of E_O is:

$$Eo = \frac{g(\rho_L - \rho_G)d_B^2}{\sigma}$$ (12)

3.3. Bubble Breakup Model

Common bubble breakup models are: the Luo model [32], the Lehr model [33], the Ghadiri model, and the Laakkonen model [34]. However, the Luo bubble breakup model has the advantages of simple form, high prediction accuracy and wide application. Thus, in this study, the Luo bubble breakup model was adopted. The specific expression of the Luo model is shown in Equation (13):

$$\Omega_{br}(V, V') = K \int_{\zeta_{min}}^{1} \frac{(1+\zeta)^2}{\zeta^n} \exp(-b\zeta^m)d\zeta$$ (13)

where K, n, m, β, b can be specifically expressed as:

$$\begin{aligned} K &= 0.9238\varepsilon^{1/3}d^{-2/3}\alpha \\ n &= 11/3, m = -11/3, \beta = 2.047 \\ b &= 12\left[f^{2/3} + (1-f)^{2/3} - 1\right]\sigma\rho^{-1}\varepsilon^{-2/3}d^{-5/3}\beta^{-1} \end{aligned}$$ (14)

3.4. Bubble Coalescence Model

Common bubble coalescence models include the Luo model, the free molecular model, and the turbulent-model. The bubble coalescence rate model can be expressed as:

$$\Omega_{ag}(V_iV_j) = \omega(V_iV_j)P(V_iV_j)$$ (15)

The collision frequency between bubbles can be expressed as:

$$\omega(V_iV_j) = \frac{\pi}{4}(d_i + d_j)n_in_j\bar{u}_{ij}$$ (16)

Based on Luo's coalescence efficiency model, the modified coefficient Ce was introduced into the bubble coalescence efficiency model. The modified bubble coalescence efficiency model is listed as follows [35]:

$$Ce = 0.319\ln(\rho/\rho_0) + 0.665$$ (17)

4. Results and Discussion

4.1. Mesh Independence

The numerical simulation was carried out by using the FLUENT 15.0 software as the platform. The CFD-PBM coupling model was presented and used to evaluate the gas holdup of the large and small bubbles influenced by the critical bubble diameter, the superficial gas velocity, pressure, surface tension and viscosity in a large-scale high pressure bubble column. In the numerical simulation, the two-dimensional axisymmetric model could optimize the time of the calculation due to the small number of meshes. Figure 3a is a two-dimensional axisymmetric geometric model taken by the numerical simulation. Figure 3b is a grid map taken from the bottom of the column at 2000 and 3200 mm. The detailed information of meshing is shown in Reference [35].

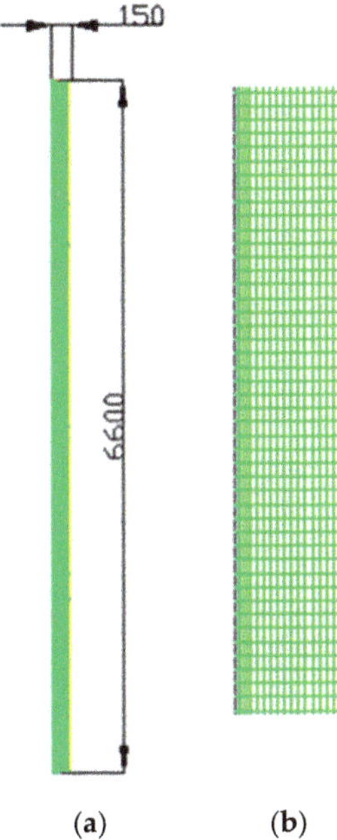

Figure 3. (a) Two-dimensional axisymmetric model. (b) A part of the meshing scheme.

The meshing of the geometric model had a great influence on the numerical simulation results. As the number of meshes increased, it not only requested improvements of the performance of the computer, it also increased the calculation time. Therefore, in order to improve the numerical simulation accuracy and the calculation efficiency, a suitable meshing method was the basis of the numerical simulation.

The mesh independence was investigated under the conditions of the superficial gas velocity of 0.317 m/s and a pressure of 0.5 MP, and, as such, four grids with grid numbers of 3960, 5940, 17160 and 47520 were used, respectively. The effects of meshing on radial gas holdup (Figure 4a), axial gas velocity (Figure 4b), and axial fluid velocity (Figure 4c) were verified. By comparison, it was found that selecting a grid with a grid number of 5940 could ensure the accuracy of calculation of each fluid mechanics parameter and could also satisfy a small calculation amount.

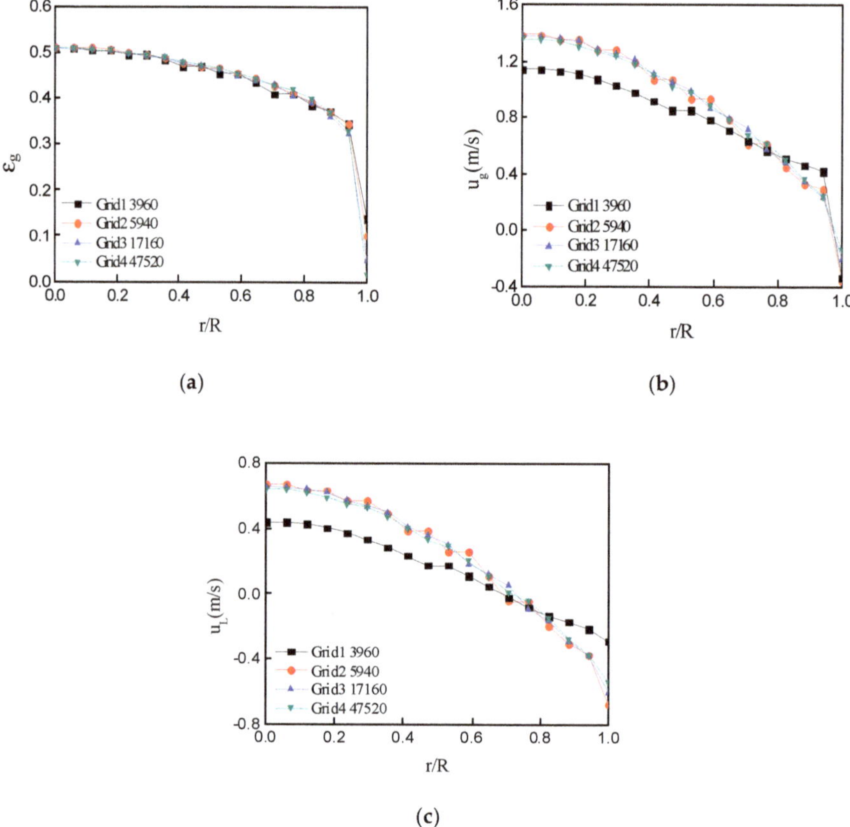

Figure 4. (**a**) Influence of grid partition on radial gas holdup. (**b**) Influence of grid partition on the axial gas velocity. (**c**) Influence of grid partition on the axial liquid velocity.

4.2. Determination of Critical Bubble Diameters

In the experiment, a bubble diameter of 3–6 mm could be regarded as small bubbles, and a bubble diameter of 10–80 mm could be regarded as large bubble [10–12]. Thus, it was very meaningful to distinguish bubble size from bubble swarm for calculating gas–liquid mass transfer characteristics. Xing et al. [20] used DGD to measure the tendency of the gas holdup of large bubbles with the superficial gas velocity in his experiment, and they gave a suggestive bubble critical value. In this study, we used 6, 7 and 8 mm as the critical bubble diameters to simulate the gas holdup, and we compared them with the experimental values obtained by DGD. From Figure 5, it can be seen that the results of simulation using different critical bubble diameters were different. It was found that the variation trend of the gas holdup of large bubbles was apparently consistent with the superficial gas velocity—that is, the gas holdup of large bubbles gradually increased with an increase of superficial gas velocity. However, when the critical bubble diameter was 6 mm, the simulation value of gas holdup of large bubbles was significantly higher than the experimental value. When the critical bubble diameter was 8 mm, the simulation value of gas holdup of the large bubble was lower than the experimental value. Therefore, the critical diameter of 7 mm of the bubble was adopted.

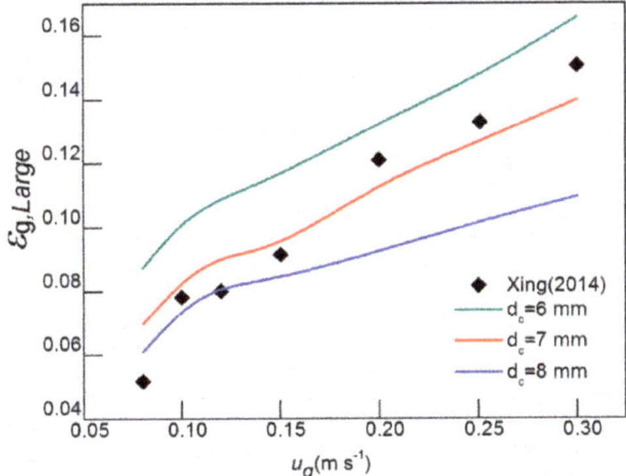

Figure 5. Effect of different critical bubble diameters on gas holdup of large bubbles.

4.3. The Gas Holdup of Large Bubbles and Small Bubbles

4.3.1. Effect of the Superficial Gas Velocity on the Gas Holdup

Figure 6 shows the effect of superficial gas velocity on the average gas holdup, large bubbles gas holdup, and small bubbles gas holdup at different pressures (0.5, 1.0, 1.5, and 2.0 MPa). The effects of superficial gas velocity with different pressures on various gas holdup were simulated by the modified CFD-PBM coupling model. At the same time, compared the date obtained by Yang [19] using DGD, the variation of the gas holdup of large and small bubbles with the change of the superficial gas velocity was analyzed. It was found that the simulation value of the gas holdup of small bubbles was in good agreement with the experimental value. However, the gas holdup of large bubbles was slightly smaller than the experimental value. With the increase of superficial gas velocity, the gas holdup of large and small bubbles increased gradually, and the increased tendency of large bubbles was smaller than small bubbles. This variation trend is consistent with the experimental results of Xing [20]. The main reason for this phenomenon is that with the increase of superficial gas velocity, the turbulence within the column was intensified, and the bubble size was relatively uniform and smaller in diameter when the breakup and coalescence between bubbles were balanced. In the case that the critical bubble diameter was determined, the number of small bubbles increased and the number of large bubbles decreased. It can also be seen from Figure 6 that with the increase of pressure, the increased rate of the average gas holdup and the gas holdup of small bubbles gradually decreased with the increase of the superficial gas velocity. This is mainly because as the pressure got higher and higher, the bubble size got smaller and narrower. However, the increasing pressure had a slight influence on the bubble size. Under the determination of critical bubble diameter, the gas holdup of the small bubble was higher, but the small bubble gradually slowed down with the increase of superficial gas velocity.

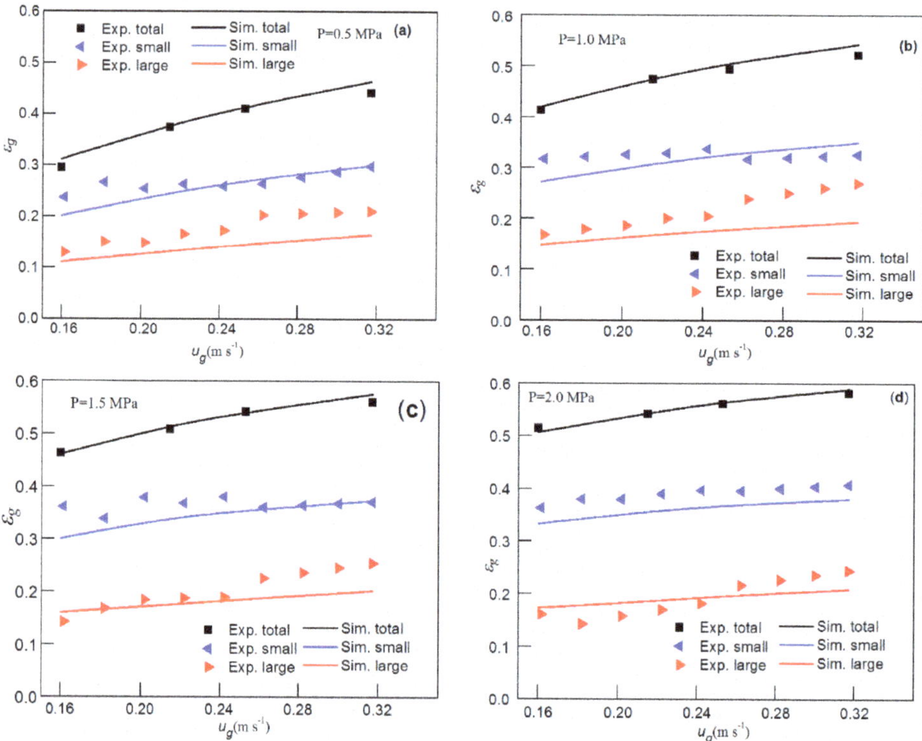

Figure 6. Effect of different superficial gas velocities on the gas holdup of large and small bubbles in different pressure ((**a**) P = 0.5 Mpa; (**b**) P = 1.0 Mpa; (**c**) P = 1.5 Mpa; (**d**) P = 2.0 Mpa).

4.3.2. Effect of the Different Pressure on the Gas Holdup

Figure 7 shows the effect of operating pressure on the gas holdup of large and small bubbles under different superficial gas velocities (0.160, 0.215, 0.253, and 0.317 m/s). The variation trend of the average gas holdup, large bubble gas holdup and small bubble gas holdup was investigated by numerical simulation. From Figure 7, it can be seen that the experimental values of Yang [19] and the values obtained through the CFD-PBM simulation are very consistent, especially the values of average gas holdup and small bubble gas holdup. For large bubble gas holdup, it can be seen that the simulated values were lower than the experimental values, especially when the pressure was higher, which makes the difference slightly obvious. This better illustrates that the modified CFD-PBM coupling model can predict the variation of gas holdup in the bubble column at different superficial gas velocities and different pressures. In addition, the gas holdup of large bubbles increased slowly with the increase of pressure. Moreover, compared with the gas holdup of large bubbles, the gas holdup of small bubbles increased significantly. The variation tendency of the gas holdup of large and small bubbles is consistent with the experimental results of Jordan et al. [36] and Krishna and Ellenberger et al. [10]. From Figure 7, it can be seen that when P ≤ 1 MPa, the average gas holdup and the gas holdup of large bubbles increased rapidly. When P ≥ 1 MPa, both increased slowly. Therefore, with the increase of pressure, the effect of pressure on the average gas holdup and the gas holdup of large bubbles gradually decreased.

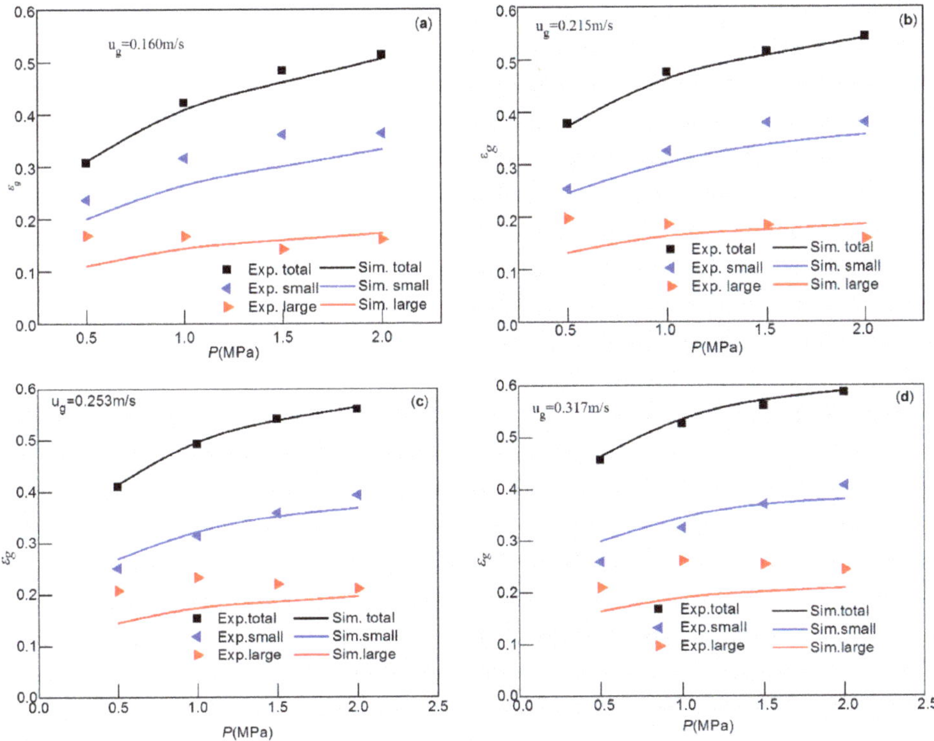

Figure 7. Effect of different pressure on the gas holdup of large bubbles and small bubbles at different superficial velocities ((**a**) u_g = 0.160 m/s; (**b**) u_g = 0.215 m/s; (**c**) u_g = 0.253 m/s; (**d**) u_g = 0.317 m/s).

4.3.3. Effect of Surface Tension on the Gas Holdup of Large and Small Bubbles

Figures 8 and 9 show the influence of different surface tensions (σ = 49.9 × 10^{-3}, 60.7 × 10^{-3}, 66.7 × 10^{-3}, 70.0 × 10^{-3} N/m) on the gas holdup of large and small bubbles at different superficial gas velocities under high-pressure conditions. From Figure 8, it can be seen that the gas holdup of small bubbles increased with the superficial gas velocity under different surface tensions. However, the gas holdup of small bubbles under a low surface tension was significantly higher than that under a high surface tension, which shows that the gas holdup of small bubbles gradually decreased with the increase of surface tension. Under the large surface tension (σ = 70.0 × 10^{-3} N/m), the simulated values agreed well with the experimental values. Under other surface tensions (σ = 49.9 × 10^{-3}, 60.7 × 10^{-3}, 66.7 × 10^{-3} N/m), the simulated value was consistent with the small bubble gas holdup measured by DGD, and there was a certain error. This may be because the critical bubble diameter was set too small. Thus, in the case of low surface tension, it helped to form small bubbles, and small bubbles rarely coalesced in the liquid phase. As such, the experimental gas holdup increased. From Figure 9, it can be seen that under different surface tensions, the gas holdup of large bubbles increased with the increase of superficial gas velocity, and the gas holdup of large bubbles increased with the increase of surface tension. The simulated value was in good agreement with the experimental value under a large surface tension. However, the simulated value of the large bubble under the small surface tension was larger than the experimental value, and, under the large surface tension, the experimental value was slightly higher than the simulated value.

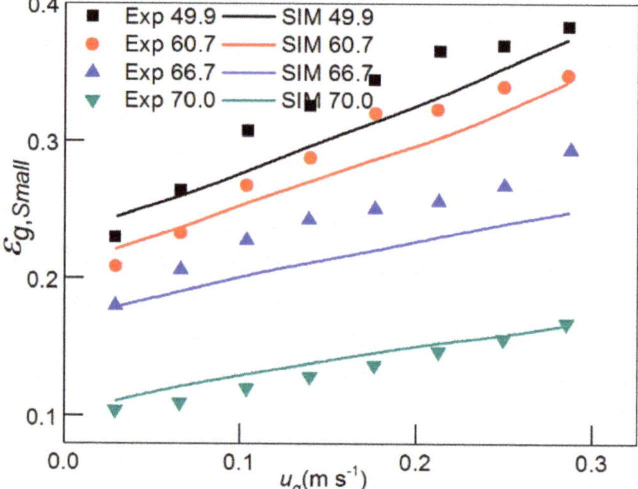

Figure 8. Effect of surface tension on the gas holdup of small bubbles.

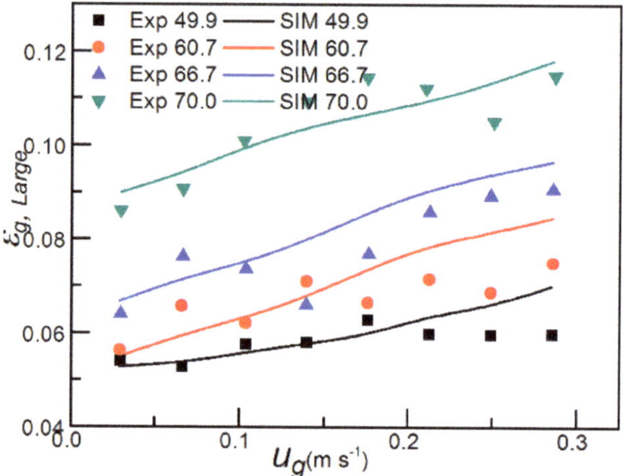

Figure 9. Effect of surface tension on the gas holdup of large bubbles.

By comparing the experimental and simulated values of the gas holdup of large and small bubbles, the modified CFD-PBM coupling model could basically investigate the influence of surface tension on the gas holdup of small bubbles in a high-pressure gas–liquid two-phase flow.

4.3.4. Effect of the Viscosity on the Gas Holdup

As can be seen from Figures 10 and 11, the modified CFD-PBM coupling model under high pressure was used to investigate the effect of different viscosities ($\mu = 1.41 \times 10^{-3}$, 1.91×10^{-3}, 2.35×10^{-3}, 3.54×10^{-3} Pas) on the gas holdup of large and small bubbles.

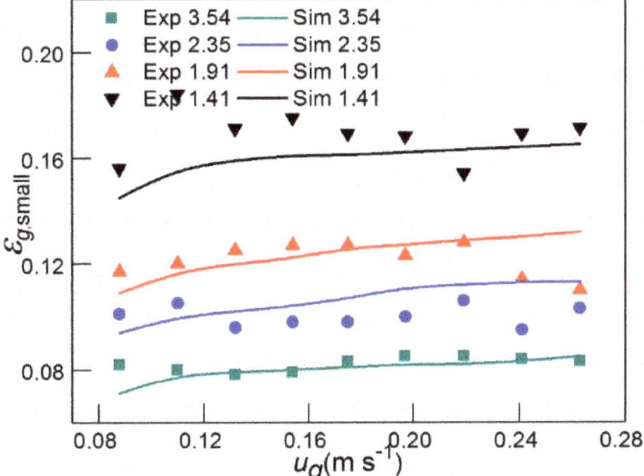

Figure 10. Effect of the viscosity on the gas holdup of small bubbles.

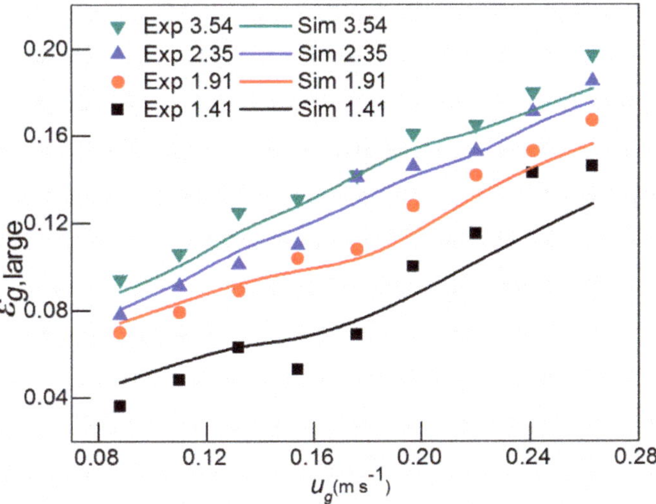

Figure 11. Effect of the viscosity on the gas holdup of large bubbles.

From Figures 10 and 11, it can be seen the gas holdup of small bubbles decreased with the increase of viscosity. With the increase of superficial gas velocity, the gas holdup first increased and then remained unchanged. The experimental results are consistent with results in Reference [20], mainly because the viscosity had little effect on the gas holdup in the case of low viscosity. As the viscosity gradually increased, the effect of viscosity on the gas holdup increased, resulting in a decrease of small bubbles in the column. The gas holdup of large bubbles increased with the increase of viscosity, and with the increase of superficial gas velocity, the gas holdup also increased. This was mainly because, with the increase of liquid viscosity, small bubbles in the column coalesced and formed large bubbles, which increased the bubble diameter and increased the gas holdup of large bubbles in the column. Yang [19] and Khare [37] also gave a reasonable explanation of the influence of viscosity on the gas holdup of large bubbles. The modified CFD-PBM coupling model was used to basically investigate the effect of viscosity on the gas holdup of small bubbles and to have a better prediction of the gas holdup in a high-pressure gas–liquid two-phase flow.

5. Conclusions

In this paper, the modified CFD-PBM coupling model by FLUENT 15.0 was used to simulate the high pressure gas–liquid two-phase flow in a bubble column, and the simulated values were compared with the experimental values. The main results were obtained as follows:

(1)   Using 6, 7 and 8 mm as critical bubble diameters, the variation trend of the gas holdup of the large bubbles with the superficial gas velocity was obtained from the simulation results, and it was compared with the gas holdup of Xing [19] in the water–air system. It was finally determined the critical bubble diameter that divided the bubble into large and small bubbles was 7 mm.

(2)   Using the modified CFD-PBM coupling model, the effects of superficial gas velocity and operating pressure on the gas holdup of large bubbles and small bubbles were analyzed. It is found that as the superficial gas velocity increased, the gas holdup of large and small bubbles increased to varying degrees. On the other hand, with the increase of pressure, the influence of pressure on the gas holdup of large bubbles gradually weakened. In the high pressure, the gas holdup of the small bubble increased with the increase of the superficial gas velocity.

(3)   Compared with the results of the cold model experiment, it is found that the modified CFD-PBM coupling model could effectively estimate the influence of surface tension and viscosity on the gas holdup of large and small bubbles. That is, the gas holdup of the small bubbles gradually decreased as the surface tension and viscosity increased. The gas holdup of the large bubble gradually increased with the increase of the surface tension and viscosity.

Author Contributions: Conceptualization, F.T. and S.N.; methodology, B.Z.; software, B.Z.; validation, F.T., S.N. and B.Z.; formal analysis, F.T.; investigation, F.T.; resources, B.Z.; data curation, F.T.; writing—original draft preparation, S.N.; writing—review and editing, S.N.; supervision, H.J. and G.H.; project administration, H.J.

Funding: This research was funded by the National Natural Science Foundation of China, grant number 91634101 and The Project of Construction of Innovative Teams and Teacher Career Development for Universities and Colleges under Beijing Municipality, grant number IDHT20180508.

Conflicts of Interest: The authors declare no conflict of interest.

Abbreviations

ε_g	——[-]	gas phase holdup
dc	——[mm]	The critical bubble diameter, mm
u_i	——[m s^{-1}]	velocity, m·s^{-1}, $i = 1$: gas phase, $i = 2$: liquid phase
g	——[m s^{-2}]	gravitational acceleration, m s^{-2}
ε_L	——[-]	liquid phase holdup
ρ	——[kg m^{-3}]	density, kg·m^{-3}
U	——[m s^{-1}]	velocity, m·s^{-1}
τ	——[-]	effective pressure tensor
g	——[m s^2]	gravitational acceleration, m·s^2
ε	——[m^2 s^{-3}]	turbulent dissipation rate, m^2·s^{-3}
μ_t	——[Pa s]	turbulent viscosity, Pa·s
K, k_L	——[m^2 s^{-2}]	turbulent kinetic energy, m^2·s^{-2}
F_D	——[N m^{-3}]	drag, N·m^{-3}
u_G	——[m s^{-1}]	gas velocity, m·s^{-1}
u_L	——[m s^{-1}]	liquid velocity, m·s^{-1}
C_D	——[-]	drag coefficient
$C_{D,\infty}$	——[-]	ideal state drag coefficient
Eo	——[-]	parameter Eo
F_L	——[N]	transverse lift, N
C_L	——[-]	transverse lift coefficient

C_{TD}	——[-]	turbulent dispersion coefficient
$F_{TD,L}\ F_{TD,G}$	——[N]	turbulent dispersion force, N
$f_{TD,limiting}$	——[-]	turbulent diffusion force model limiting function
F_{WL}	——[N]	wall lubrication force, N
C_{WL}	——[-]	wall lubrication coefficient
Eo	——[-]	parameter Eo
d_B	——[m]	diameter of the bubble, m
$\Omega_{br}(V,V')$	——[-]	bubble breakage rate
σ	——[N s^{-1}]	surface tension, N·s^{-1}
ζ	——[-]	relative diameter of the bubble
ζ_{min}	——[-]	minimum relative diameter of the bubble
$\Omega_{ag}(V_iV_j)$	——[-]	bubble coalescence rate
$\omega(V_iV_j)$	——[m^3 s^{-1}]	collision frequency between bubbles of size d_i and d_j, m^3·s^{-1}
$P(V_iV_j)$	——[-]	bubble coalescence efficiency
u_{ij}	——[-]	characteristic velocity of bubble collision

Lower subscript

G	——	gas phase
L	——	liquid phase
i	——	referring to the gas phase or the liquid phase
b	——	bubble
i,j	——	bubble section

References

1. Parisien, V.; Farrell, A.; Pjontek, D.; McKnight, C.; Wiens, J.; Macchi, A. Bubble swarm characteristics in a bubble column under high gas holdup conditions. *Chem. Eng. Sci.* **2017**, *157*, 88–98. [CrossRef]
2. Hur, Y.; Yang, J.; Jung, H.; Lee, K. Continuous alcohol addition in vaporized form and its effect on bubble behavior in a bubble column. *Chem. Eng. Res. Des.* **2014**, *92*, 804–811. [CrossRef]
3. Hlawitschka, M.; Kováts, P.; Zähringer, K.; Bart, H. Simulation and experimental validation of reactive bubble column reactors. *Chem. Eng. Sci.* **2016**, *170*, 306–319. [CrossRef]
4. Sommerfeld, M.; Bröder, D. Analysis of Hydrodynamics and Microstructure in a Bubble Column by Planar Shadow Image Velocimetry. *Ind. Eng. Chem. Res.* **2009**, *48*, 330–340. [CrossRef]
5. Saleh, S.; Mohammed, A.; Al-Jubory, F.; Barghi, S. CFD assesment of uniform bubbly flow in a bubble column. *Petrol. Sci. Eng.* **2018**, *196*, 96–107. [CrossRef]
6. Sharaf, S.; Zednikova, M.; Ruzicka, M.; Azzopardi, M. Global and local hydrodynamics of bubble columns–Effect of gas distributor. *Chem. Eng. J.* **2016**, *288*, 489–504. [CrossRef]
7. Sasaki, S.; Uchida, K.; Hayashi, K.; Tomiyama, A. Effects of column diameter and liquid height on gas holdup in air-water bubble columns. *Exp. Therm. Fluid. Sci.* **2017**, *82*, 359–366. [CrossRef]
8. Mouza, A.; Dalakoglou, G.; Paras, S. Effect of liquid properties on the performance of bubble column reactors with fine pore spargers. *Chem. Eng. Sci.* **2005**, *60*, 1465–1475. [CrossRef]
9. Jin, H.; Yang, S.; Zhang, T.; Tong, Z. Bubble behavior of a large-scale bubble column with elevated pressure. *Chem. Eng. Technol.* **2004**, *27*, 1007–1013. [CrossRef]
10. Krishna, R.; Ellenberger, J. Gas holdup in bubble column reactors operating in the churn-turbulent flow regime. *AIChE J.* **1996**, *42*, 2627–2634. [CrossRef]
11. Shetty, S.A.; Kantak, M.V.; Kelkar, B.G. Gas-phase backmixing in bubble-column reactors. *AIChE J.* **1992**, *38*, 1013–1026. [CrossRef]
12. De Swart, J.W.A.; van Vliet, R.E.; Krishna, R. Size, structure and dynamics of large bubbles in a two-dimensional slurry bubble column. *Chem. Eng. Sci.* **1996**, *51*, 4619–4629. [CrossRef]
13. Pourtousi, M.; Ganesan, P.; Sahu, J. Effect of bubble diameter size on prediction of flow pattern in Euler–Euler simulation of homogeneous bubble column regime. *Measurement* **2015**, *76*, 255–270. [CrossRef]
14. Guan, X.; Yang, N. Bubble Properties Measurement in Bubble Columns: From Homogeneous to Heterogeneous Regime. *Chem. Eng. Res. Des.* **2017**, *107*, 103–122. [CrossRef]
15. Zhang, B. Numerical Simulation of Gas-Liquid Flow in a Pressurized Bubble Column Using the CFD-PBM Coupled Model. Master's Thesis, Beijing Institute of Petrochemical Technology, Beijing, China, 2018.

16. Besagni, G.; Inzoli, F. Bubble size distributions and shapes in annular gap bubble column. *Exp. Therm. Fluid. Sci.* **2016**, *74*, 27–48. [CrossRef]

17. Gemello, L.; Plais, C.; Augier, F.; Cloupet, A.; Marchisio, D. Hydrodynamics and bubbe size in bubble columns: Effects of contaminants and spargers. *Chem. Eng. Sci.* **2018**, *184*, 93–102. [CrossRef]

18. Zhang, T.; Jin, H.; He, G.; Yang, S.; Tong, Z. Application of Pressure Transducing Technology to Measurement of Hydrodynamics in Bubble Column. *Chin. J. Chem. Eng.* **2004**, *55*, 476–480. [CrossRef]

19. Yang, S. The Hydrodynamic Characteristics of Slurry Bubble Column Reactors with Elevated Pressure. Master's Thesis, Beijing University of Chemical Technology, Beijing, China, 2004.

20. Xing, C. Experimental Study and Numerical Simulations of Bubble Column with a CFD-PBM Coupled Model. Ph.D. Thesis, Tsinghua University, Beijing, China, 2014.

21. Shaikh, A.; Al-Dahhan, M. Scale-up of Bubble Column Reactors: A Review of Current State-of-the-Art. *Ind. Eng. Chem. Res.* **2013**, *52*, 8091–8108. [CrossRef]

22. Sarhan, A.; Naser, J.; Brooks, G. CFD analysis of solid particles properties effect in three-phase flotation column. *Sep. Purif. Technol.* **2017**, *185*, 1–9. [CrossRef]

23. Yang, G.; Guo, K.; Wang, T. Numerical simulation of the bubble column at elevated pressure with a CFD-PBM coupled model. *Chem. Eng. Sci.* **2017**, *170*, 251–262. [CrossRef]

24. Roghair, I.; Van, S.; Kuipers, H. Drag force and clustering in bubble swarms. *AIChE J.* **2013**, *59*, 1791–1800. [CrossRef]

25. Qin, Y. The Measurement of Hydrodynamic Parameters and CFD Simulation of Gas-Liquid Flow Behaviors in a Pressurized Bubble Column. Master's Thesis, Beijing University of Chemical Technology, Beijing, China, 2012.

26. Lopez, M. *Turbulent Bubbly Two-Phase Flow in a Triangular Duct*; Rensselaer Rolytechnic Institute: New York, NY, USA, 1992.

27. Ueyama, K.; Miyauchi, T. Properties of recirculating turbulent two phase flow in gas bubble columns. *AIChE J.* **2010**, *25*, 258–266. [CrossRef]

28. Drew, D. Mathematical modeling of two-phase flow. *Annu. Rev. Fluid. Mech.* **1983**, *15*, 261–291. [CrossRef]

29. Zhang, Y. Hydrodynamics of Turbulent Bubble Column with and without Internals in Well-Developed Flow Region. Ph.D. Thesis, Zhejiang University, Hangzhou, China, 2011.

30. Nguyen, V.; Song, C.; Bae, B.; Euh, D. The dependence of wall lubrication force on liquid velocity in turbulent bubbly two-phase flows. *J. Nucl. Sci. Technol.* **2013**, *50*, 781–798. [CrossRef]

31. Tomiyama, A. Struggle with Computational Bubble Dynamics. *Multiph. Sci. Technol.* **1998**, *10*, 369–405. [CrossRef]

32. Luo, H.; Svendsen, H.F. Theoretical Model for drop and bubble breakup in turbulent dispersions. *AIChE J.* **1996**, *42*, 1225–1233. [CrossRef]

33. Lehr, F.; Millies, M.; Mewes, D. Bubble size distributions and flow fields in bubble column. *AIChE J.* **2002**, *48*, 2426–2443. [CrossRef]

34. Laakkonen, M.; Alopaeus, V.; Aittamaa, J. Validation of bubble breakage, coalescence and mass transfer models for gas–liquid dispersion in agitated vessel. *Chem. Eng. Sci.* **2006**, *61*, 218–228. [CrossRef]

35. Zhang, B.; Kong, L.; Jin, H.; He, G.; Yang, S.; Guo, X. CFD simulation of gas–liquid flow in a high-pressure bubble column with a modified population balance model. *Chin. J. Chem. Eng.* **2018**, *26*, 125–133. [CrossRef]

36. Jordan, U.; Schumpe, A. The gas gensity effect on mass transfer in bubble columns with organic liquids. *Chem. Eng. Sci.* **2001**, *56*, 6267–6272. [CrossRef]

37. Khare, A.; Joshi, J. Effect of fine particles on gas hold-up in three-phase sparged reactors. *Chem. Eng. J.* **1990**, *44*, 11–25. [CrossRef]

Article

Predicting By-Product Gradients of Baker's Yeast Production at Industrial Scale: A Practical Simulation Approach

Christopher Sarkizi Shams Hajian [1], Cees Haringa [2], Henk Noorman [2,3] and Ralf Takors [1,*]

[1]   Institute of Biochemical Engineering, University of Stuttgart, 70569 Stuttgart, Germany; c.sarkizi@ibvt.uni-stuttgart.de

[2]   DSM Biotechnology Center, 2613 AX Delft, The Netherlands; cees.haringa@dsm.com (C.H.); henk.noorman@dsm.com (H.N.)

[3]   Department of Biotechnology, Delft University of Technology, 2628 CD Delft, The Netherlands

[*]   Correspondence: takors@ibvt.uni-stuttgart.de

Received: 21 October 2020; Accepted: 25 November 2020; Published: 27 November 2020

Abstract: Scaling up bioprocesses is one of the most crucial steps in the commercialization of bioproducts. While it is known that concentration and shear rate gradients occur at larger scales, it is often too risky, if feasible at all, to conduct validation experiments at such scales. Using computational fluid dynamics equipped with mechanistic biochemical engineering knowledge of the process, it is possible to simulate such gradients. In this work, concentration profiles for the by-products of baker's yeast production are investigated. By applying a mechanistic black-box model, concentration heterogeneities for oxygen, glucose, ethanol, and carbon dioxide are evaluated. The results suggest that, although at low concentrations, ethanol is consumed in more than 90% of the tank volume, which prevents cell starvation, even when glucose is virtually depleted. Moreover, long exposure to high dissolved carbon dioxide levels is predicted. Two biomass concentrations, i.e., 10 and 25 g/L, are considered where, in the former, ethanol production is solely because of overflow metabolism while, in the latter, 10% of the ethanol formation is due to dissolved oxygen limitation. This method facilitates the prediction of the living conditions of the microorganism and its utilization to address the limitations via change of strain or bioreactor design or operation conditions. The outcome can also be of value to design a representative scale-down reactor to facilitate strain studies.

Keywords: scale-up; scale-down; computational fluid dynamics; *Saccharomyces cerevisiae*; mechanistic kinetic model; bioreactor; concentration gradients; digital twin; bioprocess engineering

---

1. Introduction

Bioprocesses are applied for the production of a vast spectrum of commodities, from food and pharmaceuticals to bioplastic and biofuel. Although different from their chemical counterparts, transferring promising lab approaches to industrial applications is a major challenge too. The problem lies within the different scales of lab, pilot, and industrial bioreactors. Whereas, ideally, mixed homogenous conditions are easily realized at lab scale, economic and physical constraints prevent the establishment of such ideal conditions in industrial tanks. As a result, gradients of limiting substrate concentrations, by-products, pH, temperature, and shear rates are formed inevitably. Circulating microorganisms in stirred and gassed large-scale tanks respond to the permanently changing microenvironmental conditions, finally causing uncertainty of process performance, possibly deteriorating key TRY criteria (titer, rate, yield) [1–6].

Different approaches to addressing such issues have been studied by bioreactor design experts [7–12]. The investigation of cellular interaction with substrate gradients has been the basis of numerous studies. Often, deteriorating TRY values have been reported [13,14] but there are also occasional observations of improved performance values [15]. In particular, *Saccharomyces cerevisiae* was found to respond to fluctuating conditions with improved viability [16], adapted cAMP-mediated metabolism [17–19], and global regulation [20,21]. However, most of these investigations were performed at laboratory scale, mimicking industrial-scale conditions. Real industrial-scale data, important for validation, are rare. Accordingly, researchers have been employing computational fluid dynamics (CFD) combined with metabolic models with different resolutions to shed light on gradients in the bioreactor that take place at the interface of various physical and biological phenomena [6,22–25]. It is worth mentioning that shear gradients may have a significant effect on shear sensitive hosts. Simulating shear fields and calculating the frequencies of cellular exposure may be a highly valuable tool for future applications to investigate this particular large-scale impact [26].

Saccharomyces cerevisiae strains are applied for a wide range of processes, from the food industry [27–31] and bulk chemical production [32–37] to the pharmaceutical industry [15]. These products are manufactured in large bioreactor scales, where gradients cannot be avoided. As one of the workhorses of industrial biotechnology, this yeast is well investigated [38,39], and key traits of the "Crabtree positive" strains [40–45] are thoroughly studied. One feature is the overflow production of ethanol under aerobic conditions, mirroring how *S. cerevisiae* consumes more glucose than it can metabolize. [46–48]. Under anaerobic conditions *S. cerevisiae* is known to consume ethanol [49]. Apparently, such traits may gain importance under varying microenvironmental conditions affecting TRY values and effecting population differentiation [50]. Interesting short- and long-term Crabtree effects have been observed [44] that create population responses at different timescales. Furthermore, stress exposure may cause growth phenotypes different from the well-reported Monod-type kinetics using substrate supply as the growth limiting impact. To this end, obtaining information regarding the surroundings of the cell will help to identify the triggers that could set off the stress responses [51,52], finally yielding further improved prediction of large-scale performance of the yeast.

The mathematical framework requires the joint use of CFD with microbial kinetics. Whereas the first is applied to predict hydrodynamics and mass transfer in large tanks, the latter describes the cellular phenotype which is dominated by metabolic models [53–57]. The effects interact and are both needed to predict gradients in large-scale reactors. However, recently, efforts have been made to integrate multiple levels of cellular regulation linking metabolism with enzyme activities [9] and gene expression [6]. The integration of these hierarchical control levels expands the timescales of cellular response severely [6,21,58–61], which causes extra computational burden, challenging conventional simulation capacities. With this in mind, such biokinetic models should be linked to CFD that describe the targeted phenotype with the least computational effort.

Nowadays, CFD is one of the must-have tools for process development and troubleshooting [5,62–65]. It provides the possibility to incorporate the main aspects of the process and yield further insights into the conditions inside the bioreactor. Because of the dynamic environment faced by the cells in large-scale bioreactors, different responses of cells will give rise to a heterogeneous population, which will result in heterogeneities in substrate/by-product gradients and/or in the cell population [46,48,66,67]. Setting up a simulation by itself needs to be purpose-driven for engineering applications. The computational resources and strategies should be allocated in a way that contributes the most to the goal of the project at hand.

Currently, Reynolds-averaged Navier–Stokes (RANS) methods are the most common way to model hydrodynamics including turbulence. However, more delicate methods are surfacing [68–70], with drastic changes in software and hardware requirements. Most of the bioproducts are produced in aerated fermenters. Moreover, multiphase problems are addressed with numerous methods. On the one hand, Euler–Euler approaches are used for considering the impact of the ever-changing environment on cells [71]. On the other hand, one may implement the Euler–Lagrange approach [72], for instance, if individual microorganisms are incorporated, typically as massless particles.

In this work, CFD is coupled to a mechanistic biokinetic model to predict concentration gradients in a large-scale bioreactor. The process of interest is the production of baker's yeast, *Saccharomyces cerevisiae*. A detailed assessment in terms of concentration gradients is undertaken. Aspects of biokinetics, mass transfer, and thermodynamics of the bioreactor are well characterized, and fundamentals are well established. Together with experimental validation, they prove to be an effective tool to shed light on the gradients within a bioreactor at industrial scale. By efficiently combining CFD with simple yet practical models, the assumption of non-limiting dissolved O_2 (dO_2) is evaluated. In addition to this, dissolved CO_2 (dCO_2) inhibition is known to occur at large scales [4,73] and is addressed within this work. Large-scale processes are suspected to lack real starvation in the case of overflow metabolism products that are formed near the feeding point and which are then re-consumed at the other end of the tank [4].

2. Materials and Methods

All pre-processing, solution, and post-processing were carried out with Ansys® Academic Research Workbench (2019 R1, Ansys, Canonsburg, PA, USA).

2.1. Geometry and Mesh

The reactor dimensions were based on the schematics available in the literature [25,74], as illustrated in Figure 1. For CFD simulation, a mesh with ~2.5 million hexahedron cells was generated and evaluated for turbulent kinetic energy (k) and dissipation of turbulent kinetic energy (ε) and velocity profile at the impeller heights, as discussed in the work of Haringa et al. [25].

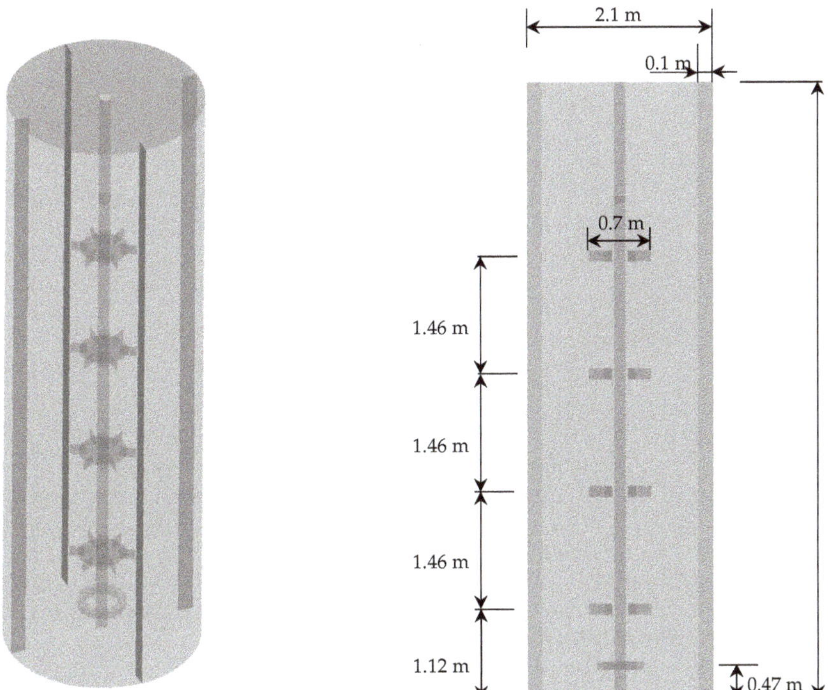

Figure 1. Stavanger bioreactor configuration and dimensions.

2.2. CFD Setup

A Eulerian model with two phases was used. Turbulence was modeled with realizable k-ε using mixture formulation for calculating turbulence of the gas phase [75]. To investigate the concentration of chemical species, namely glucose, O_2, CO_2, ethanol, every phase was defined as a mixture. The physical properties of the liquid phase were approximated as water with volume weighted mixing law for density calculation. The gas phase was assumed as ideal gas considering an average bubble diameter 0.009 m as calculated by Haringa et al. [25]. With this assumption of single bubble size, the swarm coefficient needed to be set to −1.2 in the grace drag model [76] to reproduce the gas flow regime at the bottom impeller. Surface tension was set to that of the air–water system and was 0.072 N/m. Other interphase forces were assumed not to impose significant impacts according to Scargiali et al. [77].

Boundary conditions for walls were set with no-slip conditions for liquid phase and free slip for gas phase other than the impellers, where the no-slip conditions also apply for the gas phase to improve the reproduction of the vortices behind the impellers, as suggested by Haringa et al. [25]. Gas enters the bioreactor via a sparger through a mass flow inlet with 0.231 kg/s and leaves the system on top via a degassing boundary condition. Operating pressure was set close to the boundary to 130,710 Pa [4,25,74] and operating density was set to 0, as suggested by the Fluent manual [78]. The rotation of the impeller was modeled using multiple reference frames (MRF) at 2.22 1/s, as mentioned in previous works on this bioreactor [25]. The operating conditions are summarized in Table 1.

Table 1. Operating conditions for the fermentation process.

Operation Conditions	Description	Ref.
Primary phase	~water • $\rho = 1000$ kg/m^3 • $v = 0.001$ kg/m/s	[25]
Secondary phase	~air • $\rho = 1.225$ kg/m^3 • ideal gas law • $d_b = 0.009$ m	[25]
Inter-phase forces	Drag • grace • swarm coefficient −1.2	[76,77]
Aeration rate	0.231 kg/s	[74]
Headspace pressure	130,710 Pa	[74]
Agitation rate	2.22 s^{-1}	[74]
Glucose feed	52 kg/h	[25]

After setting up the phenomena describing the system, the solution methods and strategies were included. Phase coupled SIMPLE were chosen for pressure-velocity coupling and temporal (transient formulation) and spatial discretization scheme were set to first-order upwind for the first few hundred iterations to achieve solution stability and then were set to QUICK for velocity, turbulence, and volume fraction, and temporal discretization was set to bounded second-order implicit. The residuals were set to 10^{-6} and a time step of 0.001 s with 50 iterations was chosen.

Simulations were qualified as "accomplished" once flow velocities and turbulent kinetic energies converged to pseudo-steady states with ±5% variation. Further validation was achieved by comparing mixing time (τ_{95}) and integral mass transfer coefficient ($k_l a$) with published values. Flow fields served as basis for implementing mass transfer and biokinetics in subsequent steps.

2.3. Biokinetics

Sonnleitner and Käppeli [39] introduced a black-box model to describe the substrate uptake, growth, and by-product formation of *S.cerevisiae*. Glucose is considered as substrate whereas ethanol may serve as substrate or product depending on metabolic and environmental conditions. The model assumes the respiratory capacity of the yeast as key metabolic bottleneck. If glucose uptake exceeds respiratory limits, remaining electrons are channeled via reductive pathways, leading to the secretion of ethanol. Notably, the model also allows ethanol uptake under aerobic conditions. Under anaerobic conditions, ethanol is considered as dominating product. Details are as follows:

1. Aerobic growth on glucose (indexed *sae*)
2. Anaerobic growth on glucose (indexed *san*)
3. Aerobic growth on ethanol (indexed *eae*)

Uptake rates are assumed to follow Monod kinetics (1)–(3).

$$q_s = q_{s,max} \frac{C_s}{K_s + C_s} \tag{1}$$

$$q_o = q_{o,max} \frac{C_o}{K_o + C_o} \tag{2}$$

$$q_e = q_{e,max} \frac{C_e}{K_e + C_e} \tag{3}$$

To distinguish between these cases, the catabolic capacity to metabolize glucose aerobically serves as the threshold. In essence, if biomass specific glucose uptake q_s exceeds the related oxygen demand for oxidation, $Y_{\frac{s}{o}} \cdot q_o$ meaning (4):

$$q_s > Y_{\frac{s}{o}} \cdot q_o \tag{4}$$

Aerobic ethanol formation starts. Acetaldehyde, upstream of ethanol in the fermentation pathway, serves as electron acceptor under such conditions. Accordingly, "anaerobic" carbon flux occurs and equals the remainder of the total substrate uptake (5) and (6), which will be metabolized to ethanol.

$$q_{sae} = \min\left(Y_{\frac{s}{o}} \times q_o, q_s \right) \tag{5}$$

$$q_{san} = q_s - q_{sae} \tag{6}$$

To shed light on ethanol dynamics, its consumption under aerobic conditions is also considered, prioritizing glucose [79]. Given that (7) holds true, ethanol uptake rates q_{eae} are calculated as shown in (8). In essence, the min modulator compares whether oxygen demands for ethanol oxidation after glucose consumption $\frac{(q_o - Y_{\frac{o}{s}} \cdot q_{sae})}{Y_{\frac{o}{e}}}$ exceed the Monod-type ethanol uptake kinetics q_e.

$$q_s < Y_{\frac{s}{o}} q_o \tag{7}$$

$$q_{eae} = \min\left(\frac{(q_o - Y_{\frac{o}{s}} q_{sae})}{Y_{\frac{o}{e}}}, q_e \right) \tag{8}$$

A graphical illustration of respiratory bottleneck is shown in Figure 2. The concentrations are calculated and, using Equations (1)–(3), the uptake rates are calculated, upon which the rest of the model is based. Yields for by-products are stoichiometrically approximated for every reaction.

To reveal the stoichiometry between substrates and products, elemental balances are applied to the process reaction (9), which then results in three different scenarios (10)–(12).

Process reaction

$$sCH_2O + oO_2 + nNH_3 \rightarrow CH_{1.79}O_{0.56}N_{0.15} + eCH_3O_{0.5} + cCO_2 + wH_2O \tag{9}$$

Process reaction for aerobic growth on glucose (r_{sae})

$$sCH_2O + oO_2 + nNH_3 \rightarrow CH_{1.79}O_{0.56}N_{0.15} + cCO_2 + wH_2O \tag{10}$$

Process reaction for anaerobic growth on glucose (r_{san})

$$sCH_2O + nNH_3 \rightarrow CH_{1.79}O_{0.56}N_{0.15} + eCH_3O_{0.5} + cCO_2 + wH_2O \tag{11}$$

Process reaction for aerobic growth on ethanol (r_{eae})

$$eCH_3O_{0.5} + oO_2 + nNH_3 \rightarrow CH_{1.79}O_{0.56}N_{0.15} + cCO_2 + wH_2O \tag{12}$$

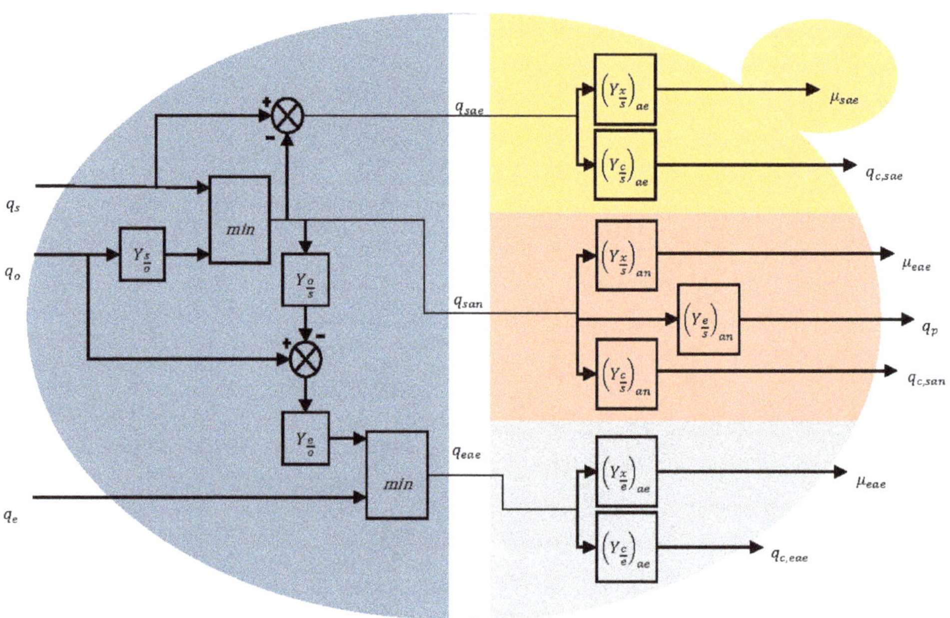

Figure 2. Graphical representation of the bottleneck concept and the interplay of substrate availability in the kinetic model for ethanol consumption under aerobic conditions and overflow metabolism based on [39].

The conservation of mass results in (13). Since all underlying phenomena do not disturb mass or elemental conservation, it is safe to conclude that there is no net conversion of elements. For this purpose, the matrices for elemental composition "E" (14), stoichiometry coefficients "S.C." (15), and reaction rates "r" (16) are set up.

$$\frac{d(E \cdot C \cdot V)}{dt} = E \cdot r \cdot V \text{ (Volume of the element)} + E \cdot MTR \text{ (Mass Transfer Rate)} \tag{13}$$

$$
E =
\begin{array}{c|ccccccc}
 & s & o & n & x & e & c & w \\
\hline
C & 1 & 0 & 0 & 1 & 1 & 1 & 0 \\
H & 2 & 0 & 3 & 1.79 & 3 & 0 & 2 \\
O & 1 & 2 & 0 & 0.56 & 0.5 & 2 & 1 \\
N & 0 & 0 & 1 & 0.15 & 0 & 0 & 0 \\
\end{array}
\tag{14}
$$

$$r = \begin{bmatrix} r_{s_i} \\ r_{o_i} \\ r_{n_i} \\ r_{x_i} \\ r_{e_i} \\ r_{c_i} \\ r_{w_i} \end{bmatrix} \tag{15}$$

$$S.C = \begin{array}{c|ccc} & S.C_{sae} & S.C_{san} & S.C_{eae} \\ \hline s & -s_{sae} & -s_{san} & 0 \\ o & -o_{sae} & 0 & -o_{eae} \\ n & -n_{sae} & -n_{san} & -n_{eae} \\ x & x_{sae} & x_{san} & x_{eae} \\ e & 0 & e_{san} & -e_{eae} \\ c & c_{sae} & c_{san} & c_{eae} \\ w & w_{sae} & w_{san} & w_{eae} \end{array} \tag{16}$$

Elemental conservation results in the following system of Equation (17):

$$E \cdot r = 0 \tag{17}$$

By solving this system based on growth rate and carbon source uptake rate, other rates are calculated for each reaction in the models (18)–(21).

$$\text{Aerobic growth on glucose :} \quad \begin{cases} r_{c_{sae}} = 6r_{s_{sae}} - r_{x_{sae}} \\ r_{o_{sae}} = 6r_{s_{sae}} - 1.05r_{x_{sae}} \end{cases} \tag{18}$$

$$\text{Anaerobic growth on glucose :} \quad \begin{cases} r_{c_{san}} = 2r_{s_{san}} - 0.3r_{x_{san}} \\ r_{e_{san}} = r_{s_{san}} - 0.7r_{x_{san}} \end{cases} \tag{19}$$

$$\text{Aerobic growth on ethanol :} \quad \begin{cases} r_{c_{eae}} = 2r_{e_{eae}} - r_{x_{eae}} \\ r_{o_{eae}} = 3r_{e_{eae}} - 1.05r_{x_{eae}} \end{cases} \tag{20}$$

The carbon source consumption rate is calculated using Equations (5)–(8) to solve the remaining equation for a known biomass concentration. The growth rate is calculated using the yield coefficient taken from the literature [80].

2.4. Mass Transfer

The mass transfer coefficient between the two phases was considered for O_2, CO_2, and ethanol (21). A common assumption is to estimate the mass transfer close to the equilibrium state. The gas phase is considered well mixed (zero resistance); hence, using the film theory, one can assume the mass transfer resistance to be on the liquid side of the interface. Moreover, for dilute gases in liquid, Henry's law (22) is implemented to calculate the equilibrium concentration "C^*" on the interface [81,82].

$$MTR = k_l a \, \Delta C \tag{21}$$

$$C_i^* = \frac{P_i}{H} \tag{22}$$

H is Henry's constant for the gas component at fermentation temperature (30 °C). k_l is modeled by the surface renewal approach [83], a is the bubble surface, and both are known to be dependent on flow characteristics [84]. Mass transfer driving force (ΔC) differs for O_2 and CO_2 since the direction of the transport is different, which leads to (23) and (24).

$$\Delta C_c = \left(C_{c_{liq}} - C_c^* \right) \tag{23}$$

$$\Delta C_o = \left(C_o^* - C_{o_{liq}} \right) \tag{24}$$

For ethanol stripping, an approach by Löser et al. [85] is implemented in which ethanol transfer to gas phase is investigated. In this way, a partition coefficient (25) linking ethanol concentration in both phases to each other is available, which allows calculation of the mass transfer driving force for ethanol.

$$K_{\frac{L}{G}} = \frac{C_{e_{liq}}}{C_{e_{gas}}} \tag{25}$$

3. Results

3.1. Flow Field Validation

For the validation of the represented flow field, several criteria were evaluated. τ_{95} was reproduced with a virtual pulse of glucose above the top impeller and by reading its concentration at the probe location, as disclosed in [74], at 0.97 m distance from the bottom. The estimated value in the current work is 186 s, which is in agreement with the results of previous investigations [4,25,74].

The simulated gas holdup of 19% slightly overpredicts experimental measurements (17.1%) [86] and previous numerical studies (17.6%) [25] but still falls within an acceptable range. In addition to the average holdup, gas distribution of the gas phase plays a crucial role in $k_l a$ calculations (Figure 3).

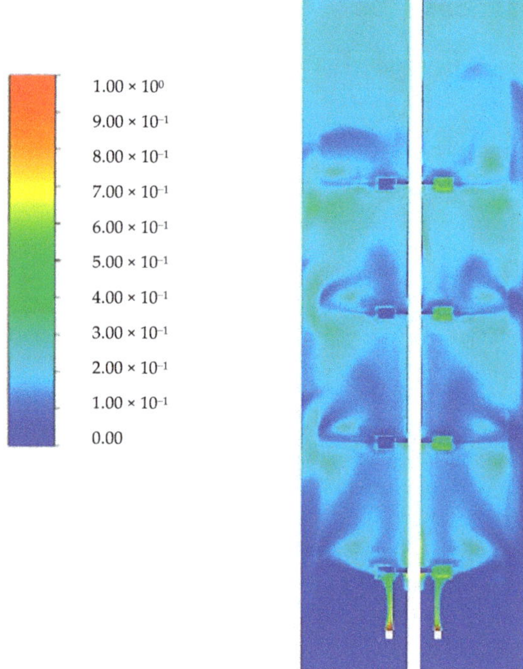

Figure 3. Gas holdup distribution can reproduce the loading regime at the bottom impeller.

Under said conditions, $k_l a$ of 190 h^{-1} is estimated, which agrees well with the experimental measurements [86].

3.2. Scenario I: Experimental Fermentation

For this scenario, conditions are considered as explained by previous investigations [25,86]. Accordingly, model suitability could be checked. dO_2 concentrations are expected to show high values at the bottom of the tank for mainly three reasons: first, higher hydrostatic pressure increases the solubility of dO_2. Second, lower metabolic activity of cells should occur due to substrate scarcity, and third, higher fraction of O_2 in gas phase close to the bottom should be observed. In contrast, opposite trends are found close to the feeding point, where the lowest dO_2 of ~3.7×10^{-5} M (~1.2 ppm) is estimated (Figure 4a). This is an order of magnitude larger than the critical value of ~4.6×10^{-6} M [87]. Furthermore, the related volume is only a negligible fraction of the entire stirred tank reactor, which gives rise to the fair assumption of "no oxygen limitation" in the bioreactor for given conditions [4,25,86].

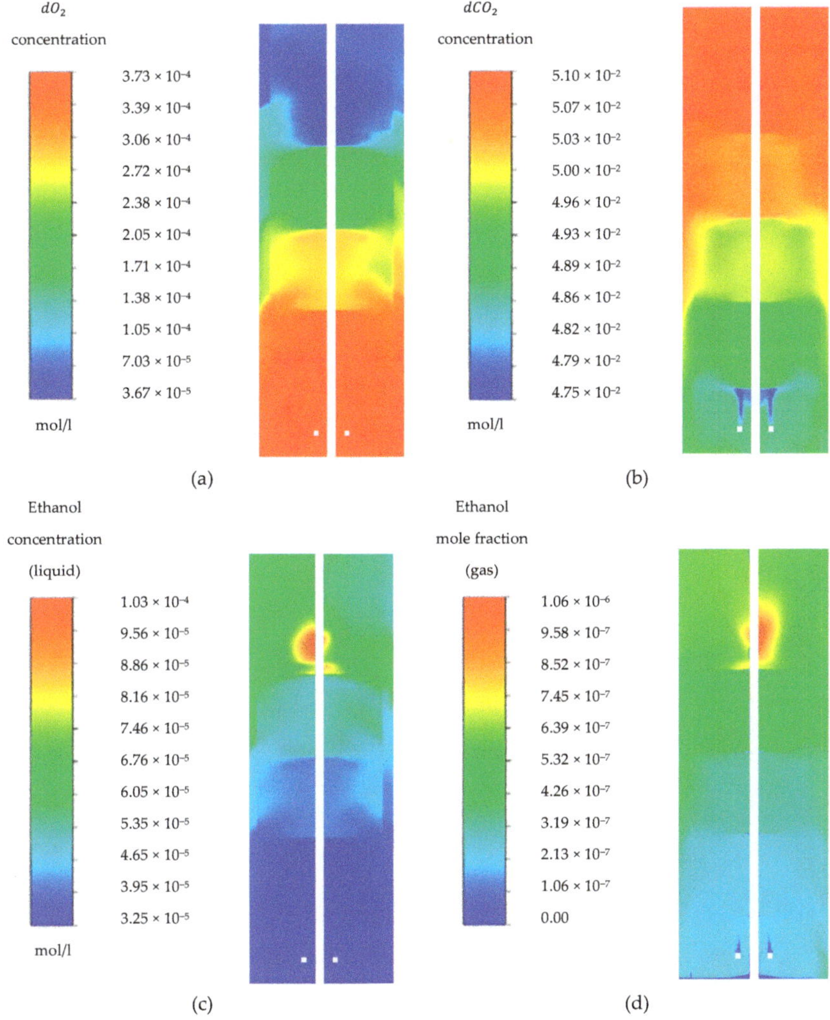

Figure 4. Concentration profiles for scenario I with 10 g/L biomass for (**a**) dO_2, (**b**) dCO_2, (**c**) ethanol in liquid phase, (**d**) stripped ethanol mole fraction in the gas phase.

dCO_2 concentrations are represented in Figure 4b. Interestingly, only minor dCO_2 gradients occur, varying no more than ±5% from average. Notably, CO_2/HCO_3^- creates a buffering system that consists of around 99% CO_2 at the operational pH 5 [73,88]. Accordingly, the simplifying assumption was made that inorganic carbon only encompasses dCO_2. The current study snapshots a pseudo-steady state of late phase yeast fermentation [82,89]. Consequently, the reasonable postulation was made that the liquid phase is saturated with respect to dCO_2. In the gas phase, CO_2 is estimated to reach mole fractions between 2.1 and 2.9%.

Ethanol gradients are more pronounced than those of dCO_2 but less so than dO_2. The highest values are observed proximate to the feeding point, reflecting the highest cellular product formation and reduced stripping. The lowest titer is found at the bottom of the tank (3.25×10^{-5} M, Figure 4c,d).

3.3. Scenario II: Protein Production

To place the model into a more industrially relevant context, biomass was increased to 25 g/L to imitate protein production [15]. For simplification, putative impacts of increased biomass concentration on the viscosity were neglected [90]. The scenario shows that a significant volume is exposed to oxygen limitation (approximately 0.37 m^3) above the top impeller, considering that 10% of saturation dO_2. dO_2 levels below 4.6×10^{-6} M were observed in a volume of 0.04 m^3, which is below the $dO_{2,crit}$ according to the available literature [87] (Figure 5a). Notably, elevated viscosity values would have even deteriorated the oxygen supply. Increasing biomass concentrations also increased microbial substrate consumption and product formation rates. As aeration and the energy input of the bioreactor remained equal, gradients for substrates and by-products became more pronounced.

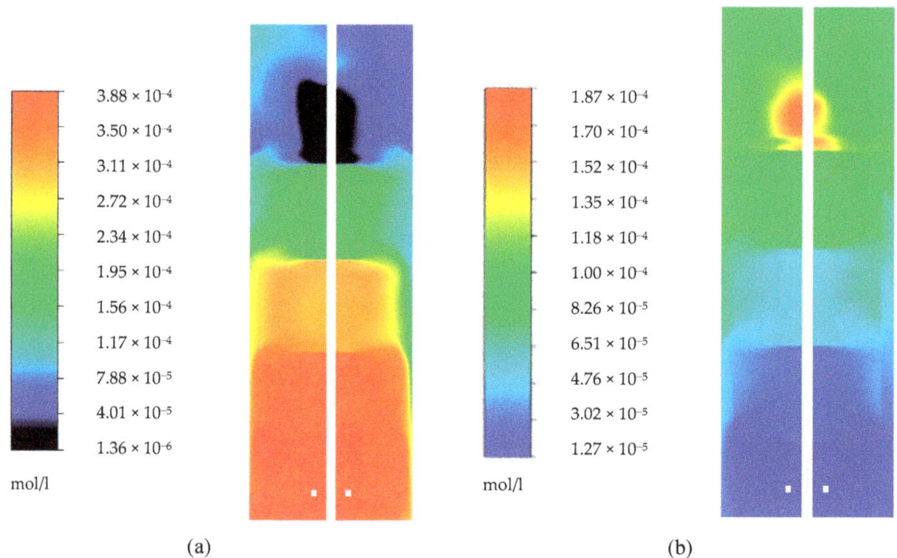

Figure 5. Concentration profiles with 25 g/L biomass concentration for (**a**) dO_2 and (**b**) ethanol.

In turn, this affected the ethanol gradient twofold. First, the drop in dO_2 resulted in higher production of ethanol around the feeding point. Second, higher biomass concentration in the tank increased the volumetric ethanol consumption, causing lower ethanol concentrations at the bottom of the tank, as observed in Figure 5b.

4. Discussion

4.1. Glucose Gradient

Figure A1 (Appendix A) shows the anticipated heterogeneous glucose distribution, disclosing a hotspot of high glucose concentrations close to the inlet and low values at the bottom of the bioreactor. As expected, the resulting gradients are more pronounced for the "25 g/L biomass" case than for 10 g/L. Interestingly, studies [4,73] provided experimental values sampled from the top, middle, and bottom regions of the bioreactor (Table A1). Notably, sampling was performed at the wall of the bioreactors, only giving a very restricted resolution of local conditions. On the contrary, simulated values of scenario I and II indicate average concentrations of total reactor slices calculated at the same height. Consequently, the comparison of simulated predictions with experimental values is intrinsically biased. Nevertheless, the comparison shows that scenario II comes closest to the measurements. At the top, high glucose levels were equally predicted by simulation and measurements. Notably, each value indicates saturated glucose uptake. The strongest deviations are found for the bottom region, where simulations overestimate the glucose consumption. Consequently, model refinements should be considered in the next generation of metabolic models by implementing the co-consumption of intracellular buffers (such as trehalose) as an additional, not yet considered, carbon source in nutrient-limiting regions.

4.2. Ethanol Gradient

To the best of our knowledge, this study represents the first example of considering ethanol formation and re-consumption in a CFD-linked large-scale bioreactor simulation. Figure 4c indicates the well-distributed presence of ethanol in the entire reactor, giving rise to the assumption that ethanol-based growth should be possible in large parts of the bioreactor. Based on the results, growth on ethanol is expected to take place in 97% bioreactor. The conclusion is in accordance with the work of Noorman [4], who anticipated that no real "starvation" zone might exist because of the occurrence of ethanol. The finding does have implications for the design of proper scale-down approaches [63,65,91] as suitable settings ideally should consider the co-substrate ethanol too. For scenario II, the average ethanol concentration was approximately 26% lower (3.17×10^{-5} M) compared to scenario I (5×10^{-5} M). Nevertheless, more than 90% of the tank may offer sufficient ethanol uptake within seconds according to a radiocarbon study [92]. Such levels might be enough to prevent an actual starvation scenario.

4.3. Oxygen Gradient

A conventional approach for estimating the occurrence of gradients is the comparison of critical timescales τ for substrate supply $\tau_{S_{supply}}$ versus substrate consumption $\tau_{S_{cons}}$. Whereas the first may be approximated by the mixing time τ_{mix} or circulation time $\tau_{circulation}$, the latter resembles the quotient of average substrate concentration divided by volumetric substrate consumption rates (26). Regarding dO_2, scenarios 1 and 2 anticipate the occurrence of gradients because $\tau_{O_{cons,1}}$ and $\tau_{O_{cons,2}}$ showing ~28 s and ~30 s are smaller than $\tau_{circulation} \approx 47$ s ($\tau_{mix} = 186$ s). Indeed, the expectation is met by the CFD simulations.

$$\tau_{O_{cons}} = \frac{C_o}{\left(Y_{\frac{o}{s}} \cdot q_{sae} + Y_{\frac{o}{e}} \cdot q_{eae}\right) \times C_x} \tag{26}$$

Assuming average values, this approach theoretically indicates that assuming the non-limiting role of dO_2 for scenario I is a reasonable approximation for the whole tank, other than a small region around the feeding point, where the dO_2 concentration is slightly above 3.67×10^{-5} M. The threshold for aerobic growth $dO_{2,crit}$ for *S. cerevisiae* is given as 4.6×10^{-6} M [87]. This allows us to make a distinction between the ethanol production caused by overflow metabolism or dO_2 limitation. Accordingly, no dO_2 limitation is observed in scenario I as mentioned; hence, all ethanol production in this case is attributed to overflow metabolism, which occurs in 1.63 m^3 of the fermenter. However, this is not the case for scenario II, where 10% of the ethanol production takes place in regions with dO_2 below the critical value. The total volume associated with ethanol production in scenario II is 0.3 m^3.

4.4. Carbon Dioxide Gradient

Although the dCO_2 gradient is practically absent when compared to those for dO_2, ethanol, and glucose (Appendix A), the key observation is the generally high level inside the bioreactor due to overpressure applied in the headspace plus the hydrostatic pressure from the liquid column. This should be accounted for in experimental scale-down. It should be noticed that by using a black-box model, some inherent flaws of such models affect the results. While such models could prove to be insightful for a specific case, the assumptions upon which the model is founded limit the generalization. In this case, the process reaction (6) considers dCO_2 only as a product of a single reaction. However, multiple decarboxylating reactions exist in the cellular metabolism, showing variable activity [93,94]. This intrinsic feature needs to be included if one wishes to reproduce the respiratory quotient (RQ). Despite this, the average ethanol consumption rate is an order of magnitude smaller than the average glucose consumption rate and, as a result, an RQ value of around 1.1 is achieved. Adding another layer of detail to the biokinetic model by including lumped reactions and metabolite pools [95] might be an interesting step forward. This might be possible by combining multi-reaction models like the one used in this work with lumped metabolic models [95,96]. From another perspective, dCO_2 creates a carbonate system in the fluid and within the cell which alters the cytosolic pH and hence induces stress and increases the cellular maintenance [64,73,97] or alters the metabolism based on gas composition [98]. Based on the actual process, one can decide to include some or all of the equilibrium reaction, but this does not fall within the scope of this work since, at pH 5, more than 99% is in the form of dCO_2. Nevertheless, using a comparatively simple approach, the results indicate that the dCO_2 gradient is rather weak compared to other species. At such levels, dCO_2 inhibition inevitably takes place at industrial scale and impacts the transcription according to recent findings [37]. Our results indicate that while fluctuations in other concentrations might be experienced by cells on short timescales, the same does not hold true for dCO_2, where cells are exposed to high dCO_2 for long timescales. The latter requires different experimental set-ups for scale-down tests.

5. Conclusions

This work suggests that in the case of baker's yeast production, ethanol production is inevitable around the feeding point—in this case, positioned at the top of the vessel. This causes lower growth rates above the top impeller and hence hinders the overall growth rate over the bioreactor volume and is not desirable when the final product is the biomass itself. It is possible to distinguish the ethanol production due to overflow metabolism (Crabtree effect) from dO_2 limitation (Pasteur effect). Such information can prove crucial for process optimization. dCO_2 gradients might not be as pronounced as the other species, but the fact that, in both scenarios, it reaches saturation levels hints at CO_2 stripping under real industrial conditions [4]. This points out the fact that, unlike glucose, ethanol, and dO_2, where fluctuations might trigger a stress response, dCO_2 stress is different in nature and should be evaluated by long-term scale-down experiments. The results further suggest that a real starvation region in the lower parts of the tank might not exist because of the presence of ethanol compensating for glucose shortage. Accordingly, scale-down experiments should consider this impact, even investigating the putative benefits for long-term protein formation [15].

Author Contributions: Funding acquisition, R.T.; Investigation, C.S.S.H.; Methodology, C.S.S.H.; Resources, R.T.; Software, R.T.; Supervision, R.T.; Writing—original draft, C.S.S.H.; Writing—review & editing, C.S.S.H., C.H., H.N. and R.T. All authors have read and agreed to the published version of the manuscript.

Funding: Authors are thankful to Höchstleistungsrechenzentrum Stuttgart (HLRS) for large-scale computations and to the University of Stuttgart for providing the required funds. This work was supported by the German Federal Ministry of Education and Research (BMBF), grant number: FKZ 031B0629. C.S.S.H. is supported by ERA CoBioTech/EU H2020 project (grant 722361) "ComRaDes", a public–private partnership between the University of

Stuttgart, TU Delft, University of Liege, DSM, Centrient Pharmaceuticals and Syngulon. C.H. and H.N. are paid employees of the DSM Biotechnology Center.

Acknowledgments: Authors acknowledge the insightful comments of Wouter van Winden in the process of manuscript preparation.

Conflicts of Interest: The authors declare no conflict of interest.

Nomenclature

C	Concentration matrix	
C_i^*	Concentration at the interface	mol/L
C_e	Ethanol concentration	mol/L
$C_{i_{liq}}$	Concentration of component "i" in liquid phase	mol/L
C_o	Oxygen concentration	mol/L
C_s	Glucose concentration	mol/L
C_x	Biomass concentration	g/L
d_b	Bubble diameter	m
dCO_2	Concentration of dissolved carbon dioxide	mol/L
dO_2	Concentration of dissolved oxygen	mol/L
$dO_{2_{crit}}$	Critical concentration of dissolved oxygen	mol/L
E	Elemental composition matrix	
H	Henry coefficient	l·atm/mol
k	Turbulent kinetic energy	m^2/s^2
$K_{L/G}$	Partition coefficient	-
K_e	Monod constant for ethanol	mol/L
$k_l a$	Mass transfer coefficient	1/h
K_o	Monod constant for oxygen	mol/L
K_s	Monod constant for glucose	mol/L
MTR	Mass transfer rate	mol/L/s
P_i	Partial pressure of the component "i" in gas phase	atm
q_e	Specific ethanol uptake rate	mol/g$_x$/s
q_{eae}	Specific ethanol uptake rate under aerobic conditions	mol/g$_x$/s
$q_{e_{max}}$	Maximum specific ethanol uptake rate	mol/g$_x$/s
q_o	Specific oxygen uptake rate	mol/g$_x$/s
$q_{o_{max}}$	Maximum specific oxygen uptake rate	mol/g$_x$/s
q_s	Specific glucose uptake rate	mol/g$_x$/s
q_{sae}	Specific glucose uptake rate for aerobic metabolism	mol/g$_x$/s
q_{san}	Specific glucose uptake rate for anaerobic metabolism	mol/g$_x$/s
$q_{s_{max}}$	Maximum specific glucose uptake rate	mol/g$_x$/s
r	Reaction rate matrix	
$r_{i_{eae}}$	Reaction rate of component "i" when growing aerobically on ethanol	mol/L/s
$r_{i_{sae}}$	Reaction rate of component "i" when growing aerobically on glucose	mol/L/s
$r_{i_{san}}$	Reaction rate of component "i" when growing anaerobically on glucose	mol/L/s
S.C.	Stoichiometry coefficients matrix	
τ_{circ}	Circulation time	s
τ_{mix}	Mixing time	s
$\tau_{i_{cons}}$	Consumption timescale of component "i"	s
$\tau_{i_{supply}}$	Supply timescale of component "i"	s
V	Volume	m^3
$Y_{\frac{c}{s}ae}$	Carbon dioxide yield (aerobic) per mole substrate	mol/mol
$Y_{\frac{c}{e}ae}$	Carbon dioxide yield (aerobic) per mole ethanol	mol/mol
$Y_{\frac{c}{s}an}$	Carbon dioxide yield (anaerobic) per mole substrate	mol/mol
$Y_{\frac{e}{s}an}$	Ethanol yield (anaerobic) per mole substrate	mol/mol
$Y_{\frac{o}{e}}$	Oxygen yield per mole ethanol	mol/mol
$Y_{\frac{o}{s}}$	Oxygen yield per mole glucose	mol/mol
$Y_{\frac{s}{o}}$	Substrate yield per mole oxygen	mol/mol
$Y_{\frac{x}{e}ae}$	Biomass yield (aerobic) per mole ethanol	mol/mol
$Y_{\frac{x}{s}ae}$	Biomass yield (aerobic) per mole glucose	mol/mol
$Y_{\frac{x}{s}an}$	Biomass yield (anaerobic) per mole glucose	mol/mol
ΔC	concentration driving force	mol/L
ε	dissipation rate of turbulent kinetic energy	m^2/s^3
ν	Kinematic viscosity	m^2/s
ρ	Density	kg/m^3

Abbreviations

cAMP	Cyclic adenosine monophosphate
CFD	Computational fluid dynamics
MRF	Multiple reference frames
QUICK	Quadratic Upstream Interpolation for Convective Kinematics
RANS	Reynolds-averaged Navier–Stokes
RQ	Respiratory quotient
SIMPLE	Semi-Implicit Method for Pressure Linked Equations
TRY	Titer, rate, yield

Appendix A. Glucose Gradients

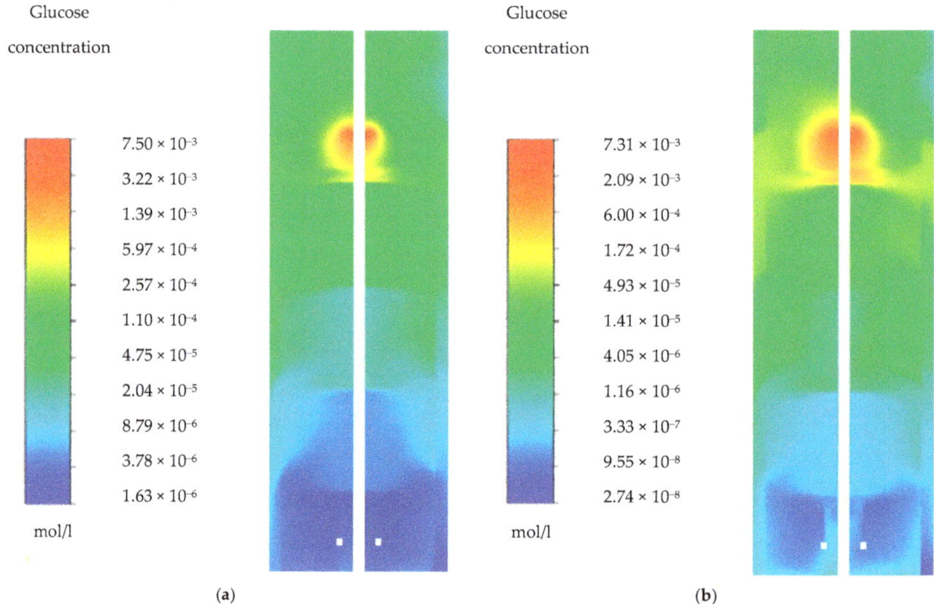

Figure A1. Glucose gradients with (**a**) 10 g/L and (**b**) 25 g/L biomass (logarithmic colormap).

As expected, a strong gradient for glucose as the main substrate exists in both cases, as shown in Figure A1, and also aligned with previous efforts [4,25,74]. Using a similar approach to Section 4.2. for glucose (A1) gives a timescale for substrate consumption of 7 and 17 s for 25 /L and 10 g/L biomass concentration, respectively. This still falls short of the circulation time $\tau_{circulation}$ of approximately 47 s, meaning that supply is slower than the demand.

$$\tau_{S_{cons}} = \frac{C_s}{(q_{sae} + q_{san}) \times C_x} \tag{A1}$$

It is worth noticing the glucose concentration predicted in this work is in the same order of magnitude (Table A1), but it drops to concentrations that are below measured quantities. This could be an interesting topic for further investigations.

Table A1. Comparison of glucose concentration at probe location (Top: 6.35 m, Mid.: 3.9 m, Bot.: 0.97 m from the bottom) from experimental (light blue 10 g/L biomass concentration) and simulation (light orange 10 g/L biomass concentration, orange 25 g/L biomass concentration) results from the literature and this work.

| | | Glucose Concentration (μmol/L) | | | | |
| | | [4] | [74] | [25] | This Work | |
					Scenario I	Scenario II
Probe location	Top	199	222	455	356	250
	Mid.	91.6	62.2	97	32.6	56
	Bot.	40	28.3	23.8	4.15	2.55

References

1. Suarez-Mendez, C.; Sousa, A.; Heijnen, J.; Wahl, A. Fast "Feast/Famine" Cycles for Studying Microbial Physiology Under Dynamic Conditions: A Case Study with *Saccharomyces cerevisiae*. *Metabolites* **2014**, *4*, 347–372. [CrossRef]
2. Suarez-Mendez, C.A.; Ras, C.; Wahl, S.A. Metabolic adjustment upon repetitive substrate perturbations using dynamic 13C-tracing in yeast. *Microb. Cell Fact.* **2017**, *16*, 1–14. [CrossRef]
3. Kresnowati, M.T.A.P.; Van Winden, W.A.; Van Gulik, W.M.; Heijnen, J.J. Energetic and metabolic transient response of *Saccharomyces cerevisiae* to benzoic acid. *FEBS J.* **2008**, *275*, 5527–5541. [CrossRef]
4. Noorman, H. An industrial perspective on bioreactor scale-down: What we can learn from combined large-scale bioprocess and model fluid studies. *Biotechnol. J.* **2011**, *6*, 934–943. [CrossRef]
5. Wang, G.; Haringa, C.; Noorman, H.; Chu, J.; Zhuang, Y. Developing a Computational Framework To Advance Bioprocess Scale-Up. *Trends Biotechnol.* **2020**, 1–11. [CrossRef]
6. Zieringer, J.; Wild, M.; Takors, R. Data-Driven In-silico Prediction of Regulation Heterogeneity and ATP Demands of Escherichia coli in Large-scale Bioreactors. *Biotechnol. Bioeng.* **2020**, bit.27568. [CrossRef] [PubMed]
7. Morchain, J.; Pigou, M.; Lebaz, N. A population balance model for bioreactors combining interdivision time distributions and micromixing concepts. *Biochem. Eng. J.* **2017**, *126*, 135–145. [CrossRef]
8. Linkès, M.; Fede, P.; Morchain, J.Ô.; Schmitz, P. Numerical investigation of subgrid mixing effects on the calculation of biological reaction rates. *Chem. Eng. Sci.* **2014**, *116*, 473–485. [CrossRef]
9. Haringa, C.; Tang, W.; Wang, G.; Deshmukh, A.T.; Van Winden, W.A.; Chu, J.; Van Gulik, W.M.; Heijnen, J.J.; Mudde, R.; Noorman, H. Computational fluid dynamics simulation of an industrial, P. chrysogenum fermentation with a coupled 9-pool metabolic model: Towards rational scale-down and design optimization. *Chem. Eng. Sci.* **2018**, *175*, 12–24. [CrossRef]
10. Haringa, C.; Mudde, R.F.; Noorman, H.J. From industrial fermentor to CFD-guided downscaling: What have we learned? *Biochem. Eng. J.* **2018**, *140*, 57–71. [CrossRef]
11. Lapin, A.; Müller, D.; Reuss, M. Dynamic behavior of microbial populations in stirred bioreactors simulated with Euler-Lagrange methods: Traveling along the lifelines of single cells. *Ind. Eng. Chem. Res.* **2004**, *43*, 4647–4656. [CrossRef]
12. Wright, M.R.; Bach, C.; Gernaey, K.V.; Krühne, U. Investigation of the effect of uncertain growth kinetics on a CFD based model for the growth of, S. cerevisiae in an industrial bioreactor. *Chem. Eng. Res. Des.* **2018**, *140*, 12–22. [CrossRef]
13. Wang, G.; Haringa, C.; Tang, W.; Noorman, H.; Chu, J.; Zhuang, Y.; Zhang, S. Coupled metabolic-hydrodynamic modeling enabling rational scale-up of industrial bioprocesses. *Biotechnol. Bioeng.* **2020**, *117*, 844–867. [CrossRef]
14. Lara, A.R.; Galindo, E.; Ramírez, O.T.; Palomares, L.A. Living with heterogeneities in bioreactors: Understanding the effects of environmental gradients on cells. *Mol. Biotechnol.* **2006**, *34*, 355–381. [CrossRef]
15. Wright, N.R.; Wulff, T.; Palmqvist, E.A.; Jørgensen, T.R.; Workman, C.T.; Sonnenschein, N.; Rønnest, N.P.; Herrgård, M.J. Fluctuations in glucose availability prevent global proteome changes and physiological transition during prolonged chemostat cultivations of *Saccharomyces cerevisiae*. *Biotechnol. Bioeng.* **2020**, *117*, 2074–2088. [CrossRef] [PubMed]

16. Binai, N.A.; Bisschops, M.M.M.; Van Breukelen, B.; Mohammed, S.; Loeff, L.; Pronk, J.T.; Heck, A.J.R.; Daran-Lapujade, P.; Slijper, M. Proteome adaptation of *Saccharomyces cerevisiae* to severe calorie restriction in retentostat cultures. *J. Proteome Res.* **2014**, *13*, 3542–3553. [CrossRef] [PubMed]

17. Kraakman, L.S.; Winderickx, J.; Thevelein, J.M.; De Winde, J.H. Structure-function analysis of yeast hexokinase: Structural requirements for triggering cAMP signalling and catabolite repression. *Biochem. J.* **1999**, *343*, 159–168. [PubMed]

18. François, J.; Parrou, J.L. Reserve carbohydrates metabolism in the yeast *Saccharomyces cerevisiae*. *FEMS Microbiol. Rev.* **2001**, *25*, 125–145. [CrossRef]

19. De Winde, J.H.; Crauwels, M.; Hohmann, S.; Thevelein, J.M.; Winderickx, J. Differential requirement of the yeast sugar kinases for sugar sensing in establishing the catabolite-repressed state. *Eur. J. Biochem.* **1996**, *241*, 633–643. [CrossRef]

20. Mans, R.; Daran, J.M.G.; Pronk, J.T. Under pressure: Evolutionary engineering of yeast strains for improved performance in fuels and chemicals production. *Curr. Opin. Biotechnol.* **2018**, *50*, 47–56. [CrossRef]

21. Zhang, J.; Martinez-Gomez, K.; Heinzle, E.; Wahl, S.A. Metabolic switches from quiescence to growth in synchronized *Saccharomyces cerevisiae*. *Metabolomics* **2019**, *15*, 1–13. [CrossRef] [PubMed]

22. Rizzi, M.; Baltes, M.; Theobald, U.; Reuss, M. In vivo analysis of metabolic dynamics in *Saccharomyces cerevisiae*: II. Mathematical model. *Biotechnol. Bioeng.* **1997**, *55*, 592–608. [CrossRef]

23. Link, H.; Fuhrer, T.; Gerosa, L.; Zamboni, N.; Sauer, U. Real-time metabolome profiling of the metabolic switch between starvation and growth. *Nat. Methods* **2015**, *12*, 1091–1097. [CrossRef] [PubMed]

24. Siebler, F.; Lapin, A.; Hermann, M.; Takors, R. The impact of CO gradients on, C. ljungdahlii in a 125 m^3 bubble column: Mass transfer, circulation time and lifeline analysis. *Chem. Eng. Sci.* **2019**, *207*, 410–423. [CrossRef]

25. Haringa, C.; Deshmukh, A.T.; Mudde, R.F.; Noorman, H.J. Euler-Lagrange analysis towards representative down-scaling of a 22 m^3 aerobic, *S. cerevisiae* fermentation. *Chem. Eng. Sci.* **2017**, *170*, 653–669. [CrossRef]

26. Kordas, M.; Konopacki, M.; Grygorcewicz, B.; Augustyniak, A.; Musik, D.; Wójcik, K.; Jędrzejczak-Silicka, M.; Rakoczy, R. Hydrodynamics and mass transfer analysis in bioflow® bioreactor systems. *Processes* **2020**, *8*, 1311. [CrossRef]

27. Di Serio, M.; Aramo, P.; De Alteriis, E.; Tesser, R.; Santacesaria, E. Quantitative analysis of the key factors affecting yeast growth. *Ind. Eng. Chem. Res.* **2003**, *42*, 5109–5116. [CrossRef]

28. Sweere, A.P.J.; Mesters, J.R.; Janse, L.; Luyben, K.C.A.M.; Kossen, N.W.F. Experimental simulation of oxygen profiles and their influence on baker's yeast production: I. One-fermentor system. *Biotechnol. Bioeng.* **1988**, *31*, 567–578. [CrossRef]

29. Sweere, A.P.J.; Matla, Y.A.; Zandvliet, J.; Luyben, K.C.A.M.; Kossen, N.W.F. Experimental simulation of glucose fluctuations—The influence of continually changing glucose concentrations on the fed-batch baker's yeast production. *Appl. Microbiol. Biotechnol.* **1988**, *28*, 109–115. [CrossRef]

30. Enfors, S.; Hedenberg, J.; Olsson, K. Bi0pr0cess Engineering Simulation of the dynamics in the Baker ' s yeast process. *Bioprocess Eng.* **1990**, *5*, 191–198. [CrossRef]

31. Sweere, A.P.J.; van Dalen, J.P.; Kishoni, E.; Luyben, K.C.A.M.; Kossen, N.W.F. Theoretical analysis of the baker's yeast production: An experimental verification at a laboratory scale—Part 2: Fed-batch fermentations. *Bioprocess Eng.* **1989**, *4*, 11–17. [CrossRef]

32. Raab, A.M.; Lang, C. Oxidative versus reductive succinic acid production in the yeast *Saccharomyces cerevisiae*. *Bioeng. Bugs.* **2011**, *2*, 120–123. [CrossRef] [PubMed]

33. Arikawa, Y.; Kuroyanagi, T.; Shimosaka, M.; Muratsubaki, H.; Enomoto, K.; Kodaira, R.; Okazaki, M. Effect of gene disruptions of the TCA cycle on production of succinic acid in *Saccharomyces cerevisiae*. *J. Biosci. Bioeng.* **1999**, *87*, 28–36. [CrossRef]

34. Otero, J.M.; Cimini, D.; Patil, K.R.; Poulsen, S.G.; Olsson, L.; Nielsen, J. Industrial Systems Biology of *Saccharomyces cerevisiae* Enables Novel Succinic Acid Cell Factory. *PLoS ONE* **2013**, *8*, e54144. [CrossRef] [PubMed]

35. Yan, D.; Wang, C.; Zhou, J.; Liu, Y.; Yang, M.; Xing, J. Construction of reductive pathway in *Saccharomyces cerevisiae* for effective succinic acid fermentation at low pH value. *Bioresour. Technol.* **2014**, *156*, 232–239. [CrossRef]

36. Nissen, T.L.; Schulze, U.; Nielsen, J.; Villadsen, J. Flux distributions in anaerobic, glucose-limited continuous cultures of *Saccharomyces cerevisiae*. *Microbiology* **1997**, *143*, 203–218. [CrossRef]

37. Hakkaart, X.; Liu, Y.; Hulst, M.; El Masoudi, A.; Peuscher, E.; Pronk, J.; Van Gulik, W.M.; Daran-Lapujade, P. Physiological responses of *Saccharomyces cerevisiae* to industrially relevant conditions: Slow growth, low pH, and high CO_2 levels. *Biotechnol. Bioeng.* **2020**, *117*, 721–735. [CrossRef]

38. Pham, H.T.B.; Larsson, G.; Enfors, S.O. Growth and energy metabolism in aerobic fed-batch cultures of *Saccharomyces cerevisiae*: Simulation and model verification. *Biotechnol. Bioeng.* **1998**, *60*, 474–482. [CrossRef]

39. Sonnleitner, B.; Käppeli, O. Growth of *Saccharomyces cerevisiae* is controlled by its limited respiratory capacity: Formulation and verification of a hypothesis. *Biotechnol. Bioeng.* **1986**, *28*, 927–937. [CrossRef]

40. Postma, E.; Alexander Scheffers, W.; Van Dijken, J.P. Kinetics of growth and glucose transport in glucose-limited chemostat cultures of *Saccharomyces cerevisiae* CBS 8066. *Yeast* **1989**, *5*, 159–165. [CrossRef]

41. De Deken, R.H. The Crabtree effect: A regulatory system in yeast. *J. Gen. Microbiol.* **1966**, *44*, 149–156. [CrossRef] [PubMed]

42. Van Urk, H.; Voll, W.S.L.; Scheffers, W.A.; Van Dijken, J.P. Transient-state analysis of metabolic fluxes in Crabtree-positive and crabtree-negative yeasts. *Appl. Environ. Microbiol.* **1990**, *56*, 281–287. [CrossRef] [PubMed]

43. Pfeiffer, T.; Morley, A. An evolutionary perspective on the Crabtree effect. *Front. Mol. Biosci.* **2014**, *1*, 1–6. [CrossRef] [PubMed]

44. Hagman, A.; Piškur, J. A study on the fundamental mechanism and the evolutionary driving forces behind aerobic fermentation in yeast. *PLoS ONE* **2015**, *10*, e0116942. [CrossRef]

45. Van Urk, H.; Postma, E.; Scheffers, W.A.; van Dijken, J.P. Glucose Transport in Crabtree-positive and Crabtree-negative. *Yeasts* **1989**, 2399–2406. [CrossRef]

46. Vrábel, P.; Van der Lans, R.G.J.M.; Van der Schot, F.N.; Luyben , K.C.A.M.; Xu, B.; Enfors, S.O. CMA: Integration of fluid dynamics and microbial kinetics in modelling of large-scale fermentations. *Chem. Eng. J.* **2001**, *84*, 463–474.

47. Mazzoleni, S.; Landi, C.; Cartenì, F.; De Alteriis, E.; Giannino, F.; Paciello, L.; Parascandola, P. A novel process-based model of microbial growth: Self-inhibition in *Saccharomyces cerevisiae* aerobic fed-batch cultures. *Microb. Cell Factories* **2015**, *14*, 1–14. [CrossRef]

48. Enfors, S.-O.; Jahic, M.; Rozkov, A.; Xu, B.; Hecker, M.; Jürgen, B.; Krüger, E.; Schweder, T.; Hamer, G.; O'Beirne, D.; et al. Physiological responses to mixing in large scale bioreactors. *J. Biotechnol.* **2001**, *85*, 175–185. [CrossRef]

49. Kozak, B.U.; van Rossum, H.M.; Niemeijer, M.S.; van Dijk, M.; Benjamin, K.; Wu, L.; Daran, J.G.; Pronk, J.T.; van Maris, A.J.A. Replacement of the initial steps of ethanol metabolism in *Saccharomyces cerevisiae* by ATP-independent acetylating acetaldehyde dehydrogenase. *FEMS Yeast Res.* **2016**, *16*, fow006. [CrossRef]

50. Delafosse, A.; Collignon, M.-L.; Calvo, S.; Delvigne, F.; Crine, M.; Thonart, P.; Toye, D. CFD-based compartment model for description of mixing in bioreactors. *Chem. Eng. Sci.* **2014**, *106*, 76–85. [CrossRef]

51. Nieß, A.; Löffler, M.; Simen, J.D.; Takors, R. Repetitive Short-Term Stimuli Imposed in Poor Mixing Zones Induce Long-Term Adaptation of *E. coli* Cultures in Large-Scale Bioreactors: Experimental Evidence and Mathematical Model. *Front. Microbiol.* **2017**, *8*, 1195. [CrossRef] [PubMed]

52. Smets, B.; Ghillebert, R.; De Snijder, P.; Binda, M.; Swinnen, E.; De Virgilio, C.; Winderickx, J. Life in the midst of scarcity: Adaptations to nutrient availability in *Saccharomyces cerevisiae*. *Curr. Genet.* **2010**, *56*, 1–32. [CrossRef] [PubMed]

53. Larsson, C.; Påhlman, I.L.; Gustafsson, L. The importance of ATP as a regulator of glycolytic flux in *Saccharomyces cerevisiae*. *Yeast* **2000**, *16*, 797–809. [CrossRef]

54. Larsson, C.; Nilsson, A.; Blomberg, A.; Gustafsson, L. Glycolytic flux is conditionally correlated with ATP concentration in *Saccharomyces cerevisiae*: A chemostat study under carbon or nitrogen-limiting conditions. *J. Bacteriol.* **1997**, *179*, 7243–7250. [CrossRef] [PubMed]

55. Somsen, O.J.G.; Hoeben, M.A.; Esgalhado, E.; Snoep, J.L.; Visser, D.; Van der Heijden, R.T.J.M.; Heijnen, J.J.; Westerhoff, H.V. Glucose and the ATP paradox in yeast. *Biochem. J.* **2000**, *352*, 593–599. [CrossRef]

56. Verma, M.; Zakhartsev, M.; Reuss, M.; Westerhoff, H.V. 'Domino' systems biology and the 'A' of ATP. *Biochim. Biophys. Acta (BBA) Gen. Subj.* **2013**, *1827*, 19–29. [CrossRef]

57. Kerkhoven, E.J.; Lahtvee, P.J.; Nielsen, J. Applications of computational modeling in metabolic engineering of yeast. *FEMS Yeast Res.* **2015**, *15*, 1–13. [CrossRef]

58. Almquist, J.; Cvijovic, M.; Hatzimanikatis, V.; Nielsen, J.; Jirstrand, M. Kinetic models in industrial biotechnology—Improving cell factory performance. *Metab. Eng.* **2014**, *24*, 38–60. [CrossRef]

59. Campbell, K.; Xia, J.; Nielsen, J. The Impact of Systems Biology on Bioprocessing. *Trends Biotechnol.* **2017**, *35*, 1156–1168. [CrossRef]

60. Campbell, K.; Westholm, J.; Kasvandik, S.; Di Bartolomeo, F.; Mormino, M.; Nielsen, J. Building blocks are synthesized on demand during the yeast cell cycle. *Proc. Natl. Acad. Sci. USA* **2020**, *117*, 7575–7583. [CrossRef]

61. Szatkowska, R.; Albornoz, M.G.; Roszkowska, K.; Holman, S.W.; Furmanek, E.; Hubbard, S.J.; Beynon, R.J.; Adamczyk, M. Glycolytic flux in *Saccharomyces cerevisiae* is dependent on RNA polymerase III and its negative regulator Maf1. *Biochem. J.* **2019**, *476*, 1053–1082. [CrossRef] [PubMed]

62. Delvigne, F.; Takors, R.; Mudde, R.; van Gulik, W.; Noorman, H. Bioprocess scale-up/down as integrative enabling technology: From fluid mechanics to systems biology and beyond. *Microb. Biotechnol.* **2017**, *10*, 1267–1274. [CrossRef] [PubMed]

63. Takors, R. Scale-up of microbial processes: Impacts, tools and open questions. *J. Biotechnol.* **2012**, *160*, 3–9. [CrossRef]

64. Takors, R. Biochemical engineering provides mindset, tools and solutions for the driving questions of a sustainable future. *Eng. Life Sci.* **2020**, *20*, 5–6. [CrossRef] [PubMed]

65. Noorman, H.J.; Heijnen, J.J. Biochemical engineering's grand adventure. *Chem. Eng. Sci.* **2017**, *170*, 677–693. [CrossRef]

66. Morchain, J.; Gabelle, J.C.; Cockx, A. A coupled population balance model and CFD approach for the simulation of mixing issues in lab-scale and industrial bioreactors. *AIChE J.* **2014**, *60*, 27–40. [CrossRef]

67. Lemoine, A.; Delvigne, F.; Bockisch, A.; Neubauer, P.; Junne, S. Tools for the determination of population heterogeneity caused by inhomogeneous cultivation conditions. *J. Biotechnol.* **2017**, *251*, 84–93. [CrossRef]

68. Murthy, B.N.; Joshi, J.B. Assessment of standard k-ε{lunate}, RSM and LES turbulence models in a baffled stirred vessel agitated by various impeller designs. *Chem. Eng. Sci.* **2008**, *63*, 5468–5495. [CrossRef]

69. Derksen, J.J. Direct simulations of mixing of liquids with density and viscosity differences. *Ind. Eng. Chem. Res.* **2012**, *51*, 6948–6957. [CrossRef]

70. Witz, C.; Treffer, D.; Hardiman, T.; Khinast, J. Local gas holdup simulation and validation of industrial-scale aerated bioreactors. *Chem. Eng. Sci.* **2016**, *152*, 636–648. [CrossRef]

71. Pigou, M.; Morchain, J. Investigating the interactions between physical and biological heterogeneities in bioreactors using compartment, population balance and metabolic models. *Chem. Eng. Sci.* **2015**, *126*, 267–282. [CrossRef]

72. Sarkizi Shams Hajian, C.; Zieringer, J.; Takors, R. Chapter 15: Euler-Lagrangian simulations – a proper tool for predicting cellular performance in industrial scale bioreactors. In *Digital Twins—Applications to the Design and Optimization of Bioprocesses*; Herwig, C., Möller, J., Pörtner, R., Eds.; Springer Nature Switzerland AG: Cham, Switzerland, 2020.

73. Eigenstetter, G.; Takors, R. Dynamic modeling reveals a three-step response of *Saccharomyces cerevisiae* to high CO_2 levels accompanied by increasing ATP demands. *FEMS Yeast Res.* **2017**, *17*, 1–11. [CrossRef]

74. Larsson, G.; Törnkvist, M.; Ståhl Wernersson, E.; Trägårdh, C.; Noorman, H.; Enfors, S.O. Substrate gradients in bioreactors: Origin and consequences. *Bioprocess Eng.* **1996**, *14*, 281–289. [CrossRef]

75. Marshall, E.M.; Bakker, A. Computational Fluid Mixing. In *Handbook of Industrial Mixing*; Atiemo-Obeng, V.A., Calabrese, R.V., Eds.; John Wiley & Sons, Inc.: Hoboken, NJ, USA, 2004; pp. 257–343. [CrossRef]

76. Clift, R.; Grace, J.R.; Weber, M.E. *Bubbles, Drops, and Particles*; (Dover Books on Engineering); Clift, R., Grace, J.R., Weber, M.E., Eds.; Dover Publ.: Mineola, NY, USA, 2005; Volume XIII, p. 381.

77. Scargiali, F.; D'Orazio, A.; Grisafi, F.; Brucato, A. Modelling and simulation of gas—Liquid hydrodynamics in mechanically stirred tanks. *Chem Eng Res Des.* **2007**, *85*, 637–646. [CrossRef]

78. *Ansys® Academic Research Fluent, Release 2019 R1, Fluent user's Guide*; ANSYS, Inc.: Canonsburg, PA, USA, 2019.

79. Weusthuis, R.A.; Visser, W.; Pronk, J.T.; Scheffers, W.A.; Van Dijken, J.P. Effects of oxygen limitation on sugar metabolism in yeasts: A continuous-culture study of the Kluyver effect. *Microbiology* **1994**, *140*, 703–715. [CrossRef] [PubMed]

80. Sweere, A.P.J.; Giesselbach, J.; Barendse, R.; de Krieger, R.; Honderd, G.; Luyben, K.C.A.M. Modelling the dynamic behaviour of *Saccharomyces cerevisiae* and its application in control experiments. *Appl. Microbiol. Biotechnol.* **1988**, *28*, 116–127. [CrossRef]

81. Kuschel, M.; Siebler, F.; Takors, R. Lagrangian Trajectories to Predict the Formation of Population Heterogeneity in Large-Scale Bioreactors. *Bioengineering* **2017**, *4*, 27. [CrossRef]

82. Zieringer, J.; Takors, R. In Silico Prediction of Large-Scale Microbial Production Performance: Constraints for Getting Proper Data-Driven Models. *Comput. Struct. Biotechnol. J.* **2018**, *16*, 246–256. [CrossRef]

83. Alves, S.S.; Vasconcelos, J.M.T.; Orvalho, S.P. Mass transfer to clean bubbles at low turbulent energy dissipation. *Chem. Eng. Sci.* **2006**, *61*, 1334–1337. [CrossRef]

84. Sieblist, C.; Hägeholz, O.; Aehle, M.; Jenzsch, M.; Pohlscheidt, M.; Lübbert, A. Insights into large-scale cell-culture reactors: II. Gas-phase mixing and CO 2 stripping. *Biotechnol. J.* **2011**, *6*, 1547–1556. [CrossRef]

85. Löser, C.; Schröder, A.; Deponte, S.; Bley, T. Balancing the ethanol formation in continuous bioreactors with ethanol stripping. *Eng. Life Sci.* **2005**, *5*, 325–332. [CrossRef]

86. Vrábel, P.; Van Der Lans, R.G.J.M.; Cui, Y.Q.; Luyben, K.C.A.M. Compartment model approach: Mixing in large scale aerated reactors with multiple impellers. *Chem. Eng. Res. Des.* **1999**, *77*, 291–302. [CrossRef]

87. Bailey, J.E.; Ollis, D.F. *Biochemical Engineering Fundamentals*. McGraw-Hill Chemical Engineering Series); McGraw-Hill. 1986. Available online: https://books.google.de/books?id=KM9TAAAAMAAJ (accessed on 26 September 2020).

88. Blombach, B.; Takors, R. CO_2—Intrinsic product, essential substrate, and regulatory trigger of microbial and mammalian production processes. *Front. Bioeng. Biotechnol.* **2015**, *3*, 1–11. [CrossRef] [PubMed]

89. Kuschel, M.; Takors, R. Simulated oxygen and glucose gradients as a prerequisite for predicting industrial scale performance a priori. *Biotechnol. Bioeng.* **2020**, *117*, 2760–2770. [CrossRef]

90. Mancini, M.; Moresi, M. Rheological behaviour of baker's yeast suspensions. *J. Food Eng.* **2000**, *44*, 225–231. [CrossRef]

91. Delvigne, F.; Destain, J.; Thonart, P. A methodology for the design of scale-down bioreactors by the use of mixing and circulation stochastic models. *Biochem. Eng. J.* **2006**, *28*, 256–268. [CrossRef]

92. Guijarro, J.M.; Lagunas, R. *Saccharomyces cerevisiae* does not accumulate ethanol against a concentration gradient. *J. Bacteriol.* **1984**, *160*, 874–878. [CrossRef] [PubMed]

93. Piškur, J.; Rozpedowska, E.; Polakova, S.; Merico, A.; Compagno, C. How did Saccharomyces evolve to become a good brewer? *Trends Genet.* **2006**, *22*, 183–186. [CrossRef]

94. Thomson, J.M.; A Gaucher, E.; Burgan, M.F.; De Kee, D.W.; Li, T.; Aris, J.P.; A Benner, S. Resurrecting ancestral alcohol dehydrogenases from yeast. *Nat. Genet.* **2005**, *37*, 630–635. [CrossRef]

95. Tang, W.; Deshmukh, A.T.; Haringa, C.; Wang, G.; Van Gulik, W.; Van Winden, W.; Reuss, M.; Heijnen, J.J.; Xia, J.; Chu, J.; et al. A 9-pool metabolic structured kinetic model describing days to seconds dynamics of growth and product formation by Penicillium chrysogenum. *Biotechnol. Bioeng.* **2017**, *114*, 1733–1743. [CrossRef]

96. La, A.; Du, H.; Taidi, B.; Perré, P. A predictive dynamic yeast model based on component, energy, and electron carrier balances. *Biotechnol. Bioeng.* **2020**, *117*, 2728–2740. [CrossRef] [PubMed]

97. Orij, R.; Brul, S.; Smits, G.J. Intracellular pH is a tightly controlled signal in yeast. *Biochim. Biophys. Acta (BBA) Gen. Subj.* **2011**, *1810*, 933–944. [CrossRef] [PubMed]

98. Pham, T.H.; Mauvais, G.; Vergoignan, C.; De Coninck, J.; Dumont, F.; Lherminier, J.; Cachon, R.; Feron, G. Gaseous environments modify physiology in the brewing yeast *Saccharomyces cerevisiae* during batch alcoholic fermentation. *J. Appl. Microbiol.* **2008**, *105*, 858–874. [CrossRef] [PubMed]

Publisher's Note: MDPI stays neutral with regard to jurisdictional claims in published maps and institutional affiliations.

Article

Droplet Characteristics of Rotating Packed Bed in H_2S Absorption: A Computational Fluid Dynamics Analysis

Zhihong Wang *, Xuxiang Wu, Tao Yang *, Shicheng Wang, Zhixi Liu and Xiaodong Dan

School of Chemistry and Chemical Engineering, Southwest Petroleum University, Chengdu 610500, China; Wxx2alley@163.com (X.W.); 18200351814@163.com (S.W.); 18780273398@163.com (Z.L.); dongceswpu@163.com (X.D.)

* Correspondence: wzhswpu@swpu.edu.cn (Z.W.); yangtao9264826@outlook.com (T.Y.); Tel.: +86-1390-8215-506 (Z.W.); +86-1354-1167-685 (T.Y.)

Received: 7 August 2019; Accepted: 8 October 2019; Published: 11 October 2019

Abstract: Rotating packed bed (RPB) has been demonstrated as a significant and emerging technology to be applied in natural gas desulfurization. However, droplet characteristics and principle in H_2S selective absorption with N-methyldiethanolamine (MDEA) solution have seldom been fully investigated by experimental method. Therefore, a 3D Eulerian–Lagrangian approach has been established to investigate the droplet characteristics. The discrete phase model (DPM) is implemented to track the behavior of droplets, meanwhile the collision model and breakup model are employed to describe the coalescence and breakup of droplets. The simulation results indicate that rotating speed and radial position have a dominant impact on droplet velocity, average residence time and average diameter rather than initial droplet velocity. A short residence time of 0.039–0.085 s is credited in this study for faster mass transfer and reaction rate in RPB. The average droplet diameter decreases when the initial droplet velocity and rotating speed enhances. Restriction of minimum droplet diameter for it to be broken and an appropriate rotating speed have also been elaborated. Additional correlations on droplet velocity and diameter have been obtained mainly considering the rotating speed and radial position in RPB. This proposed formula leads to a much better understanding of droplet characteristics in RPB.

Keywords: rotating packed bed; natural gas desulfurization; droplet characteristic; Eulerian–Lagrangian approach; computational fluid dynamics

---

1. Introduction

Rotating packed bed (RPB) has been demonstrated as a novel and efficient equipment in process intensification of mass transfer and reaction [1], as displayed in Figure 1. Gas phase and liquid phase contact counter-currently in the packing by centrifugal force, and the height of mass transfer unit (HTU) in RPB can dramatically decrease by 1–3 orders of magnitude compared with that in a conventional column. Therefore, RPB can evidently enhance productivity while also remarkably reducing the device size and equipment investments. RPB has been maturely applied to the traditional chemical industry [2] in processes such as distillation, absorption and extraction. It has gradually begun to play a crucial role in gas purification, especially for H_2S removal in natural gas.

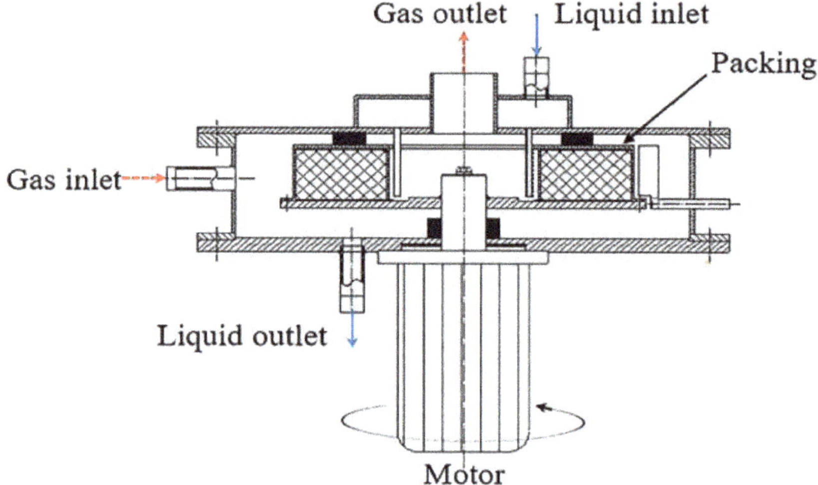

Figure 1. Schematic diagram of rotating packed bed (RPB) (counter-currently).

Various investigations have been carried out to reveal the gas–liquid behavior in RPB by experimental method. Burns and Ramshaw [3] exhibited a visual liquid patterns, shown in Figure 2, in the way that a video camera was arranged rotating synchronously with the packing. Liquid patterns can be distinguished into pore flow, droplet flow and film flow in different stages and positions. At a lower rotating speed, the rivulet flow was dominant while droplet flow mainly existed at a higher rotating speed. Guo [4] found that liquid impinged and deformed intensively within the initial 7–10 mm from the inner diameter and that liquid performed two states in the packing: Films on the packing surface and films flying in the gap of packing. Moreover, Li [5] investigated the effect of a various operating conditions especially for the layer numbers on the liquid patterns and drew a conclusion that the droplet was the main pattern in packing. As for recent research, Sang [6] indicated a criterion to distinguish two typical liquid patterns: Ligament flow and droplet flow. The Computational fluid dynamics is called CFD for short, which is a combination of computer and numerical techniques. The nature of CFD is that some physical phenomena can be obtained by solving related partial differential equations in computer. Due to the lower cost and visualization of process and results of CFD technology, the application of CFD has become a powerful approach to depict the liquid patterns and behaviors in RPB. Shi [7] verified that the flow patterns were varied under different rotating speeds in CFD simulation. Ouyang emphasized that a higher viscosity of liquid led a liquid line predominantly [8]. In another simulation article, Ouyang et al. pointed out that droplet diameter increased with decreasing rotating speed, liquid initial velocity and number of layers in a Rotor–Stator reactor (RSR) [9]. Conclusively, Xie [10] figured out that the collision and merging of droplets occurred between droplets and packing and that the liquid flow tended to form droplets under the influence of surface tension (larger contact angles) and high rotating speeds (1000–1500 rpm).

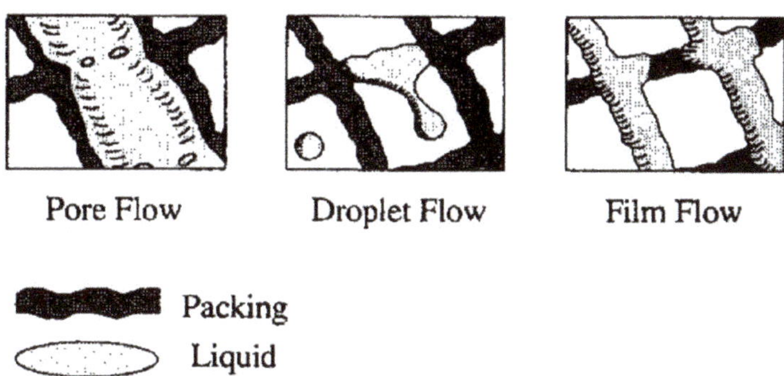

Figure 2. Three types of liquid pattern in RPB with a video camera, reproduced with permission from [3]. Copyright Elsevier, 1996.

Aforementioned visual studies through experimental method and CFD method elaborate that typical flow pattern in RPB is liquid droplet. Although characteristics of droplets such as velocity, average residence time and average diameter exert great influence in design and improvement of RPB, there are still few studies on them. Consequently the present work investigated the influence of rotating speed, initial droplet velocity and initial droplet diameter on droplet velocity, average droplet residence time as well as average droplet diameter by CFD method. Through introducing the collision and breakup model, correlations of droplet velocity and average diameter were proposed. Additionally, a phenomenon called "end effect" in the end zone and a principle of processing intensification contributing to H_2S selective absorption were also presented.

2. Simulation

2.1. Physical Model and Grid Refinement of RPB

The structure of RPB in this study is based on a pilot installation built in a natural gas purification plant, and Table 1 displays the main dimensions of RPB. It has a 24 mm height, 45 mm inner diameter and 160 mm outer diameter. Packing is regarded as a crucial part of RPB and it generally consists of uniform thickness wire, which has a 20 mm height, 48 mm inner diameter and 92 mm outer diameter.

Table 1. The main geometric dimensions of RPB.

	Inner Diameter (mm)	Outer Diameter (mm)	Height (mm)
RPB	45	160	24
Packing	48	92	20

In order to make the 3D model as consistent as possible with the actual structure, there are three parts in 3D model: End zone, packing zone and cavity zone (Figure 3a). The packing zone is built with 21 concentric layers, and the foursquare obstacles model is employed to simulate the wire behavior in RPB. Consequently, the side length of each foursquare is 0.8 mm and the center distances between two foursquares in radial and tangential directions are 2 mm and 5 mm, respectively. Four annulus distributors set uniformly in the inner radius of RPB are set as liquid inlets and an outer ring is the outlet.

A computational 3D model grid is built by ICEM 19.0. In order to describe more details, a grid refinement has been made, namely different numbers of cells are meshed in different parts. At the same time, the packing zone and the end zone are meshed by tetrahedral cell, and the cavity zone is meshed by mixed cell. The meshing result is shown in Figure 3b.

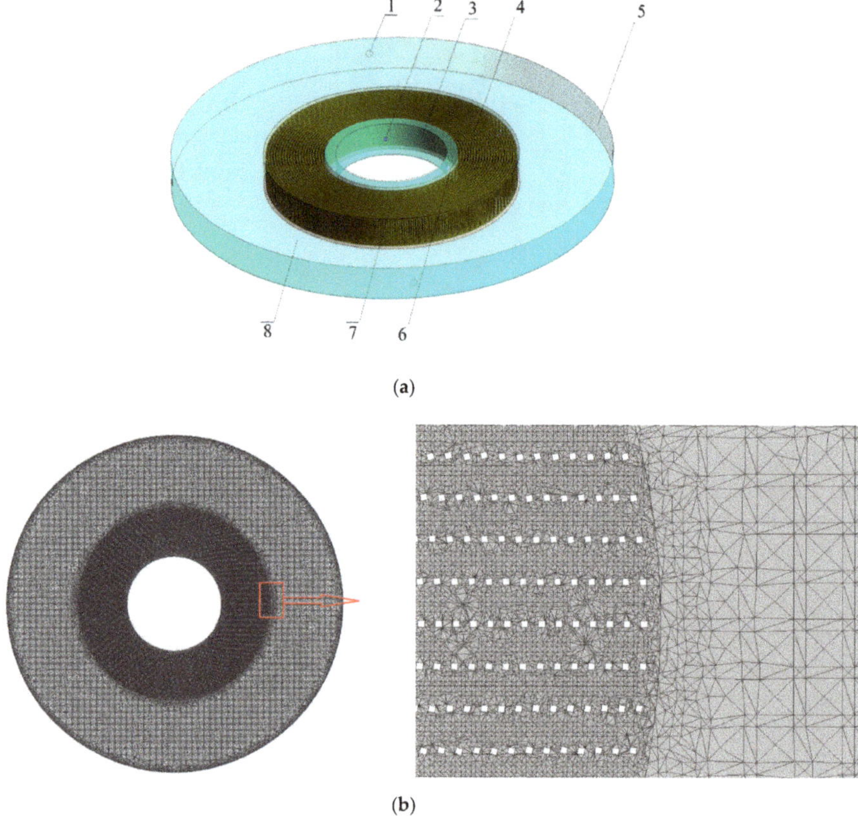

Figure 3. (**a**) 3D physical model diagram (1, gas inlet; 2, liquid inlet; 3, gas outlet; 4, foursquare obstacles; 5, liquid outlet; 6, end zone; 7, packing zone; 8, cavity zone). (**b**) Grid refinement.

2.2. Mathematical Modelling

In this study, based on an assumption that the second phase as a dispersed part occupies a low volume fraction (below 10%), the Eulerian–Lagrangian approach is used in which the natural gas is considered as a continuous phase while the MDEA solution is regarded as a discrete dispersed phase. Forces, like drag force, buoyancy force and virtual mass force, are taken into account to describe the particle force balance. A turbulence model is introduced to reflect the turbulent process when packing shears the MDEA solution intensely under high rotation speed. Models, such as Taylor analogy breakup (TAB) model and O'Rourke model, are carried out to predict the breakup and coalescence of droplets.

2.2.1. Governing Equations

The basic equations coupled in CFD models for solving mass and momentum conservation equations are listed below for an incompressible, laminar flow, non-accelerating and unsteady state process:

Mass conservation equation

$$\frac{\partial \rho}{\partial t} + \nabla \cdot \left(\rho \vec{v} \right) = 0, \tag{1}$$

Momentum conservation equation

$$\frac{\partial(\rho\vec{v})}{\partial t} + \nabla\cdot(\rho\vec{v}\vec{v}) = -\nabla p + \rho\vec{g} + \nabla\cdot\left(\mu_{eff}\left(\nabla\vec{v} + (\nabla\vec{v})^T\right)\right) + \vec{F},\tag{2}$$

where p is the static pressure, ρ is the fluid density, $\vec{g}$ is the gravitational vector, μ_{eff} is the molecular viscosity and $\vec{F}$ is the external body force representing the interactions between the dispersed phase and the continuous phase. As for a flow in a rotating reference frame, like the rotation in RPB, the origin of the non-accelerating system is determined by a centrifugal acceleration $\vec{\omega}$ in a rotating system, hence the governing equations of flow in a rotating reference frame are as follows:

$$\frac{\partial\rho}{\partial t} + \nabla\cdot(\rho\vec{v}_r) = 0,\tag{3}$$

$$\frac{\partial(\rho\vec{v}_r)}{\partial t} + \nabla\cdot(\rho\vec{v}_r\vec{v}_r) = -\nabla p + \rho\left(2\vec{\omega}\times\vec{v}_r + \vec{\omega}\times\vec{u}_r\right) + \nabla\cdot\left(\mu_{eff}\left(\nabla\vec{v}_r + (\nabla\vec{v}_r)^T\right)\right) + \vec{F},\tag{4}$$

$$\vec{v}_r = \vec{v} - \vec{u}_r,\tag{5}$$

$$\vec{u}_r = \vec{\omega}\times\vec{r},\tag{6}$$

where $\vec{v}_r$ is the relative velocity, $\vec{\omega}$ is the centrifugal acceleration, $\vec{\omega}\times\vec{v}_r$ is the Coriolis acceleration, $\vec{u}_r$ is the whirl velocity and $\vec{\omega}\times\vec{u}_r$ is the centripetal acceleration.

Although the Navier–Stokes equation can accurately describe the mass and momentum of the laminar flow, the turbulence caused by intensive movement can evidently influence the transport process. Therefore, the turbulence model determines additional variables as time-averaged or ensemble-averaged in the modified governing equations.

2.2.2. Turbulence Model

The standard k–ε model is presented by molecular viscosity and dynamic viscosity due to the turbulence and it is a two-equation model. It is a semi-empirical model, and the solution of the k equation allows the turbulent kinetic energy to be determined while the result of the ε equation can be considered relying on phenomenological considerations and empiricism.

$$\frac{\partial}{\partial t}(\rho k) + \nabla\cdot(\rho k\vec{v}_r) = \nabla\cdot\left(\left(\mu + \frac{\mu_t}{\sigma_k}\right)\cdot\nabla k\right) + G_k + G_b - \rho\varepsilon,\tag{7}$$

$$\frac{\partial}{\partial t}(\rho\varepsilon) + \nabla\cdot(\rho\varepsilon\vec{v}_r) = \nabla\cdot\left(\left(\mu + \frac{\mu_t}{\sigma_\varepsilon}\right)\cdot\nabla\varepsilon\right) + c_{\varepsilon1}\frac{\varepsilon}{k}(G_k + c_{\varepsilon3}G_b) - c_{\varepsilon2}\rho\frac{\varepsilon^2}{k},\tag{8}$$

$$G_k = \mu_t\nabla\vec{v}\cdot\left(\nabla\vec{v} + (\nabla\vec{v})^T\right),\tag{9}$$

$$G_b = -\frac{\mu_t}{\rho Pr_t}g\nabla\rho,\tag{10}$$

$$\mu_t = C_\mu\rho\frac{k^2}{\varepsilon},\tag{11}$$

where G_k represents the influence of the mean velocity gradients, G_b represents the influence of the buoyancy force, μ_t is the turbulent viscosity, Pr_t is the turbulent Prandtl number for energy, k is the turbulence kinetic energy and ε is the dissipation rate. $c_{\varepsilon1}$, $c_{\varepsilon2}$, $c_{\varepsilon3}$, C_μ, σ_k, σ_ε are the model constants and the default values are: $c_{\varepsilon1} = 1.44$, $c_{\varepsilon2} = 1.92$, $C_\mu = 0.09$, $\sigma_k = 1.0$, $\sigma_\varepsilon = 1.3$.

2.2.3. Droplet Force Balance

In a Lagrangian reference frame, the force balance on the particle which drives the particle acceleration is due to the difference in velocity between fluid and particles as well as various influences of force, like drag force, gravity, virtual mass force and centrifugal force in RPB. Hence, a particle force balance can be written as:

$$\frac{du_p}{dt} = F_D(u - u_p) + \frac{(\rho_p - \rho)}{\rho_p} g + F_x,$$ (12)

Drag force

$$F_D = \frac{18\mu}{\rho_p d_p^2} \frac{C_D Re}{24},$$ (13)

$$Re = \frac{\rho d_p |u_p - u|}{\mu},$$ (14)

where $F_D(u - u_p)$ is the drag force per unit particle mass, $\frac{(\rho_p - \rho)}{\rho_p} g$ is the gravity per unit particle mass, F_x is the additional acceleration term (force per unit particle mass), C_D is the drag coefficient, Re is the relative Reynolds number, u_p is the particle velocity, u is the fluid velocity and ρ_p is the particle density.

Virtual mass force

$$F_m = \frac{1}{2} \frac{\rho}{\rho_p} \frac{d}{dt}(u - u_p),$$ (15)

Forces in rotating reference frames

$$F_{r,x} = \left(1 - \frac{\rho}{\rho_p}\right)\omega^2 x + 2\omega\left(u_{y,p} - \frac{\rho}{\rho_p}u_y\right),$$ (16)

$$F_{r,y} = \left(1 - \frac{\rho}{\rho_p}\right)\omega^2 y + 2\omega\left(u_{x,p} - \frac{\rho}{\rho_p}u_x\right),$$ (17)

where F_m is the virtual mass force, $F_{r,x}$ and $F_{r,y}$ are the forces in rotating reference frames in x and y directions, respectively and ω is the rotating speed.

2.2.4. Droplet Coalescence and Breakup Model

The coalescence model

The O'Rourke model a is model based on energy balance; the droplets will coalescence once the kinetic energy cannot overcome the surface energy of a new droplet [11]. We is defined as a ratio of inertial force to surface tension to quantify the collision and coalescence of droplets. Therefore, the critical offset b is a function of the collisional, shown in Figure 4.

$$We = \frac{\rho_p \left|\vec{u}_l - \vec{u}_s\right|^2 \bar{d}}{\sigma},$$ (18)

$$We \geq \frac{2.4(r_l + r_s)^2 f\left(\frac{r_l}{r_s}\right)}{b^2},$$ (19)

$$b_{crit} = (r_l + r_s)\sqrt{\min\left\{1.0, \frac{2.4f\left(\frac{r_l}{r_s}\right)}{We}\right\}},$$ (20)

$$f\left(\frac{r_l}{r_s}\right) = \left(\frac{r_l}{r_s}\right)^3 - 2.4\left(\frac{r_l}{r_s}\right)^2 + 2.7\frac{r_l}{r_s},$$ (21)

where We is the Weber number, $\vec{u}_l$ and $\vec{u}_s$ are the velocities of large droplets and small droplets, respectively, $\bar{d}$ is the arithmetic mean diameter of two droplets, r_l and r_s are the radii of large droplets and small droplets, respectively, b and b_{crit} are the actual collision parameter and the critical offset of collision, respectively. If $b < b_{crit}$, the result of collision is coalescence, and the new velocity based on conservation of momentum and kinetic energy is calculated below:

$$v' = \frac{m_l v_l + m_s v_s + m_s(v_l - v_s)}{m_l + m_s}\left(\frac{b - b_{crit}}{r_l + r_s - b_{crit}}\right),\tag{22}$$

where m_l and m_s are the masses of large droplets and small droplets, respectively.

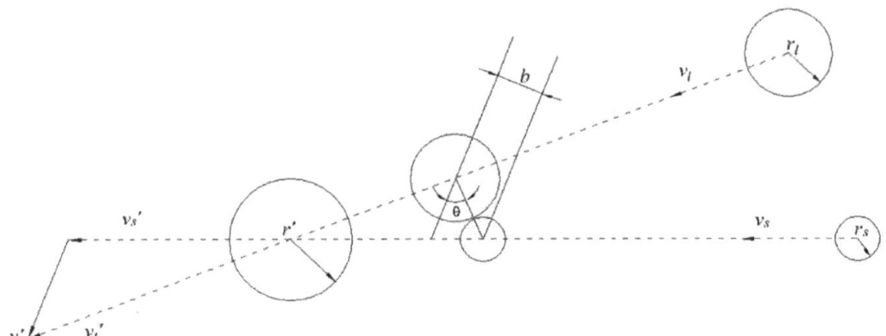

Figure 4. Schematic diagram of the coalescence process.

Taylor Analogy Breakup (TAB) model

The TAB model governs the oscillating and distorting droplet; when the droplet oscillations grow to a critical value, the large droplet will break up into small droplets. The droplet distortion governing equation is defined as follows:

$$F - kx - d\frac{dx}{dt} = m\frac{d^2x}{dt^2},\tag{23}$$

where x is the displacement of the droplet, m is the droplet mass, F is the aerodynamic force of the droplet, kx is the surface tension of the droplet and $d\frac{dx}{dt}$ is the viscous force.

$$\frac{F}{m} = C_F\frac{\rho|u - u_p|^2}{\rho_p r_p},\tag{24}$$

$$\frac{d}{m} = C_d\frac{\rho\mu_p}{\rho_p r_p^2},\tag{25}$$

$$\frac{k}{m} = C_k\frac{\sigma_p}{\rho_p r_p^3},\tag{26}$$

$$\frac{d^2y}{dt^2} = \frac{C_F}{C_b}\frac{\rho|\vec{u} - \vec{u}_p|^2}{\rho_p r_p^2} - C_k\frac{\sigma_p}{\rho_p r_p^3}y - C_d\frac{\mu_p}{\rho_p r_p^2}\frac{dy}{dt},\tag{27}$$

where C_F, C_d, C_k are dimensionless constants, $C_F = 1/3$, $C_d = 5$, $C_k = 8$. Setting $y = x/(C_b r_p)$, when $C_b = 0.5$ and $y > 1$, the non-dimensional equation for droplet breakup is shown in Equation (28).

$$y(t) = \mathrm{We}_r + e^{-\frac{t}{t_d}}\left\{(y_0 - \mathrm{We}_r)\cos(\omega t) + \frac{1}{\omega}\left(\frac{dy_0}{dt} + \frac{y_0 - \mathrm{We}_r}{t_d}\right)\sin(\omega t)\right\},\tag{28}$$

$$We_r = \frac{C_F}{C_b C_k} \frac{\rho \left| \vec{u} - \vec{u}_p \right|^2 r_p}{\sigma_p}, \tag{29}$$

$$t_d = \frac{2 \rho_p r_p^2}{C_d \mu_p}, \tag{30}$$

$$\omega^2 = C_k \frac{\sigma_p}{\rho_p r_p^3} - \frac{1}{t_d^2}. \tag{31}$$

The size of droplet is determined by the energy conservation between the parent droplet and the child droplets which satisfies the Sauter distribution and the number of child droplets can be calculated by mass conservation.

$$r' = \frac{1 + \frac{1}{20} y^2 + \frac{1}{8} \frac{\rho_p r_p^3}{\sigma_p} \left(\frac{dy}{dt} \right)^2}{r_p} \tag{32}$$

2.3. Fluid Properties

MDEA aqueous solution is one of the conventional absorbents for H_2S selective absorption. In this study, the gas mixture consists mainly of methane and the mass fraction of the aqueous solution is 35%. The properties of the gas and liquid used for the CFD simulations are shown in Table 2. The MDEA aqueous solution is assumed to operate at a constant temperature of 30 °C and 2 MPa, which are close to the real operation conditions for H_2S selective absorption in RPB. Under these temperature and pressure conditions, the density and viscosity of the gas mixture are 15.99 kg/m^3 and 1.203×10^{-5} kg/(m·s), respectively; the density and viscosity of the aqueous solution are and 1027 kg/m^3 and 1.203×10^{-5} kg/(m·s), respectively.

Table 2. Fluid properties.

	CH$_4$	C$_2$H$_6$	C$_3$H$_8$	C$_4$H$_{10}$	C$_5$H$_{12}$	CO$_2$	H$_2$S	N$_2$
Gas (mol %)	85.71	2.30	0.73	0.47	0.24	4.25	5.04	1.27
Liquid (m %)	An MDEA aqueous solution with a mass fraction of 35%							

2.4. Solution Procedure

The domain of the packing zone is defined as the moving reference frame (MFR) and the rest of RPB: End zone and cavity zone are relative to a stationary reference frame in the Fluent 19.0 software. Thus, the packing with a reflect boundary rotates in a rotating speed of 300 to 900 rpm and the wall condition with no-slip and escape boundary is set to simulate the casing of RPB. The gas inlet and outlet in RPB are set as the velocity inlet boundary and pressure outlet boundary conditions, respectively, while the liquid inlet and outlet are defined as discrete phase surface injection and trap boundary condition, respectively. The gas velocity is considered as a constant 0.5 m/s, and the droplets are injected into the RPB in this way that initial droplet velocity is specified by a range of 0.5 to 2.5 m/s, the initial droplet diameter is set to 1 to 5 mm and the number of droplets injected into the RPB is 400. Steady simulations are implemented with standard k–ε model and DPM model to investigate the droplet characteristics in RPB. The droplet discrete phase interacts with the continuous phase every five iterations and an unsteady particle tracking is applied to track the droplet trajectories individually with a time step size of 0.001 s for a maximum of 5000 steps. Drag force, gravity, virtual mass force and centrifugal force are considered in this solution and droplet coalescence and breakup models mentioned above are also introduced. The pressure–velocity coupling is resolved by SIMPLE scheme and spatial discretization methods for gradient and pressure are least squares cell-based and PRESTO, respectively. The second-order upwind scheme is employed for solving the momentum equations and turbulence equations. For the convergence absolute criteria, the residuals in mass conservation, momentum conservation and turbulence are less than 1×10^{-5}. In addition, the maximum number of iterations for steady calculation is 4000 to achieve the steady state in RPB.

2.5. Grid Independence

As is known, the number of grids has a great impact on the accuracy of the simulation result. Therefore, a grid independence study is performed to obtain a reasonable computational mesh. Fixing the droplet diameter of 2 mm, rotating speed of 300 rpm and initial droplet velocity of 0.5 m/s as the initial conditions and choosing the droplet average residence time as the only dependent variable, five different grids consisting of 1.4, 2.5, 3.3, 4.3 and 5 million unstructured cells have been implemented to analyze the effect of the cell on the droplet average residence time. As shown in Figure 5, the average residence time is almost same when the number of cells reaches 4.3 million, which may be considered as a responsible grid to predict droplet characteristics. Finally, the number of 4,349,428 cells is selected to investigate other simulation results while considering the computing resources and accuracy.

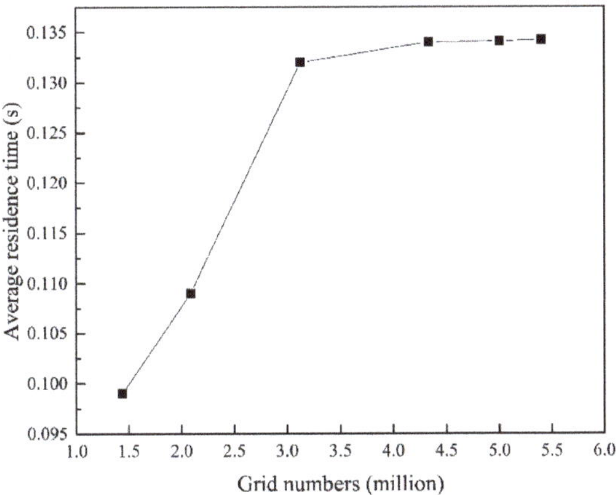

Figure 5. The grid independence checking.

3. Results and Discussions

Since droplet form in RPB is the dominant liquid pattern, its characteristics such as velocity, residence time and diameter in mass transfer and reactions will play a crucial role in novel research. Therefore, a CFD simulation on droplet characteristics will demonstrate a visual flow distribution as well as significant guide for H_2S selective absorption into MDEA solution in RPB.

3.1. Droplet Velocity in RPB

Velocity distribution, as a common and crucial characteristic of droplets, has a direct effect on residence time and diameter, therefore the velocity value and vector under various rotating speeds and initial droplet velocities are shown in Figure 6. Similarly, the velocity is distributed uniformly and symmetrically along the radial and tangential directions. The liquid is injected and dispersed uniformly into packing zone from the liquid distributor and it rapidly obtains a synchronous tangential speed as it is packing, finally liquid droplets are spun out from the outer packing zone into the cavity randomly.

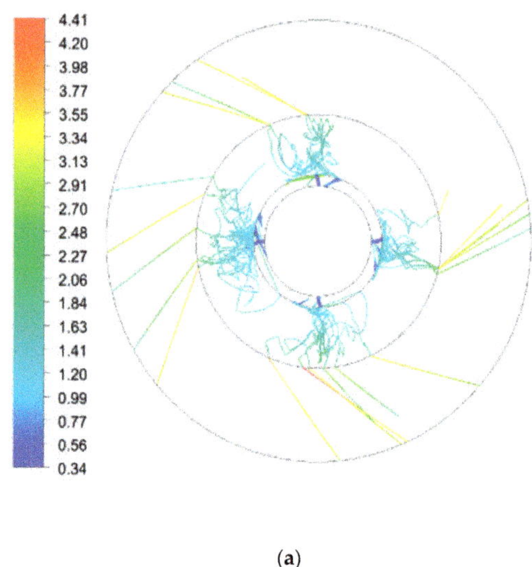

(**a**)

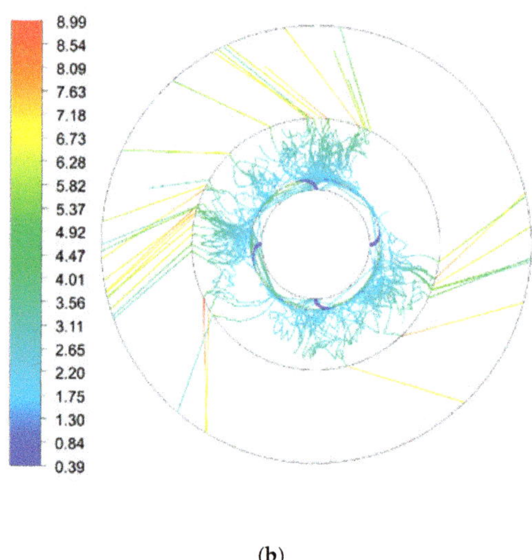

(**b**)

Figure 6. *Cont.*

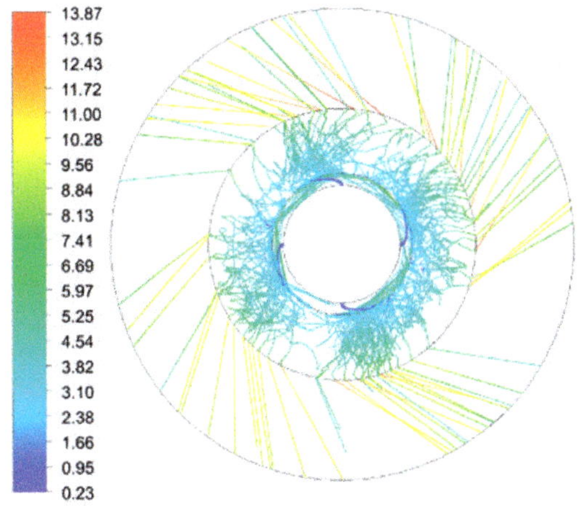

(**c**)

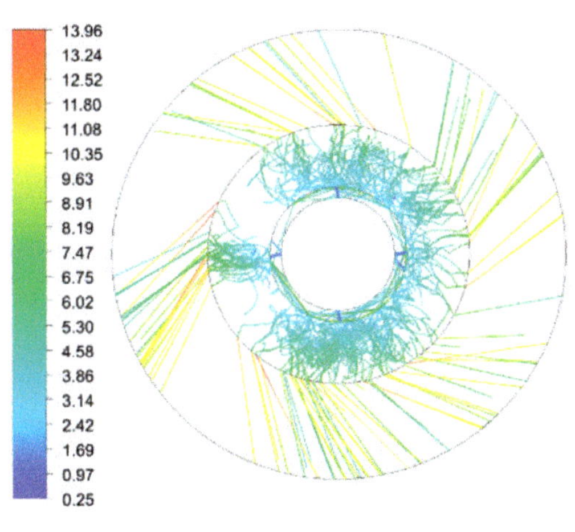

(**d**)

Figure 6. *Cont.*

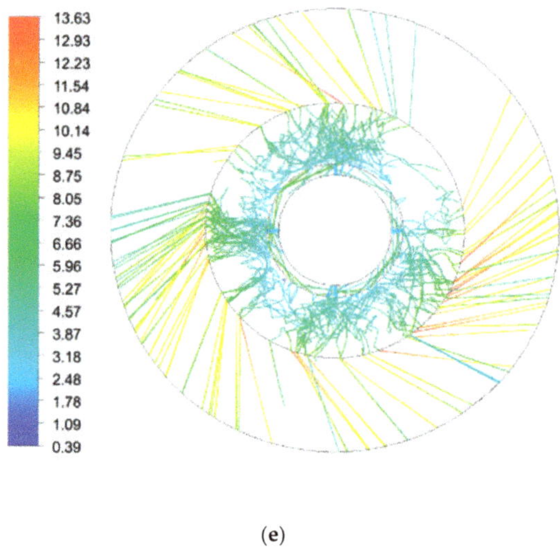

(e)

Figure 6. The velocity values and vectors under various initial droplet velocities and rotating speeds. (**a**) $u_0 = 0.5$ m/s, $\omega = 300$ rpm; (**b**) $u_0 = 0.5$ m/s, $\omega = 600$ rpm; (**c**) $u_0 = 0.5$ m/s, $\omega = 900$ rpm; (**d**) $u_0 = 1.5$ m/s, $\omega = 600$ rpm; (**e**) $u_0 = 2.5$ m/s, $\omega = 600$ rpm.

In the end zone, a parabolic droplet motion is observed with the increase of rotating speed from 300–900 rpm, since the initial droplet velocity in the radial direction is lower than that in tangential direction. Further, a phenomena of back-mixing, called "end effect" in our review [12] on mass transfer process in RPB, was also confirmed to have a superiority in processing intensification on mass transfer [13]. In the packing zone, droplets move in a radial spiral and distribute irregularly. An analysis on different rotating speeds and initial droplet velocities illustrates that the radial velocity is mainly affected by initial droplet velocity while the tangential velocity is related to rotating speed and radial position. In the cavity zone, droplets move towards the outer wall of RPB with a constant angle which decreases with the increase of rotating speed, and it manifests that rotating speed has a major impact on resultant droplet velocity.

Under the sustaining shear from the packing, a large proportion of droplets are synchronous with packing and the droplet velocity increases linearly along the radial position. Some fluctuation of droplet velocity has also been observed because the collision, breakup and coalescence of droplets take place among droplet–droplet, droplet–gas and droplet–packing interactions.

3.1.1. Effect of Initial Droplet Velocity on Droplet Velocity

Figure 7 show the effect of initial droplet velocity on droplet velocity in the packing zone under the condition that initial droplet diameter is 3 mm and rotating speed is 900 rpm. In this zone, the droplet velocity is seldom influenced by initial droplet velocity. The reason is that the initial droplet velocity makes less contribution to resultant velocity compared with the tangential velocity generated by centrifugal force. On the other hand, the inelastic collision on droplets results in the loss of kinetic energy once droplets are sheared by packing.

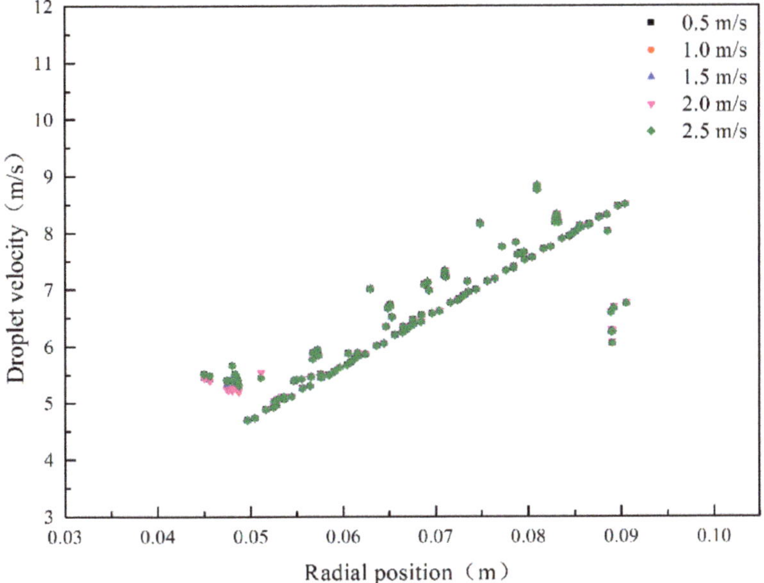

Figure 7. The effect of initial droplet velocity on droplet velocity in the packing zone.

3.1.2. Effect of Rotating Speed on Droplet Velocity

Figure 8 establishes the effect of rotating speed on droplet velocity in the packing zone under the condition that initial droplet diameter is 3 mm and initial droplet velocity is 0.5 m/s. In the packing zone, the rotating speed is dominant for droplet velocity increasing and it illustrates that the droplet velocity increases along with rotating speed increasing. As a consequence, Equation (33) for predicting droplet velocity in RPB is proposed by considering the influence of initial droplet velocity, rotating speed and radial position; validation Equation (34), predicted from Sang [14], is implemented. Finally, the mathematic expression declares that rotating speed and radial position have the main impacts on droplet velocity.

$$u_p = 0.6088\omega^{0.9819} u_0^{0.0008} R^{0.7702}, \tag{33}$$

$$u_p = 1.013\omega^{0.9994} u_0^{0.008} R^{1.005} \mu^{-0.009} \sigma^{0.036}. \tag{34}$$

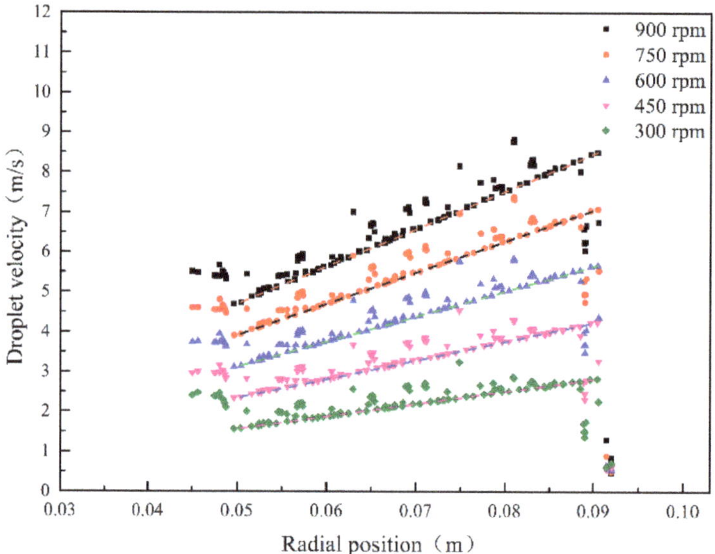

Figure 8. The effect of rotating speed on droplet velocity in the packing zone.

3.2. Average Residence Time Distribution in RPB

Analysis on selective absorption with MDEA indicated that it was kinetically selective towards H_2S while thermodynamically selective towards CO_2 [15]. According to the theory of the diffusion–reaction process for H_2S selective absorption, the reaction and mass transfer on H_2S are both instantaneously fast, while those process are restrained in CO_2 mass transfer into the liquid film. Therefore, average residence time of droplet is considered as a key parameter in the H_2S selective absorption into MDEA process in RPB. A method to obtain the residence time distribution is applied by tracking droplets, and residence time distributions in RPB under various initial droplet velocities and rotating speeds are shown in Figure 9.

Droplets interact intensely with the rotating packing and they are driven by centrifugal force, gravity, gas viscous resistance and packing frictional resistance, resulting in a frequent path change. Most of the droplets are accelerated to a synchronous motion with packing and are ejected in a radial spiral. With the increase of initial droplet velocity and rotating speed, the turbulence intensity increases with the increase of random trajectory. Table 3 summarizes the number of tracked droplets and average residence time in RPB under different rotating speeds and initial droplet velocities.

Table 3. Average residence time under various initial droplet velocities and rotating speeds.

Rotating Speed (rpm)	300	600	900	900	900
Initial droplet velocity (m/s)	0.5	0.5	0.5	1.5	2.5
Number of droplets	58	111	183	360	391
Total residence time (s)	7.7	12.7	15.5	17.2	15.0
Average residence time (s)	0.132	0.114	0.085	0.048	0.039

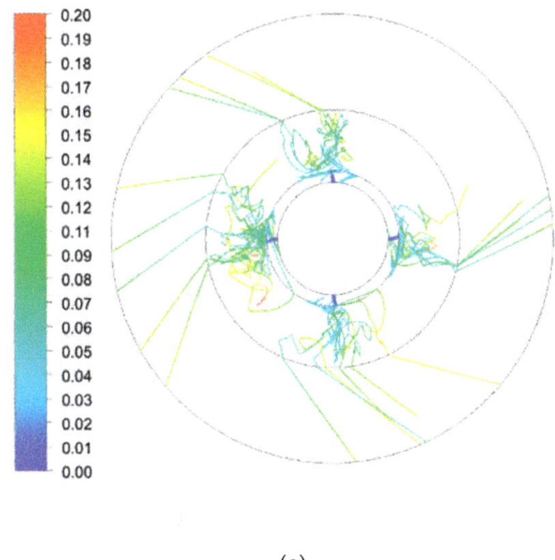

(**a**)

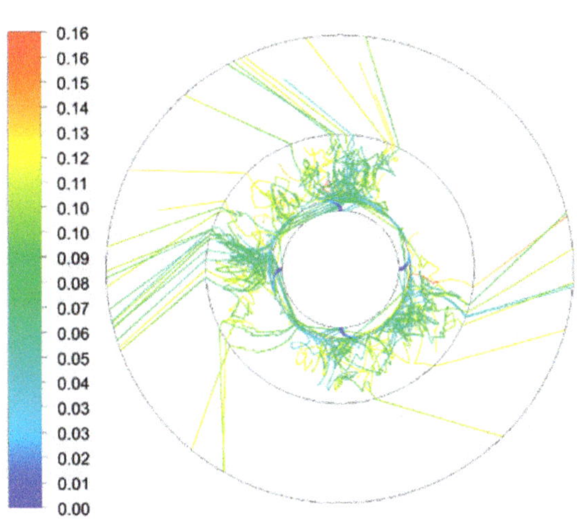

(**b**)

Figure 9. *Cont.*

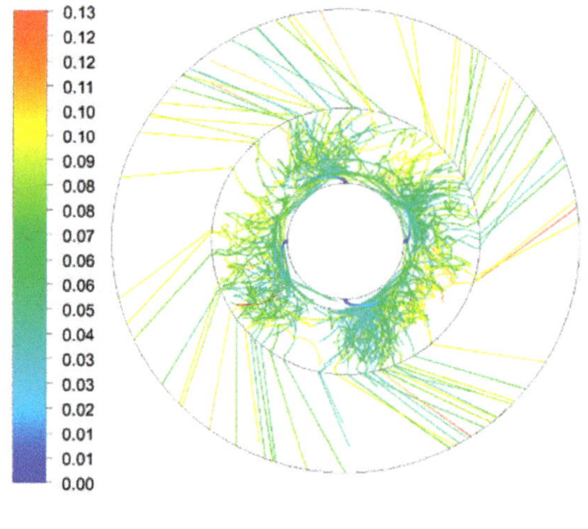

(**c**)

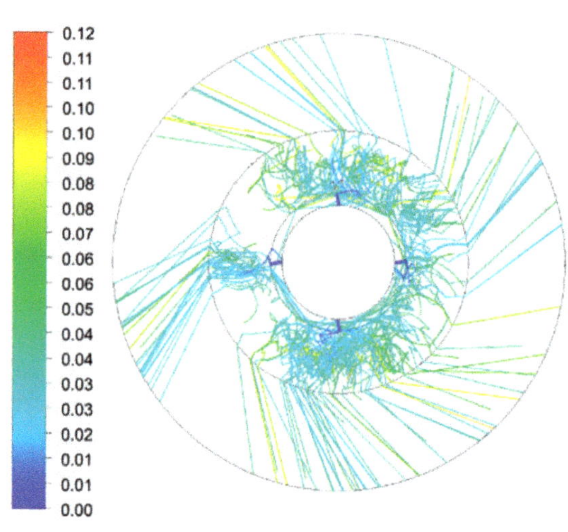

(**d**)

Figure 9. *Cont.*

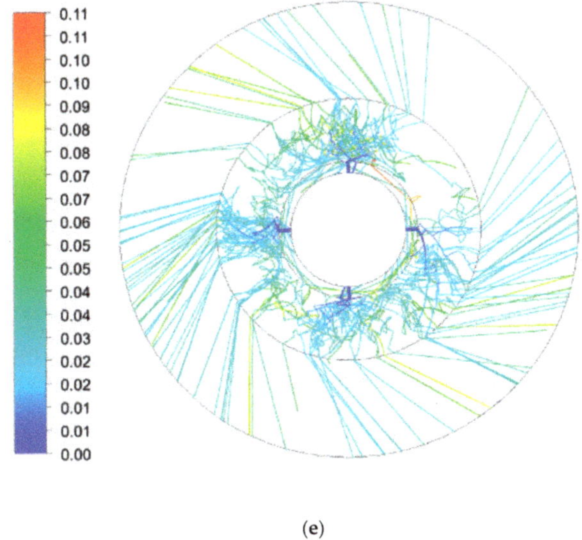

(e)

Figure 9. Residence time distribution in RPB under various initial droplet velocities and rotating speeds. (**a**) $u_0 = 0.5$ m/s, $\omega = 300$ rpm; (**b**) $u_0 = 0.5$ m/s, $\omega = 600$ rpm; (**c**) $u_0 = 0.5$ m/s, $\omega = 900$ rpm; (**d**) $u_0 = 1.5$ m/s, $\omega = 600$ rpm; (**e**) $u_0 = 2.5$ m/s, $\omega = 600$ rpm.

3.2.1. Effect of Initial Droplet Velocity on Average Residence Time

At the same rotating speed of 900 rpm, the average residence time changes drastically from a low initial velocity to a high initial velocity. Because of the increase of initial velocity, the time to pass through the whole packing will shorten correspondingly. As it is displayed in Figure 10, there is a turning point when the initial velocity reaches 1.5 m/s and the average residence time is reduced gradually.

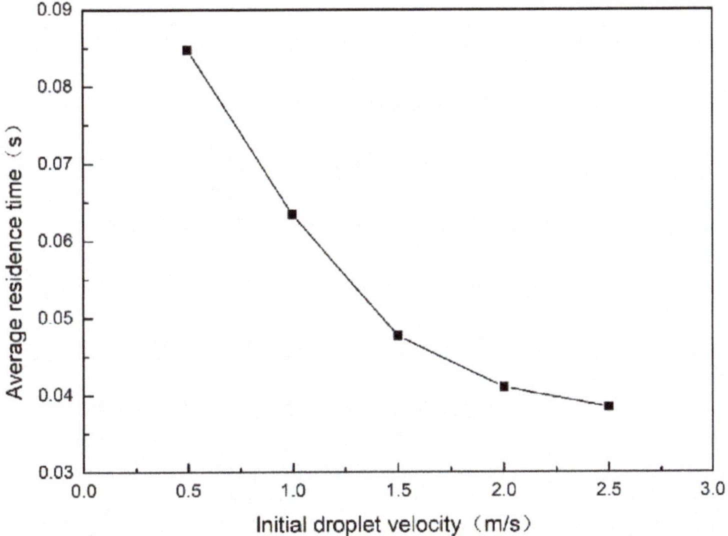

Figure 10. Effect of initial droplet velocity on average residence time.

3.2.2. Effect of Rotating Speed on Average Residence Time

As shown in Figure 11, the droplet attains a tangential velocity when it enters the packing zone and a synchronous velocity will be achieved in a short time. For the sake of reducing residence time in RPB for H_2S selective absorption, enhancing rotating speed seems to be an optimal way. Lower residence time in RPB compared with that in conventional column is ascribed to intensification process under centrifugal force. The higher the droplet velocity is, the lower the residence time obtained will be. Therefore, a commonsense calculation can be proposed to describe the residence time in RPB, and that is the ratio of radial distance to droplet velocity, calculated by Equation (35).

$$t = \frac{R}{u_p} \tag{35}$$

In the process of natural gas desulfurization and purification, H_2S selective absorption and CO_2 partial removal must meet the requirements on commercial natural gas, and the energy consumption also needs to be reduced in MDEA solution regeneration process. Therefore, higher mass transfer efficiency and lower residence time in RPB raise various concerns in natural gas purification. In the quantitative description of Qian's work [16], it only took about 2.0×10^{-9} s for H_2S to establish a steady concentration gradient while the process needed 1 s to complete for CO_2. Thus, it can be highlighted from Figure 10, that the residence time can be only 0.039–0.085 s under a rotating speed of 900 rpm. Droplets go through the packing in a short residence time, especially with high initial droplet velocity and rotating speed.

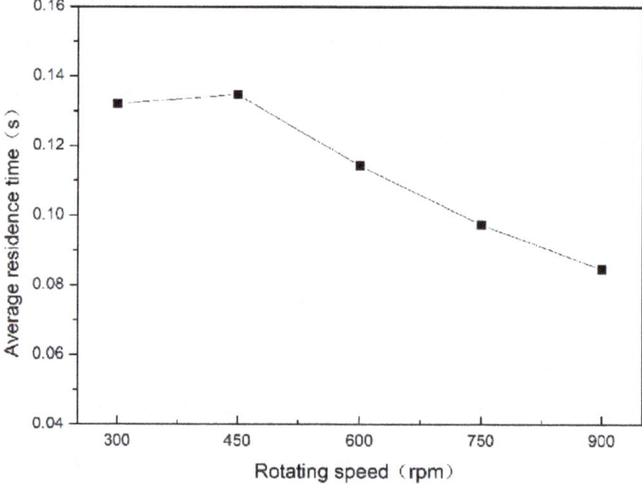

Figure 11. Effect of rotating speed on average residence time.

3.3. Droplet Diameter Distribution in RPB

Droplet surface renewal frequency, which is defined as the percentage of surface updated per unit of time and effective mass transfer area, can be investigated by the droplet diameter distribution along radial position. Figure 12 shows the 3D droplet diameter distribution with an initial droplet velocity of 1.5 m/s, rotating speed of 600 rpm and initial droplet diameter of 3 mm. The droplet diameter changes a lot along the radial position, and collision and shear between droplet and packing result in sustaining breakup and coalescence. This also indicates that frequent surface renewal makes a prominent contribution to mass transfer.

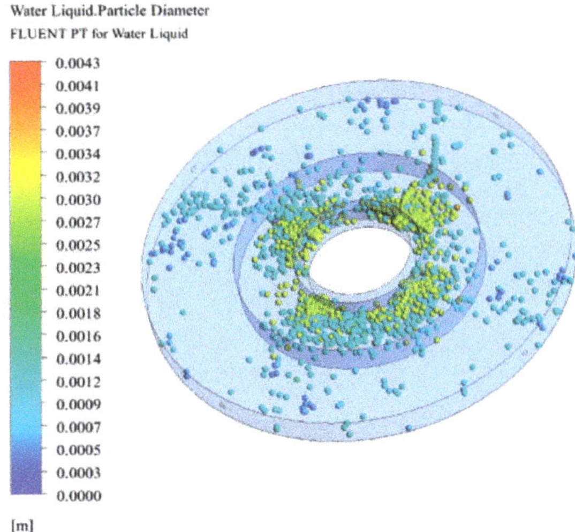

Figure 12. A 3D droplet diameter distribution in RPB.

3.3.1. Effect of Initial Droplet Diameter on Droplet Diameter Distribution

Figure 13 reveals that the droplet diameter increases with an increasing initial droplet diameter. After a larger droplet is introduced from the inlet, it tends to be broken up by packing because of lower surface tension needed to be overcome, and this results in a frequent surface renewal and an intensifying mass transfer process. Guo [17] obtained a formula by fitting relevant experimental data, $d = 0.7284 \left(\sigma / \rho \omega^2 R \right)^{0.5}$, which indicated that the droplet diameter decreases along the radial direction. Both Figures 13 and 14 show that the average droplet diameter is negatively correlated with the radius position, which is consistent with Guo's research. Therefore, it also manifests that the CFD model introduced in this study reasonably predicts the droplet diameter distribution.

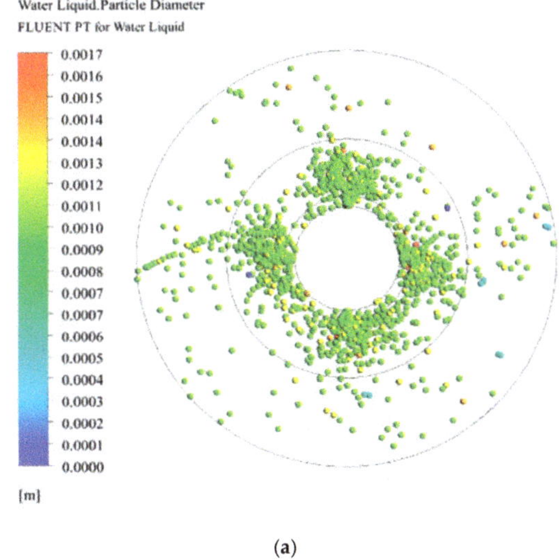

(**a**)

Figure 13. *Cont.*

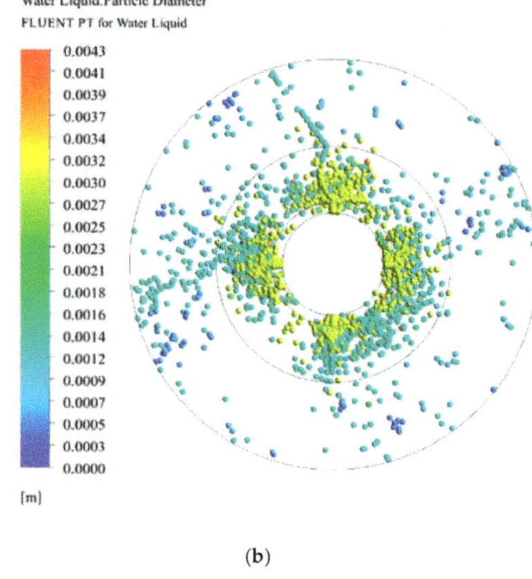

(**b**)

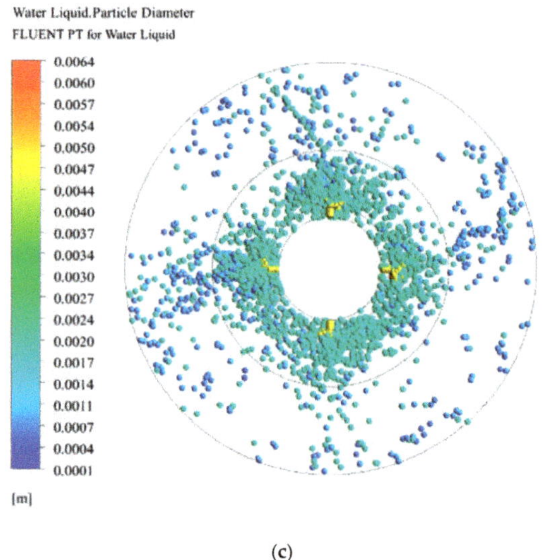

(**c**)

Figure 13. The droplet diameter distribution under different initial droplet diameters. (**a**) $d_0 = 1$ mm; (**b**) $d_0 = 3$ mm; (**c**) $d_0 = 5$ mm.

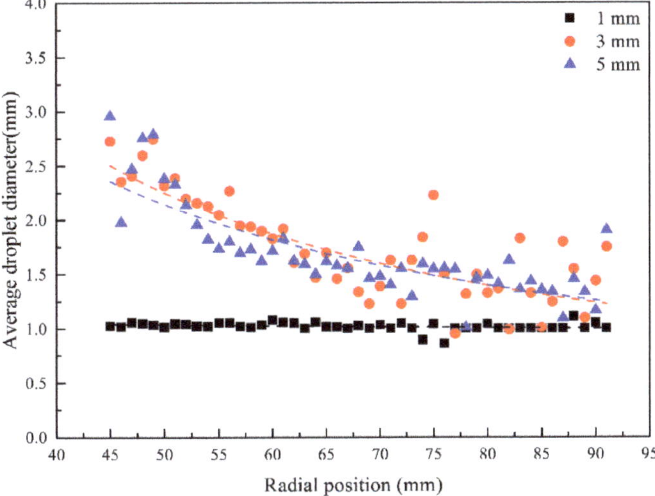

Figure 14. Effect of initial droplet diameter on average droplet diameter.

Specially, Figure 14 illuminates a particular phenomenon: There is rarely collision in the packing zone when the initial droplet diameter is 1 mm, compared with others. The reasons for the above phenomenon are as follows: (1) the packing size is larger than the initial droplet diameter; and (2) the inertial force under this condition is not enough to break the small droplet. Therefore, the study of the collision between 1-mm-diameter initial droplets and packing needs smaller packing size and larger inertial force. In a word, the effect of initial droplet diameter on droplet diameter distribution analyzed above can not only provide guidance for liquid distributor design, but also make a suitable proposal on choosing packing size.

3.3.2. Effect of Initial Droplet Velocity and Rotating Speed on Droplet Diameter

Figures 15 and 16 illustrate that the average droplet diameter along the radial position decreases with the increase of initial droplet velocity from 0.5 to 2.5 m/s and with the increase of rotating speed from 300 rpm to 900 rpm. The droplet obtains the tangential velocity from the packing immediately, resulting in a violent collision and breakup among droplet and packing. Therefore, the droplet surface renewal frequency and effective mass transfer area have been enhanced remarkably.

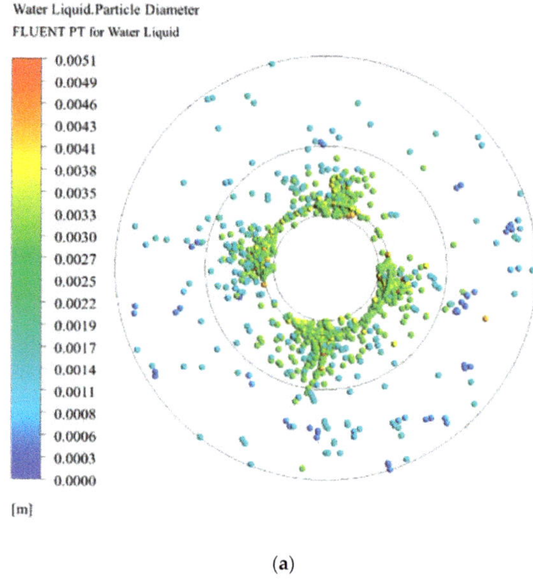

(**a**)

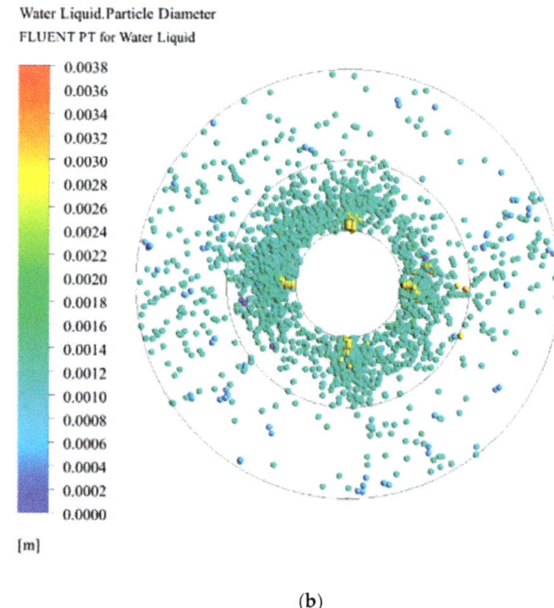

(**b**)

Figure 15. *Cont.*

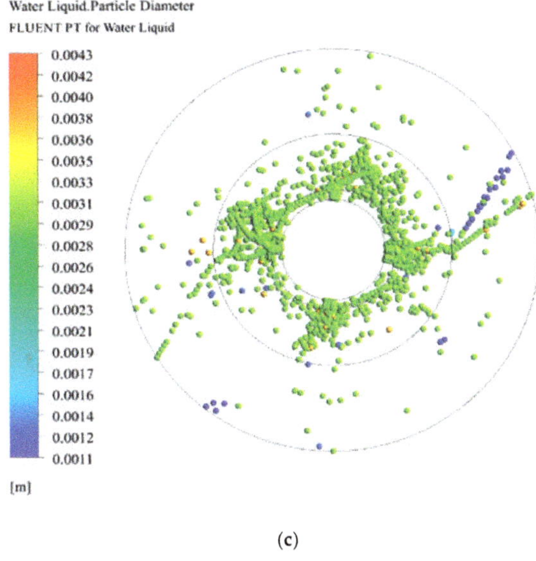

(**c**)

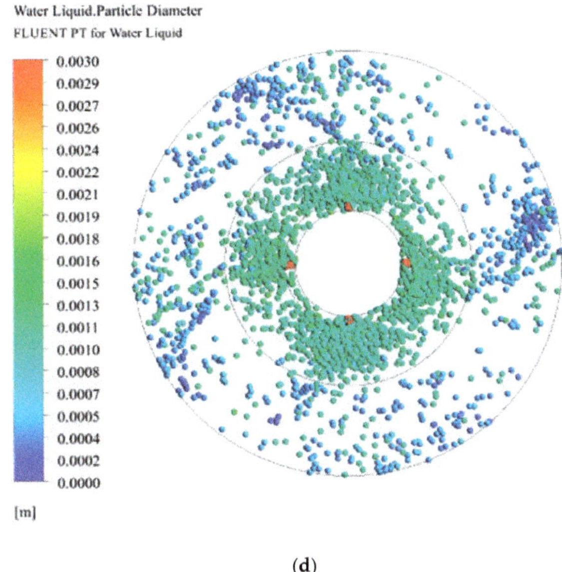

(**d**)

Figure 15. Droplet diameter under various initial droplet velocities and rotating speeds. (**a**) $u_0 = 0.5$ m/s, $\omega = 600$ rpm; (**b**) $u_0 = 2.5$ m/s, $\omega = 600$ rpm; (**c**) $u_0 = 1.5$ m/s, $\omega = 300$ rpm; (**d**) $u_0 = 1.5$ m/s, $\omega = 900$ rpm.

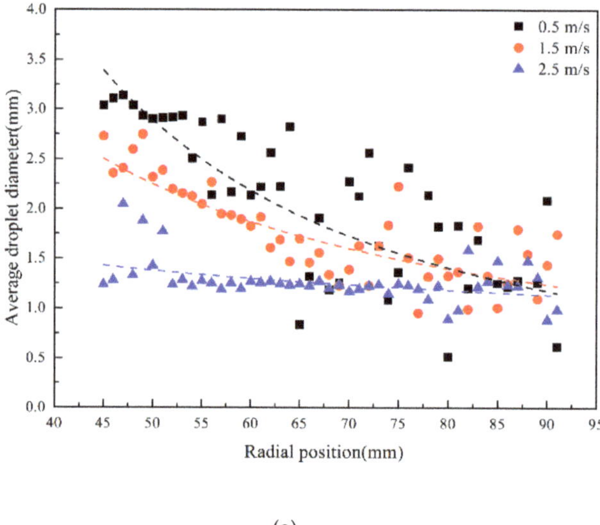

(**a**)

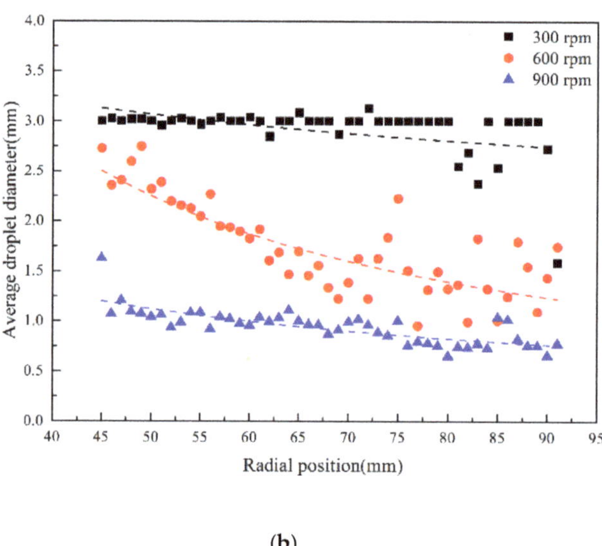

(**b**)

Figure 16. Effect of initial droplet velocity and rotating speed on average droplet diameter. (**a**) $d_0 = 3$ mm, $\omega = 600$ rpm; (**b**) $d_0 = 3$ mm, $u_0 = 1.5$ m/s.

As shown in Figure 16a, under the same conditions of initial droplet diameter of 3 mm and rotating speed of 600 rpm, the average droplet diameter decreases drastically with the increase of initial droplet velocity from 0.5 to 2.5 m/s in the packing zone. In other words, the average droplet diameter with initial velocity of 2.5 m/s is less than the average droplet diameter with initial velocity of 1.5 m/s, which is less than the average droplet diameter with initial velocity of 0.5 m/s in same radial position. A higher initial droplet velocity means a greater initial kinetic energy, which is beneficial to the breakup of droplets when kinetic energy is greater than surface energy, so the phenomenon is obtained in Figure 16a. In addition, the initial droplet diameter of 3 mm will gradually break up into smaller droplets along the radial position and finally reaches a constant diameter of 1.25 mm at the outer packing zone. Thus, conclusions can be drawn that the minimum droplet diameter in RPB only depends on the rotating speed and that there exists a minimum rotating packing radial length for droplets to fully fragmentize.

From Figure 16b, it can be observed distinctly that average droplet diameter distribution changes regularly on initial droplet diameter and velocity of 3 mm and 1.5 m/s, respectively. The effect of rotating speed on droplet diameter indicates that the droplet diameter decreases with the increasing rotating speed. The centrifugal force is linear to radial position and is proportional to the second power of rotating speed. Therefore, a higher centrifugal force results in more drastic collisions and smaller droplet diameters. Further, a balance between shear force generated by rotating packing and surface tension on droplets is proposed and discussed under an appropriate rotating speed. When the rotating speed decreases to 300 rpm, the shear force cannot meet the surface tension, resulting in only a few droplet breakups at the end of packing. On the contrary (at 900 rpm), shear force is dominant instead of surface tension, so the droplet diameter changes gently.

Compared with the effect of initial droplet velocity on droplet diameter, a more obvious impact of rotating speed is obtained. Radial velocity and tangential velocity all contribute to droplet breakup, but the tangential velocity caused by rotating speed is dominant. Therefore, considering the effect of rotating speed, radial position and fluid density on droplet diameter, a fitting expression is obtained:

$$d_p = 14.96 \left[\frac{\rho}{(\rho_p - \rho)\omega^2 R} \right]^{0.8805} \tag{36}$$

3.4. Principle of Processing Intensification in RPB

A comparison on particle diameter is conducted between RPB and conventional deposition process under gravity. Droplet force balance has been discussed in Equation (12) and the droplet diameter is calculated by two equations in Table 4, which were derived from Equations (12) and (13), considering mainly the centrifugal force in RPB (or gravity in deposition process) and drag force. Droplet diameter in different force fields is related to fluid density, droplet density, fluid velocity, droplet velocity and main acceleration. Generally speaking, the behavior in RPB can be considered as a processing intensification in force fields by replacing the term "g" with "$\omega^2 R$". Thus, greater acceleration results in smaller droplet diameter and larger droplet surface area of mass transfer. According to the theory of mass transfer, the mass transfer rate is proportional to the surface area. Therefore, the mass transfer efficiency of RPB is 1–3 orders of magnitude greater than that in conventional equipment. In addition, some intensification in principle mentioned above will be implemented by coupling multiple fields like magnetic, electric and microwaves to explore their potential in RPB.

Table 4. Droplet diameter under different force fields.

Force Field	Droplet Diameter
Rotating packed bed	$d_p = \dfrac{3C_D\rho(u-u_p)^2}{4(\rho_p-\rho)\omega^2 R}$
deposition process under gravity	$d_p = \dfrac{3C_D\rho(u-u_p)^2}{4(\rho_p-\rho)g}$

On the other hand, an arrangement should be taken as soon as possible by increasing the relative velocity between droplet and packing to enhance the collision of droplet-packing. Thus, an impinging stream distributor in an impinging stream rotating packed bed (IS-RPB) [18] has been put forward to increase the droplet initial velocity from the liquid distributor. A novel RPB called split packing RPB [19,20] (SP-RPB) can enhance the relative velocity in the packing zone by operating two separate rotating packings in opposing directions, which promotes a large surface area and makes a great contribution to surface renewal and mass transfer in the process of H_2S selective absorption into MDEA solution.

4. Conclusions

A 3D CFD Eulerian–Lagrangian approach has been built by introducing a droplet breakup and coalescence model to investigate the droplet characteristics of RPB in H_2S selective absorption into MDEA solution. Droplet characteristics such as droplet velocity, average residence time and average diameter in RPB have been analyzed by diagrams and correlations, which are compared with available experimental data in the literature [14,17]. The results show that the velocity increases with increasing rotating speed and radial position, but the opposite conclusion is made on the average residence time. Specially, in the end zone, a phenomenon called "end effect" in mass transfer intensification has been observed and can be illuminated by droplet back-mixing. A correlation on droplet velocity has been deduced in Equation (33) mainly associated with rotating speed and radial position rather than initial droplet velocity. Under the condition of 900 rpm, a short average residence time 0.039–0.085 s in RPB has been recommended for H_2S selective absorption into MDEA solution. This is because the reaction and mass transfer rate of H_2S are both instantaneously fast compared with CO_2, thus a short average residence time allows for efficient selective absorption between H_2S and CO_2. When the initial droplet velocity and rotating speed increase, the average droplet diameter decreases inordinately. However, the initial droplet diameter has a restriction (1 mm) to be captured and broken by the packing size under the simulation conditions. Furthermore, conclusions are made that the rotating speed determines the minimum droplet diameter and that a packing length in the radial direction is needed to meet droplet breakup completely. In addition, a balance between shear force and surface tension on the droplet indicates an appropriate rotating speed. A correlation (Equation (36)) on droplet diameter is obtained considering the effect of rotating speed, radial position and fluid density.

The simulation results indicate this CFD approach has the capability in describing droplet characteristics, and an investigation on principle in RPB has been conducted based on these results. Processing in RPB can be regarded as an intensification of a traditional separation device in gravity by replacing the term "g" with "$\omega^2 R$" and some coupled field with centrifugal force like magnetic, electric and microwaves, which will inspire further potential in RPB. Finally, IS-RPB and SP-RPB as novel equipment have made a great contribution to surface renewal and mass transfer in the process of H_2S selective absorption into MDEA solution through increasing the relative velocity and collision between droplets and packing.

Author Contributions: All authors contributed to the manuscript. Z.W. proposed the main idea; X.W. performed the CFD simulation; T.Y. wrote the manuscript; S.W. improved the manuscript; Z.L. provided many key suggestions; X.D. collected and analyzed data.

Funding: This research received no external funding.

Conflicts of Interest: The authors declare no conflict of interest.

Nomenclatures

CFD	Computational fluid dynamics
HTU	Height of mass transfer unit
IS-RPB	Impinging stream RPB
RPB	Rotating packed bed
RSR	Rotor-stator reactor
SP-RPB	Split packing RPB
TAB	Taylor analogy breakup

Latin symbols

C_D	Drag coefficient
F	Aerodynamic force of droplet (N)
$\vec{F}$	External body force (N)
F_D	Drag force (N)
$F_{r,x}$	Force in rotating reference frame in x direction (N)
$F_{r,y}$	Force in rotating reference frame in y direction (N)
F_m	Virtual mass force (N)
F_x	Additional acceleration term
G_b	Influence of the buoyancy force
G_k	Influence of the mean velocity gradients
Pr_t	Turbulent Prandtl number for energy
R	Radial position (m)
b	Actual collision parameter
b_{crit}	Critical offset of collision
d_0	Initial droplet diameter (mm)
$\bar{d}$	Arithmetic mean diameter of two droplet (m)
$\vec{g}$	Gravitational vector (9.8 m/s^2)
k	Turbulence kinetic energy
m	Mass of droplet (kg)
m_l	Mass of large droplet (kg)
m_s	Mass of small droplet (kg)
p	Static pressure (Pa)
r_l	Radius of large droplet (m)
r_s	Radius of small droplet (m)
u	Fluid velocity (m/s)
u_0	Initial droplet velocity (m/s)
$\vec{u}_l$	Velocity of large droplet (m/s)
u_p	Droplet velocity (m/s)
$\vec{u}_r$	Whirl velocity (m/s)
$\vec{u}_s$	Velocity of small droplet (m/s)
$\vec{v}_r$	Relative velocity (m/s)
x	Displacement of droplet (m)

Greek symbols

ρ	Fluid density (kg/m^3)
ρ_g	Gas density (kg/m^3)
ρ_l	Liquid density (kg/m^3)
ρ_p	Droplet density (kg/m^3)
ω	Rotating speed (rpm)
$\vec{\omega}$	Centrifugal acceleration (rad/s)
μ_g	Gas viscosity (mPa·s)
μ_l	Liquid viscosity (mPa·s)
μ_{eff}	Molecular viscosity (mPa·s)
μ_t	Turbulent viscosity (mPa·s)
σ	Liquid surface tension (N/m)
ε	Dissipation rate

Dimensionless groups

$Re = \dfrac{\rho d_p \lvert u_p - u \rvert}{\mu}$	Reynolds number
$We = \dfrac{\rho_p \lvert \vec{u}_l - \vec{u}_s \rvert^2 \bar{d}}{\sigma}$	Weber number

References

1. Chen, Y.-S.; Lin, F.-Y.; Lin, C.-C.; Tai, C.Y.-D.; Liu, H.-S. Packing Characteristics for Mass Transfer in a Rotating Packed Bed. *Ind. Eng. Chem. Res.* **2006**, *45*, 6846–6853. [CrossRef]
2. Neumann, K.; Gladyszewski, K.; Gross, K.; Qammar, H.; Wenzel, D.; Gorak, A.; Skiborowski, M. A guide on the industrial application of rotating packed beds. *Chem. Eng. Res. Des.* **2018**, *134*, 443–462. [CrossRef]
3. Burns, J.R.; Ramshaw, C. Process intensification: Visual study of liquid maldistribution in rotating packed beds. *Chem. Eng. Sci.* **1996**, *51*, 1347–1352. [CrossRef]
4. Guo, K.; Guo, F.; Feng, Y.; Chen, J.; Zheng, C.; Gardner, N.C. Synchronous visual and RTD study on liquid flow in rotating packed-bed contactor. *Chem. Eng. Sci.* **2000**, *55*, 1699–1706. [CrossRef]
5. Li, Y.; Wang, S.; Sun, B.; Arowo, M.; Zou, H.; Chen, J.; Shao, L. Visual study of liquid flow in a rotor-stator reactor. *Chem. Eng. Sci.* **2015**, *134*, 521–530. [CrossRef]
6. Sang, L.; Luo, Y.; Chu, G.-W.; Zhang, J.-P.; Xiang, Y.; Chen, J.-F. Liquid flow pattern transition, droplet diameter and size distribution in the cavity zone of a rotating packed bed: A visual study. *Chem. Eng. Sci.* **2017**, *158*, 429–438. [CrossRef]
7. Shi, X.; Xiang, Y.; Wen, L.-X.; Chen, J.-F. CFD analysis of liquid phase flow in a rotating packed bed reactor. *Chem. Eng. J.* **2013**, *228*, 1040–1049. [CrossRef]
8. Ouyang, Y.; Zou, H.-K.; Gao, X.-Y.; Chu, G.-W.; Xiang, Y.; Chen, J.-F. Computational fluid dynamics modeling of viscous liquid flow characteristics and end effect in rotating packed bed. *Chem. Eng. Process. Process Intensif.* **2018**, *123*, 185–194. [CrossRef]
9. Ouyang, Y.; Wang, S.; Xiang, Y.; Zhao, Z.; Wang, J.; Shao, L. CFD analyses of liquid flow characteristics in a rotor-stator reactor. *Chem. Eng. Res. Des.* **2018**, *134*, 186–197. [CrossRef]
10. Xie, P.; Lu, X.; Yang, X.; Ingham, D.; Ma, L.; Pourkashanian, M. Characteristics of liquid flow in a rotating packed bed for CO_2 capture: A CFD analysis. *Chem. Eng. Sci.* **2017**, *172*, 216–229. [CrossRef]
11. Ko, G.H.; Ryou, H.S. Modeling of droplet collision-induced breakup process. *Int. J. Multiph. Flow* **2005**, *31*, 723–738. [CrossRef]
12. Wang, Z.; Yang, T.; Liu, Z.; Wang, S.; Gao, Y.; Wu, M. Mass Transfer in a Rotating Packed Bed: A Critical Review. *Chem. Eng. Process. Process Intensif.* **2019**, *139*, 78–94. [CrossRef]
13. Liu, Y.; Luo, Y.; Chu, G.-W.; Luo, J.-Z.; Arowo, M.; Chen, J.-F. 3D numerical simulation of a rotating packed bed with structured stainless steel wire mesh packing. *Chem. Eng. Sci.* **2017**, *170*, 365–377. [CrossRef]
14. Sang, L.; Luo, Y.; Chu, G.W.; Liu, Y.Z.; Liu, X.Z.; Chen, J.F. Modeling and experimental studies of mass transfer in the cavity zone of a rotating packed bed. *Chem. Eng. Sci.* **2017**, *170*, 355–364. [CrossRef]
15. Qian, Z.; Li, Z.-H.; Guo, K. Industrial Applied and Modeling Research on Selective H_2S Removal Using a Rotating Packed Bed. *Ind. Eng. Chem. Res.* **2012**, *51*, 8108–8116. [CrossRef]
16. Qian, Z.; Xu, L.-B.; Li, Z.-H.; Li, H.; Guo, K. Selective Absorption of H_2S from a Gas Mixture with CO_2 by Aqueous N-Methyldiethanolamine in a Rotating Packed Bed. *Ind. Eng. Chem. Res.* **2010**, *49*, 6196–6203. [CrossRef]
17. Guo, F.; Zheng, C.; Guo, K.; Feng, Y.; Gardner, N.C. Hydrodynamics and mass transfer in cross-flow rotating packed bed. *Chem. Eng. Sci.* **1997**, *52*, 3853–3859. [CrossRef]
18. Yang, P.-F.; Luo, S.; Zhang, D.-S.; Yang, P.-Z.; Liu, Y.-Z.; Jiao, W.-Z. Extraction of nitrobenzene from aqueous solution in impinging stream rotating packed bed. *Chem. Eng. Process. Process Intensif.* **2018**, *124*, 255–260. [CrossRef]
19. Rajan, S.; Kumar, M.; Ansari, M.J.; Rao, D.P.; Kaistha, N. Limiting Gas Liquid Flows and Mass Transfer in a Novel Rotating Packed Bed (HiGee). *Ind. Eng. Chem. Res.* **2011**, *50*, 986–997. [CrossRef]
20. Shivhare, M.K.; Rao, D.P.; Kaistha, N. Mass transfer studies on split-packing and single-block packing rotating packed beds. *Chem. Eng. Process. Process Intensif.* **2013**, *71*, 115–124. [CrossRef]

 *processes*

Article

A Computational Fluid Dynamics Approach for the Modeling of Gas Separation in Membrane Modules

Salman Qadir, Arshad Hussain and Muhammad Ahsan *

School of Chemical and Materials Engineering (SCME), National University of Sciences and Technology, (NUST), Islamabad 44000, Pakistan
* Correspondence: ahsan@scme.nust.edu.pk; Tel.: +92-51-9085-5125

Received: 1 May 2019; Accepted: 28 June 2019; Published: 3 July 2019

Abstract: Natural gas demand has increased rapidly across the globe in the last decade, and it is set to play an important role in meeting future energy requirements. Natural gas is mainly produced from fossil fuel and is a side product of crude oil produced beneath the earth's crust. Materials hazardous to the environment, like CO_2, H_2S, and C_2H_4, are present in raw natural gas. Therefore, purification of the gaseous mixture is required for use in different industrial applications. A comprehensive computational fluid dynamics (CFD) model was proposed to perform the separation of natural gas from other gases using membrane modules. The CFD technique was utilized to estimate gas flow variations in membrane modules for gas separation. CFD was applied to different membrane modules to study gas transport through the membrane and flux, and to separate the binary gas mixtures. The different parameters of membrane modules, like feed and permeate pressure, module length, and membrane thickness, have been investigated successfully. CFD allows changing the specifications of membrane modules to better configure the simulation results. It was concluded that in a membrane module with increasing feed pressure, the pressure gradient also increased, which resulted in higher flux, higher permeation, and maximum purity of the permeate. Due to the high purity of the gaseous product in the permeate, the concentration polarization effect was determined to be negligible. The results obtained from the proposed CFD approach were verified by comparing with the values available in the literature.

Keywords: computational fluid dynamics; membrane module; gas separation; concentration polarization

---

1. Introduction

Natural gas is considered one of the significant fossil fuels. It is found in subsurface reservoirs and mostly produced as a byproduct of oil production. The demand for natural gas has seen a considerable rise in recent years [1,2]. As per the reports of the U.S. Energy Information Administration (EIA), global natural gas consumption is increasing 4% annually. It is estimated that 10% of the world power sector depends on natural gas [3]. By 2040, power production and industrial usage are expected to be 73% dependent on natural gas. Natural gas consists of several chemical species, including methane (CH_4), ethane, propane, butane, water vapor, nitrogen, and acidic gases such as carbon dioxide (CO_2) and hydrogen sulfide (H_2S). The species comprising natural gas, like CO_2, N_2, water vapor, and H_2S, are considered impurities [4]. The different concentrations of impurities in natural gas is in the range 4–50%, depending on the reservoir. The presence of these impurities can significantly affect pipelines, in terms of corrosion, and raise health and safety concerns [5]. Therefore, typical pipeline specifications usually mandate that the concentration of carbon dioxide in natural gas not exceed 2–5 volume percent, making it necessary to treat natural gas and remove the impurities before it is transported. In recent years, the membrane separation process has been widely used because it is more economical compared to other methods [6].

Membrane gas separation can remove unnecessary species from a gas mixture. The membrane allows only the desired components of a mixture to pass through because of its selectivity [7]. It is a widely used process due to its reliability, separation performance, low maintenance, and easy operation [8]. Moreover, membrane technology is used for the separation of various fuel mixtures in different industries as a result of economic competitiveness and other current demanding situations related to competitive environments [9]. It has many applications in industrial sectors, hydrogen recovery from ammonia, hydrogen recovery in refineries, air separation for oxygen purification, sour gas treatment, and carbon dioxide removal from natural gas [10]. These processes are integrated with big industrial units to perform specific industrial operations. It is estimated that the use of membrane gas separation will increase substantially in 2020 [11]. It is an essential fact that the use of this technology will decrease the unit operation cost of gas separation and reduce the environmental hazards [12].

The different mathematical models for the separation of gaseous mixtures using a membrane were developed using altered assumptions [13]. Recently, it was found that the most commonly available commercial membrane modules for gas separation were hollow fiber and spiral wound membrane modules [14,15]. Many researchers have investigated flow behavior with different module arrangements for gas separation and desalination processes [16,17]. Alrehili's study [18] showed the different arrangements of fibers with parallel feed channels that made a hybrid membrane module. The simulation results of the hybrid module gave better membrane flux for both spiral wound and hollow fiber membrane module configurations. Ahsan [19] applied computational fluid dynamics (CFD) modeling to gas separation using a finite difference method in polymeric membranes. Saeed [20] described laminar flow behavior in spacers with narrow channels, and mass transfer coefficient calculated with different wire spacing. Karode [21] determined that pressure dropped with bounding surfaces in a rectangular channel. The effect of shear stresses on both sides of the membrane was also observed.

Other researchers developed a mathematical algorithm for flow behavior and membrane surfaces, and investigated the concentration polarization phenomena in gas separation processes [22]. The transfer of CO_2 gas molecules through the membrane increased due to higher flux on the feed side, but rejected molecules of other gases that then accumulated on the membrane surface [23]. For this reason, concentration polarization occurs in membrane processes. Mourgues [24] showed the effect of concentration polarization on membrane separation processes for both counter-current and co-current patterns. The most important factors were analyzed based on feed pressure, permeability, and selectivity of the mixture.

Several researchers developed membrane processes to study the influence of concentration polarization on the feed side. Ahsan and Hussein [25], in their CFD model, studied energy transfer phenomena in the membrane using permeation flux. Coroneo [26] developed a three-dimensional single-tube membrane module to define flux, based on Sievert's law, considering both membrane sides. Recently, a non-isothermal model was developed to evaluate the effect of temperature on permeance [27]. In another study, Chen [28] considered co-current and counter-current flow patterns of the membrane process at different operating conditions using COMSOL Multiphysics 4.0a software.

Another study looked at plug flow and perfect mixing channel, commonly used for modeling membrane gas separation, and reported the fluid behavior in permeate [29]. Flat sheet membrane modules were widely used to evaluate membrane performance. The incompressible Navier–Stokes model was used to improve the fluid chamber while the solid stress–strain model was used to enhance the mechanical performance of the module [30].

In this study, the CFD technique was used to solve the model equations. The permeability of the membrane was measured by introducing the species of interest into the feed gas. The effect of gas flow profiles on gas separation in the membrane modules is reported. A three-dimensional (3D) model was established using CFD simulation in COMSOL Multiphysics software. The geometry of the flat sheet and spiral wound membrane modules were determined, and co-current and counter-current models were used to find the flow profiles. The effect of molar flux on species transported en mass through the membrane was considered. The binary gas mixtures CO_2/CH_4 and CH_4/C_2H_6 were used for separation in the flat sheet and spiral wound membrane modules, respectively. The investigated parameters, permeability, feed pressure, permeate pressure, and feed gas concentration, were compared with the data available in the literature.

2. Numerical Methods

The CFD technique (COMSOL Multiphysics®4.3, COMSOL, Inc., Burlington, MA, USA, 2012) is used in simulations looking at flow profiles in the membrane module. In this study, the flat sheet and spiral wound membrane modules were used for the simulations. The data used for the simulations were taken from the literature and are shown in Table 1.

Table 1. Properties for membranes modules. Reproduced with permission from R. Qi, M.A. Henson, Separation and Purification Technology; published by Elsevier, 1998.

Parameters	Spiral Wound Membrane Module [31]	Flat Sheet Membrane Module [32]	Units
Feed Pressure	35×10^5	106.7×10^3	Pa
Permeate Pressure	1.05×10^5	1.1×10^3	Pa
Feed gas	$0.20\ CO_2$	$0.395\ CH_4$	Mole fraction
Permeance	1.48×10^{-9}	2×10^{-7}	$mol/(m^2{\cdot}s{\cdot}pa)$
Selectivity	20	2.73	
Module Diameter	0.3	6×10^{-3}	m
Module Length	1	0.8	m
Thickness	2.84×10^{-3}	15×10^{-6}	m

The numerical simulations were performed using the COMSOL Multiphysics® package (COMSOL Multiphysics®4.3, COMSOL, Inc., Burlington, MA, USA, 2012). The spiral wound and flat sheet membranes were developed with 3D axisymmetric geometry. Fick's law of permeation was used for the main transport of diluted species through the membrane. The following assumptions were made [13].

2.1. Assumptions

1. Steady-state and ideal gas conditions;
2. Isothermal conditions;
3. Solution–diffusion mechanism for permeation;
4. Permeance not dependent on the concentration of gas or the feed pressure;
5. No axial mixing of gaseous molecules;
6. Constant pressure drop on the feed and permeate side.

2.2. Mathematical Modeling of Mass Transport

The model accounts for diffusion transport in a unit cell of the structure sketched in Figure 1. The unit cell is a small part of the membrane that is representative of the whole membrane. In this model, the initial flux in the membrane was studied, and corresponded to the largest difference in concentration between the two chambers on different sides of the membrane. The supporting structure in the membrane consisted of a mesh structure made up of a dense and rigid polymeric material. This system was used to classify the permeability properties of a membrane to certain species.

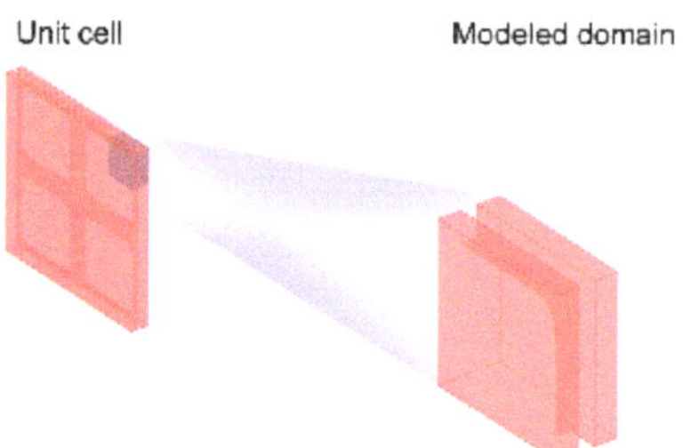

Figure 1. Membrane model unit.

2.2.1. Convection and Diffusion Model

In the convection and diffusion model, the mass transport equation contained both phenomena. In chemical engineering, the convection and diffusion model is applicable for mass transport, which is described by Fick's Law. In the 19th century, Fick gave the simplest definition of diffusion for mass transport. The mass flux of any species is directly proportional to the concentration gradient due to diffusion. The rate of change of concentration at a point in space is proportional to the second derivative of concentration with space;

$$\frac{\partial C_a}{\partial t} + v_x \frac{\partial C_a}{\partial x} + v_y \frac{\partial C_a}{\partial y} + v_z \frac{\partial C_a}{\partial z} = D_{AB}\left[\frac{\partial^2 C_a}{\partial x} + \frac{\partial^2 C_a}{\partial y} + \frac{\partial^2 C_a}{\partial z}\right] + R_a \tag{1}$$

Convective transport accounts for the bulk flow of fluid due to velocity v. This term v can be solved analytically or by solving the momentum equation with the mass balance equation. All these expressions include time (t), and spatial and velocity components are used for mass balance with convection. The v_x, v_y, and v_z are velocity fields in three dimensions. R_a describes the rate of reaction and it should be equal to zero as no reaction is involved in gas separation in membrane modules.

2.2.2. Diffusion Term

The chemical species is transferred from high to low regions due to diffusion, and becomes a mass transfer phenomenon with time and space. The chemical species is dissolved in a solvent or any gas mixture, for example as oxygen enrichment in the air. The evolution of species mass transfer depends on its concentration with respect to time and space. The gradient for diffusion occurs as a result of the motion of the molecules. Due to the kinetic energy of molecules, they can collide with each other in random directions. Then, flux N_a for diffusion can be written as;

$$N_a = -D_{AB}\left[\frac{\partial^2 C_a}{\partial x} + \frac{\partial^2 C_a}{\partial y} + \frac{\partial^2 C_a}{\partial z}\right] \tag{2}$$

where mass transfer occurs as a result of diffusion D_{AB}.

2.2.3. Membrane Model

The unit cell is a small part of the membrane module that demonstrates the whole membrane module. The steady-state diffusion equation, which shows mass transport in a model with diffusion, can be written as;

$$D_{AB}\left[\frac{\partial^2 C_a}{\partial x} + \frac{\partial^2 C_a}{\partial y} + \frac{\partial^2 C_a}{\partial z}\right] = 0 \tag{3}$$

It can also be represented in terms of a nebula operator;

$$D \cdot \nabla^2 C_a = 0 \tag{4}$$

where C_a denotes concentration (mole/m^3) and D_{AB} is the diffusion coefficient of the diffusing species (m^2/s). All boundaries are considered to be insulating

$$D_{AB}\left[\frac{\partial C_a}{\partial x} + \frac{\partial C_a}{\partial y} + \frac{\partial Ca}{\partial z}\right] = 0 \tag{5}$$

$$D\ \nabla C_a \cdot n = 0 \tag{6}$$

The two faces are applied as boundary conditions where the concentration of the two components is set from high to low. Boundary Condition 1 (B.C.1) is considered to be high concentrations on the feed side and Boundary Condition 2 (B.C.2) is represented on the reject side in low concentrations.

B.C.1

$$C = C_0 \tag{7}$$

B.C.2

$$C = C_{0,1} \tag{8}$$

2.2.4. Membrane Flux

Diffusion through the membrane was represented in terms of the initial boundary condition C_0 and the final boundary condition $C_{0,1}$.

$$N_a = \frac{D}{\delta}(C_0 - C_{0,1}) \tag{9}$$

where $\frac{D}{\delta}$ is a barrier with a corresponding thickness.

The permeability (P) of gas can be defined as a product of the diffusion coefficient (D) and solubility coefficient (S),

$$P = D \cdot S. \tag{10}$$

The relationship between partial pressure and concentration is defined as

$$C = S \cdot p. \tag{11}$$

The flux through the membrane can be presented in terms of permeability and partial pressure,

$$N_a = \frac{\frac{P}{S}}{\delta}(S \cdot pf - S \cdot ph) \tag{12}$$

where feed pressure pf and permeate pressure ph are used for calculating gradient across the membrane. Solubility is obtained using a ratio of the concentration gradient and partial pressure difference for the binary mixture. The solubility S can be calculated as

$$S = \frac{\Delta p}{\Delta C}. \tag{13}$$

This membrane model was used to study gas separation in flat sheet and spiral wound membrane modules.

2.3. Geometry

A sketch of the geometries of flat sheet and spiral wound membrane modules is shown in Figure 2.

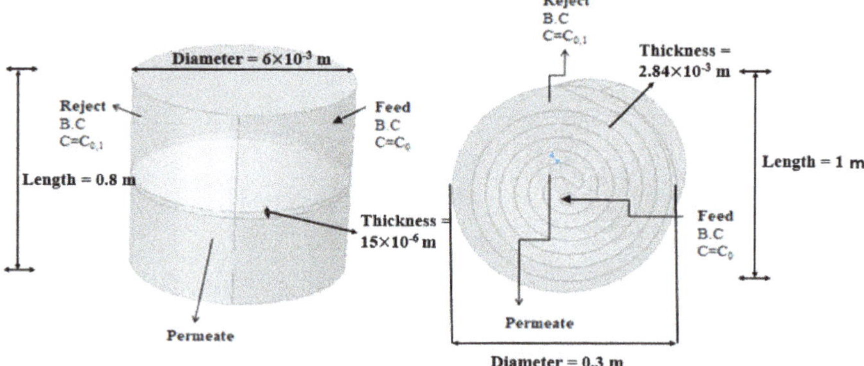

Figure 2. Schematic diagram of the flat sheet and spiral wound membrane module.

2.4. Meshing

The subdomains are often called elements or cells, and the collection of all elements or cells is called a mesh (Figure 3). In this process, a physics-controlled mesh was applied. An extremely fine grid element was used, while a further increase did not affect the model results. The mesh consisted of 1,324,604 domain elements, and 46,750 boundary elements, and 900 edge elements were used for the flat sheet membrane module. The mesh consisted of 3,257,454 domain elements, 413,420 boundary elements, and 3439 edge elements used for the spiral wound membrane module.

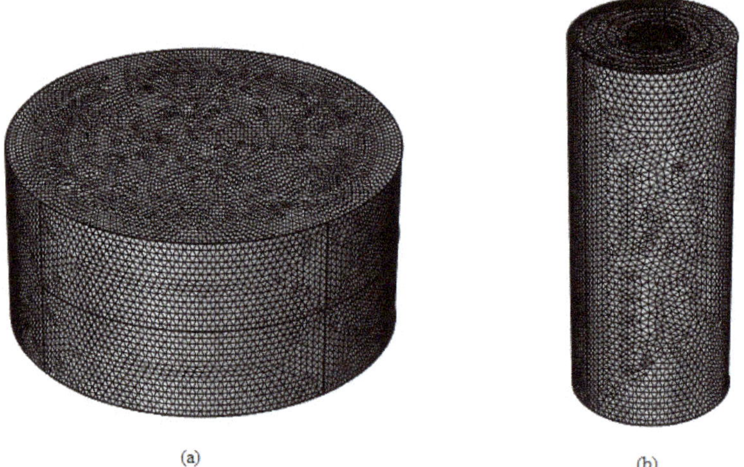

(a)

(b)

Figure 3. Meshing of membrane module. (**a**) flat sheet; (**b**) spiral wound.

3. Results

CFD analysis was performed for the flat sheet and spiral wound membrane modules. The simulation was carried out using COMSOL Multiphysics software 5.2a. Fluxes of gases and concentration variations were found across the length of membrane modules. Various parameters like feed pressure, concentrations across the length of the module, permeability, and feed were considered in this study.

3.1. CH_4/C_2H_6 Separation

The flat sheet membrane module is usually used for pilot plant testing or lab scale testing. The separation of methane and ethane was considered for the flat sheet membrane module in this study. The simulation of a flat sheet membrane module was performed to check the concentration change on the reject and permeate sides. The feed gas entered the membrane module and the permeate collected at the bottom of the module. A cross-flow model with specific boundary conditions was used. The boundary conditions included a high feed concentration on the right side and low feed concentration on the left side. Figure 4a shows the concentration variation of CH_4 on the feed and permeate sides. The contour shows the concentration variation of gas on both sides of the membrane. The slice centration shows the gas variations in the center of the module. Figure 4b shows the differing concentration variation from the feed side to the permeate side. The gas moved through the flat sheet module and permeate collected at the bottom.

Figure 4c shows the concentration gradient of CH_4 present in the flat sheet membrane module. Streamlines were used for concentration gradient representation in Figure 4c. The lines moving from high to low and passing through the membrane represent the gas diffusion through the membrane. The membrane is located in the center of the module and the permeate is shown on the bottom surface. The results verified that a gradient was present, and that gas diffused through the membrane. The flux was calculated for the given parameters using COMSOL Multiphysics software. The contour in Figure 4d indicates the flux variations in the flat sheet membrane module. The color bar shows the flux magnitude in the module. The flux calculated in the module can be explained by Fick's law.

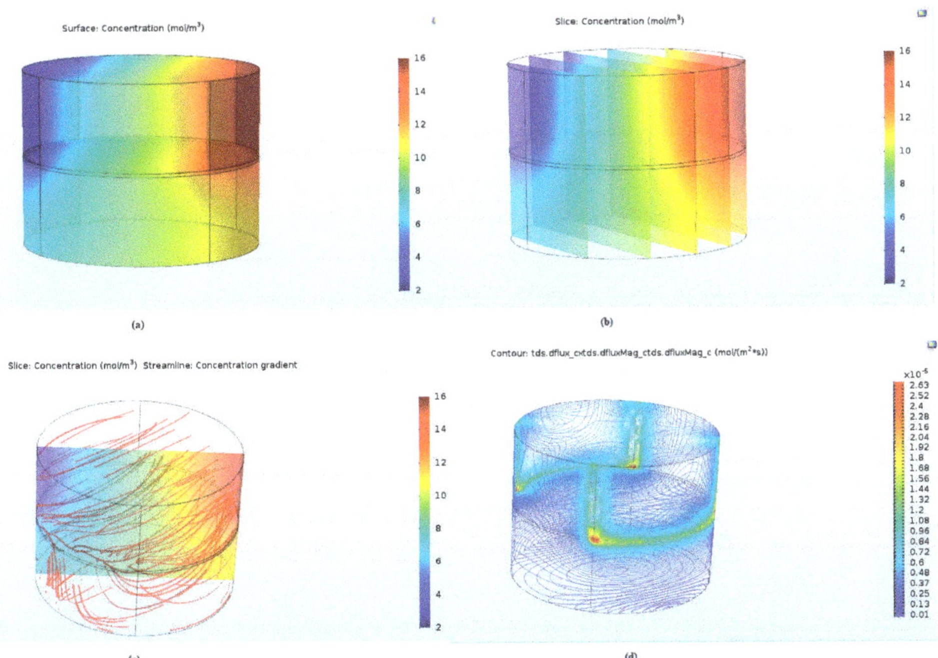

Figure 4. (**a**) CH_4 gas concentration variation in the flat sheet membrane module; (**b**) slice shows CH_4 gas variation in the center of the module; (**c**) line shows the concentration gradient variation in the membrane module and (**d**) diffusive flux variation for CH_4 in a flat sheet membrane module.

3.2. CH$_4$/CO$_2$ Separation

The simulation of a spiral wound membrane module was presented to show the concentration of feed CO_2 on the reject and permeate sides. The feed gas entered the membrane module from the central tube and moved through to the end. The module shows that the concentration of CO_2 changed throughout Figure 5a. The high concentration was applied to the first wrapped sheet as a boundary condition, permeate collected in the permeate channel, and a low concentration was applied to the second plate of the membrane module.

The contour shows that the gas concentration varied along the membrane on the permeate and reject sides. Figure 5b shows the slice concentration of CO_2 in a spiral wound membrane module. The result shows the concentration variation in the center of the module. It verifies that the gas diffused through the membrane from the feed channel and collected in the permeate channel. Figure 5c shows the concentration gradient present in the spiral wound membrane module for CH$_4$/CO$_2$ separation. The arrows show that the gas passes through the membrane from the feed side to the permeate collector of the module. The results prove that a gradient was present, and that the gas diffused through the membrane. Figure 5d shows the diffusive flux magnitude in the spiral wound membrane module for CO_2. The diffusive flux magnitude in the membrane module was obtained using concentration and permeability parameters. The flux through the membrane was calculated in the gas separation process and depended on gas diffusion in the perm-selective membrane, due to the pressure and concentration gradient. It illustated that mass transport occurred through the membrane in the spiral wound membrane module.

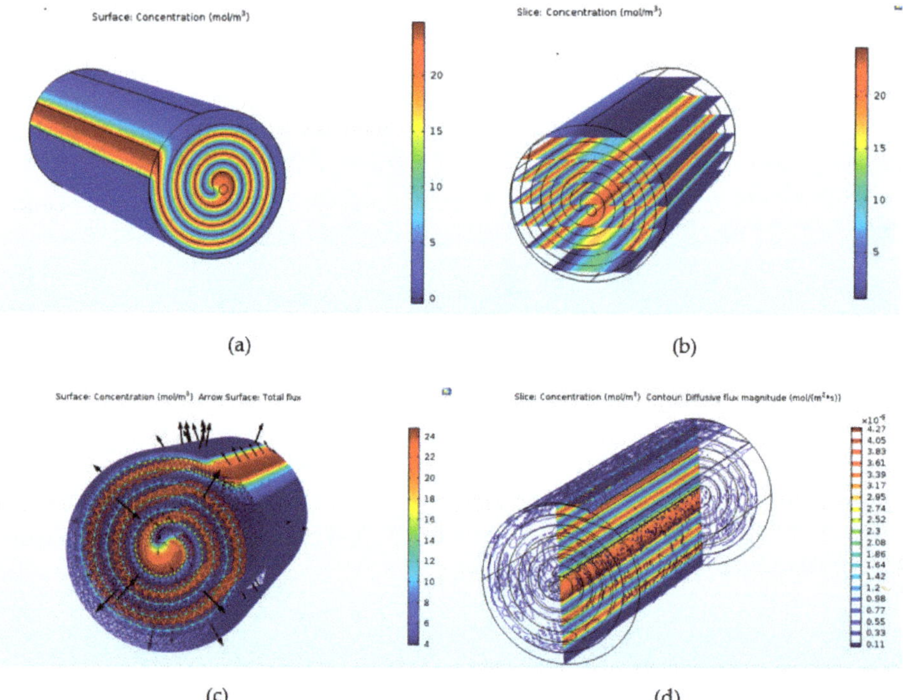

Figure 5. (**a**) CO_2 gas concentration variation in the spiral wound membrane module; (**b**) slice shows CO$_4$ gas variation in the center of the module; (**c**) line shows concentration gradient variation in the membrane module and (**d**) diffusive flux variation for CO$_4$ in the spiral wound membrane module.

3.3. Parametric Study of the Spiral Wound Membrane Module

Feed Pressure and Permeate Pressure

The membrane length was considered for the spiral wound membrane module, and the results were verified via the literature. Figure 6 shows that permeate pressure was a function of membrane length for different values. Basically, the permeate pressure was less than the feed pressure. Therefore, it was necessary to calculate the changes in permeate pressure according to length. Increasing the membrane module's length decreased the permeate pressure because of the gradient changes throughout the module from the feed side to the reject side. Therefore, less mass transfer occurred as a result of less gradient in the module. It was observed that feed pressure had a reverse effect on permeate pressure. An increase in the feed pressure produced more gradient across the membrane, which resulted in higher mass transfer through the membrane. Figure 7 represents the feed pressure variation with methane recovery in the spiral wound module. The results show that higher amounts of methane permeate were obtained in the end because a high gradient was produced, and gas diffusion through the membrane was very high.

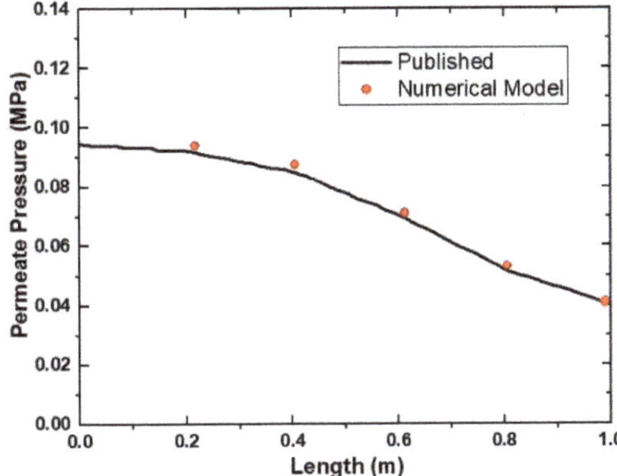

Figure 6. Permeate pressure variation with module length in a spiral wound membrane module compared with published values [31].

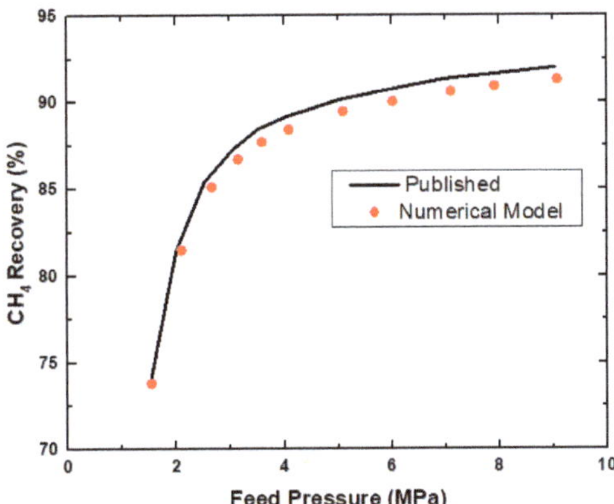

Figure 7. CH$_4$ gas variation with feed pressure in the spiral wound membrane module compared with published values [31].

3.4. Parametric Study for the Flat Sheet Membrane Module

Feed Pressure and Length

It was observed that the permeate concentration changed along the length of the module. The more permeable components passed through the membrane while other components were left as rejects. The CH_4 gas was more suitable for permeation through the membrane. Figure 8 demonstrates that the mole fraction of CH_4 increased on the permeate side across the length of the module and decreased on the reject side, with respect to length. The ethane gas increased on the reject side because the permeation of ethane through the membrane was very low. Therefore, the maximum possible purity of the gas was achieved.

Figure 9 shows that increasing the feed pressure causes the gradient to increase. Due to high feed pressure, the permeation of components through the membrane increased when the driving force increased. Therefore, it was estimated that the higher the feed pressure the better the separation. This was also true for methane recovery on the permeate side because increasing feed pressure resulted in more pressure gradient and turbulence for mass transfer. Therefore, the gas passed through the membrane. The permeate values calculated by the numerical model for the spiral wound and flat sheet membrane modules are shown in Table 2 for comparison with published data. A maximum difference of 10.8% and 8.7% was found for the spiral wound and flat sheet membrane modules, respectively.

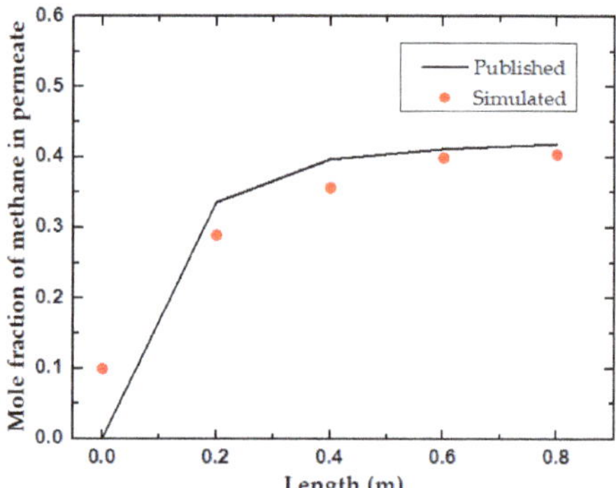

Figure 8. Mole fraction of methane with feed pressure in the flat sheet membrane module compared with published values [32].

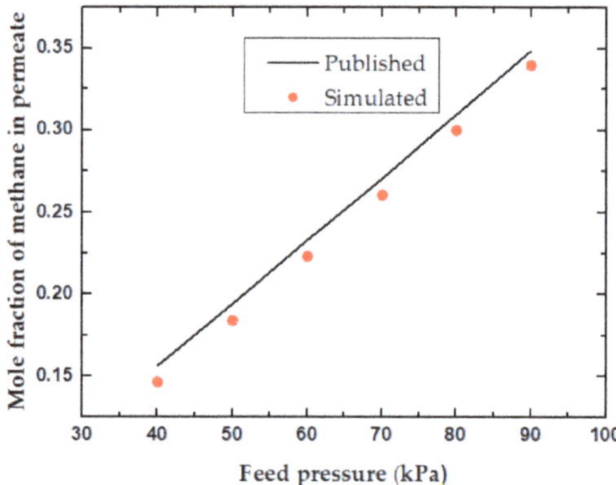

Figure 9. Mole fraction of methane change with a length of the flat sheet membrane module compared with published values [32].

Table 2. Comparison of the model results with the published data for CH_4 (permeate) at different values of selectivity.

Spiral Wound Membrane Modules				Flat Sheet Membrane Module			
Selectivity P_A/P_B	CH$_4$ (Permeate)		Difference (%)	Selectivity P_A/P_B	CH$_4$ (Permeate)		Difference (%)
	Published [31]	This Model			Published [32]	This Model	
20	0.881	0.786	10.8	2	0.457	0.417	8.7
40	0.936	0.903	3.5	4	0.672	0.635	5.5
60	0.952	0.913	4.1	6	0.749	0.718	4.2
80	0.967	0.929	3.9	8	0.891	0.829	6.9

4. Discussion

Computational fluid dynamics simulations were used to study mass transport in spiral wound and flat sheet membrane modules. Three-dimensional geometrics were considered to analyze the membrane-based gas separation for the binary gas mixture. Counter-current and cross-flow membrane models were used to verify the results. The simple Fick's law was used to explain mass flux transport of the binary gas mixture through the membrane. The membrane model was defined using COMSOL Multiphysics software for the separation of the binary mixture. The membrane model was applied as a thin diffusion barrier, which allowed certain species to pass through the membrane. The effect of molar flux for species in mass transport through a membrane was considered. In the present model, the different parameters were investigated to verify the models in the literature. Different gas mixtures, like CO_2/CH_4 and CH_4/C_2H_6, were investigated in different membrane modules.

A spiral wound membrane module was discussed for CH_4/CO_2 membrane separation. The module geometry consisted of three flat sheets that wrapped around a central tube. The membrane collector was present at the center of the feed and reject sides. The cross-flow model was applied to verify the literature results. In the spiral wound membrane module, the increase in membrane length showed a considerable decrease in concentration polarization. When the length of the membrane increased, an increase in the residual mole fraction was observed. An increase in the permeate purity was also observed, which indicated that the concentration polarization was negligible.

A flat sheet membrane module was considered for the separation of CH_4/C_2H_6. The model showed a variation of membrane results that have been discussed. This was novel work describing the flow profiles of gases in different membrane modules. The contour also showed the total flux, concentration gradient, and diffusive flux magnitude of different membrane modules. The investigated parameters were compared with published results. It was observed that in a flat sheet membrane module with increasing feed pressure, the pressure gradient also increased, which resulted in higher flux, higher permeation, and maximum purity of the permeate. The concentration polarization was observed to be negligible. Furthermore, with increasing module length in the flat sheet membrane module, a decrease in concentration polarization was observed because the increase in the module length resulted in more permeation of the desired component and an increase in permeate purity. In the spiral wound membrane module, increasing the membrane length resulted in a considerable decrease in concentration polarization. When the length of the membrane was increased, an increase in the residual mole fraction was observed. An increase in the permeate purity was also observed, which indicated that the concentration polarization was negligible.

5. Conclusions

The membrane modules were modeled using CFD to obtain the maximum possible value of the desired gas in the permeate. The effect of concentration polarization on gas separation performance was negligible. Different parameters were studied in the membrane modules, such as feed pressure, module length, permeate pressure, and feed concentration. Increasing feed pressure in the membrane modules caused the pressure gradient to increase. Therefore, maximum mass transfer occurred through the membrane. Moreover, the length was considered to show gas variation in the permeate. In the end, the contours of gas flow profiles in the membrane modules were successfully reported. Modeling predictions were compared with the published data and validated, and it was found that there was good agreement between them for the different values of selectivity. The comparison indicated that the membrane modules were very efficient in terms of the separation of the desired gas at a higher pressure gradient.

Author Contributions: A.H. proposed the main idea and methodology; S.Q. performed the CFD analysis and wrote the manuscript; M.A. and A.H. provided key suggestions and improved the manuscript.

Funding: This research received no external funding.

Conflicts of Interest: The authors declare no conflicts of interest.

List of symbols

C	Concentration of a (mol/m^3)
D_{AB}	Diffusion coefficient (m^2/s)
C_0	Initial concentration (mol/m^3)
$C_{0,1}$	Final concentration (mol/m^3)
P	Permeance ($mol/(m^2 \cdot s \cdot pa)$)
S	Solubility in the membrane ($mol/(m^3 \cdot pa)$)
pf	Feed pressure (Pa)
ph	Permeate pressure (Pa)
Δ	Membrane thickness (m)
ΔC	Difference in concentration (mol/m^3)
Δp	Gradient of the partial pressure of gases (Pa)
N_a	Diffusive flux ($mol/m^2 \cdot s$)

References

1. Alkhamis, N.; Anqi, A.E.; Oztekin, A. Computational study of gas separation using a hollow fiber membrane. *Int. J. Heat Mass Transf.* **2015**, *89*, 749–759. [CrossRef]
2. Wetenhall, B.; Race, J.; Downie, M. The effect of CO_2 purity on the development of pipeline networks for carbon capture and storage schemes. *Int. J. Greenh. Gas Control* **2014**, *30*, 197–211. [CrossRef]
3. Conti, J.; Holtberg, P.; Diefenderfer, J.; LaRose, A.; Turnure, J.T.; Westfall, L. *International Energy Outlook 2016 with Projections to 2040*; USDOE Energy Information Administration (EIA): Washington, DC, USA, 2016.
4. Hakim, A.K. Numerical Simulation of Gas Separation by Hollow Fiber Membrane. 2017. Theses and Dissertations. 2624. Available online: http://preserve.lehigh.edu/etd/2624 (accessed on 2 July 2018).
5. Forward, Y. A carbon capture and storage network for Yorkshire and Humber. In *An Introduction to Understanding the Transportation of CO_2 from Yorkshire and Humber Emitters into Offshore Storage Sites*; Yorkshire Forward Victoria House: Leeds, UK, 2008.
6. Mechleri, E.; Brown, S.; Fennell, P.S.; Mac Dowell, N. CO_2 capture and storage (CCS) cost reduction via infrastructure right-sizing. *Chem. Eng. Res. Des.* **2017**, *119*, 130–139. [CrossRef]
7. Bernardo, P.; Drioli, E.; Golemme, G. Membrane gas separation: A review/state of the art. *Ind. Eng. Chem. Res.* **2009**, *48*, 4638–4663. [CrossRef]
8. Bernardo, P.; Clarizia, G. 30 years of membrane technology for gas separation. *Chem. Eng.* **2013**, *32*, 1999–2004.
9. Shamsabadi, A.A.; Kargari, A.; Farshadpour, F.; Laki, S. Mathematical modeling of CO_2/CH_4 separation by hollow fiber membrane module using finite difference method. *J. Membr. Sep. Technol.* **2012**, *1*, 19–29.
10. Nunes, S.P.; Peinemann, K.-V. *Membrane Technology*; Wiley Online Library: Hoboken, NJ, USA, 2001.
11. Chen, W.-H.; Lin, C.-H.; Lin, Y.-L. Flow-field design for improving hydrogen recovery in a palladium membrane tube. *J. Membr. Sci.* **2014**, *472*, 45–54. [CrossRef]
12. Lock, S.; Lau, K.; Ahmad, F.; Shariff, A. Modeling, simulation and economic analysis of CO_2 capture from natural gas using cocurrent, countercurrent and radial crossflow hollow fiber membrane. *Int. J. Greenh. Gas Control* **2015**, *36*, 114–134. [CrossRef]

13. Katoh, T.; Tokumura, M.; Yoshikawa, H.; Kawase, Y. Dynamic simulation of multicomponent gas separation by hollow-fiber membrane module: Nonideal mixing flows in permeate and residue sides using the tanks-in-series model. *Sep. Purif. Technol.* **2011**, *76*, 362–372. [CrossRef]

14. Geankoplis, C.J. *Transport Processes and Separation Process Principles: (Includes Unit Operations)*; Prentice Hall Professional Technical Reference: Upper Saddle River, NJ, USA, 2003.

15. Wankat, P.C. *Separation Process Engineering*; Pearson Education: London, UK, 2006.

16. Marriott, J.; Sørensen, E.; Bogle, I. Detailed mathematical modelling of membrane modules. *Comput. Chem. Eng.* **2001**, *25*, 693–700. [CrossRef]

17. Pan, C. Gas separation by permeators with high-flux asymmetric membranes. *AIChE J.* **1983**, *29*, 545–552. [CrossRef]

18. Alrehili, M.; Usta, M.; Alkhamis, N.; Anqi, A.E.; Oztekin, A. Flows past arrays of hollow fiber membranes–Gas separation. *Int. J. Heat Mass Transf.* **2016**, *97*, 400–411. [CrossRef]

19. Ahsan, M.; Hussain, A. A computational fluid dynamics (CFD) approach for the modeling of flux in a polymeric membrane using finite volume method. *Mech. Ind.* **2017**, *18*, 406. [CrossRef]

20. Saeed, A.; Vuthaluru, R.; Yang, Y.; Vuthaluru, H.B. Effect of feed spacer arrangement on flow dynamics through spacer filled membranes. *Desalination* **2012**, *285*, 163–169. [CrossRef]

21. Karode, S.K.; Kumar, A. Flow visualization through spacer filled channels by computational fluid dynamics I.: Pressure drop and shear rate calculations for flat sheet geometry. *J. Membr. Sci.* **2001**, *193*, 69–84. [CrossRef]

22. Thundyil, M.J.; Koros, W.J. Mathematical modeling of gas separation permeators—For radial crossflow, countercurrent, and cocurrent hollow fiber membrane modules. *J. Membr. Sci.* **1997**, *125*, 275–291. [CrossRef]

23. Alkhamis, N.; Oztekin, D.E.; Anqi, A.E.; Alsaiari, A.; Oztekin, A. Numerical study of gas separation using a membrane. *Int. J. Heat Mass Transf.* **2015**, *80*, 835–843. [CrossRef]

24. Mourgues, A.; Sanchez, J. Theoretical analysis of concentration polarization in membrane modules for gas separation with feed inside the hollow-fibers. *J. Membr. Sci.* **2005**, *252*, 133–144. [CrossRef]

25. Ahsan, M.; Hussain, A. Computational fluid dynamics (CFD) modeling of heat transfer in a polymeric membrane using finite volume method. *J. Therm. Sci.* **2016**, *25*, 564–570. [CrossRef]

26. Coroneo, M.; Montante, G.; Baschetti, M.G.; Paglianti, A. CFD modelling of inorganic membrane modules for gas mixture separation. *Chem. Eng. Sci.* **2009**, *64*, 1085–1094. [CrossRef]

27. Coroneo, M.; Montante, G.; Catalano, J.; Paglianti, A. Modelling the effect of operating conditions on hydrodynamics and mass transfer in a Pd–Ag membrane module for H_2 purification. *J. Membr. Sci.* **2009**, *343*, 34–41. [CrossRef]

28. Chen, W.-H.; Syu, W.-Z.; Hung, C.-I.; Lin, Y.-L.; Yang, C.-C. A numerical approach of conjugate hydrogen permeation and polarization in a Pd membrane tube. *Int. J. Hydrog. Energy* **2012**, *37*, 12666–12679. [CrossRef]

29. Szwast, M. Modelling the gas flow in permeate channel in membrane gas separation process. *Chem. Process Eng.* **2018**, *39*, 271–280.

30. Santafé-Moros, A.; Gozálvez-Zafrilla, J. Design of a Flat Membrane Module for Fouling and Permselectivity Studies. In Proceedings of the COMSOL Conference, Paris, France, 15–17 November 2010; pp. 1–7.

31. Qi, R.; Henson, M. Optimization-based design of spiral-wound membrane systems for CO_2/CH_4 separations. *Sep. Purif. Technol.* **1998**, *13*, 209–225. [CrossRef]

32. Gholami, G.; Soleimani, M.; Takht Ravanchi, M. Mathematical Modeling of Gas Separation Process with Flat Carbon Membrane. *J. Membr. Sci. Res.* **2015**, *1*, 90–95.

 *processes*

Article

Influence of Particle Charge and Size Distribution on Triboelectric Separation—New Evidence Revealed by In Situ Particle Size Measurements

Johann Landauer * and Petra Foerst

Chair of Process Systems Engineering, TUM School of Life Sciences Weihenstephan, Technical University of Munich, Gregor-Mendel-Straße 4, 85354 Freising, Germany; petra.foerst@tum.de
* Correspondence: johann.landauer@tum.de; Tel.: +49-8161-71-5172

Received: 20 May 2019; Accepted: 15 June 2019; Published: 19 June 2019

Abstract: Triboelectric charging is a potentially suitable tool for separating fine dry powders, but the charging process is not yet completely understood. Although physical descriptions of triboelectric charging have been proposed, these proposals generally assume the standard conditions of particles and surfaces without considering dispersity. To better understand the influence of particle charge on particle size distribution, we determined the in situ particle size in a protein–starch mixture injected into a separation chamber. The particle size distribution of the mixture was determined near the electrodes at different distances from the separation chamber inlet. The particle size decreased along both electrodes, indicating a higher protein than starch content near the electrodes. Moreover, the height distribution of the powder deposition and protein content along the electrodes were determined in further experiments, and the minimum charge of a particle that ensures its separation in a given region of the separation chamber was determined in a computational fluid dynamics simulation. According to the results, the charge on the particles is distributed and apparently independent of particle size.

Keywords: triboelectric separation; particle size distribution; particle charge; binary mixture; in situ particle size measurement; charge estimation

---

1. Introduction

Electrostatic effects were first recognized by the ancient Greek philosophers, who generated electricity by rubbing amber with fur. Thales of Miletus is often called the discoverer of the triboelectric effect [1,2]; however, this ancient observation has not been completely understood. Triboelectric charging of conductive materials is described by work function [3,4]. At conductor–insulator and insulator–insulator contacts, triboelectric charging has been described with "effective work function" [5,6], electron transfer [7], ion transfer [8,9], and material transfer [10,11]. Furthermore, contact charging is affected by environmental conditions such as humidity [12–14] and physical impact [15–17].

Triboelectric charging is undesired in process engineering because it interferes with pneumatic conveying [18,19], fluidized beds [20,21], mixing [22,23], and tablet pressing [24]. Moreover, it is a surface effect, indicating that particle surface plays a critical role. The known factors affecting triboelectric charging are particle area (indicated by particle size) [25–30], surface roughness [31–33], chemical composition [34], and elasticity (indicated by contact area) [35–38]; however, most experimental studies assumed uniform particles or contact surfaces. In most applications, the particles are not monodispersed and have no defined surface; however, the particles are dispersed in size, surface area, elasticity, crystallinity, and morphology.

To use triboelectric charging and subsequent separation as a tool to separate particles due to their chemical composition, surface morphology, crystallinity, or particle morphology, it is necessary to understand the influence of particle size distribution or powder composition as well as the influence of non-ideal conditions. All these factors show the necessity of triboelectric separation experiments with real, but defined, powders (like starch and protein) in order to use triboelectric separation to enrich, e.g., protein in lupine flour [39] and to take into account further influencing factors. Furthermore, the use of a starch–protein mixture as a model substrate for triboelectric charging is anticipated to have a possible application to enrich protein out of cereals or legumes. This ability of triboelectric separation has been demonstrated [39–46].

Hitherto, lots of studies have been carried out with well-defined particles or with inhomogeneous and undefined organic systems. Supplementary to these findings, real powders with a defined chemical composition and dispersity in particle size should be investigated. The discussion of influencing factors, such as particle morphology or further particle properties, suggest that particle surface charge is affected by the particle size distribution, in turn influencing the triboelectric charging effect. As the particle charge strongly affects the subsequent separation step, particles with different charges become separated at different regions on the electrodes, depending on the flow profile in the separation chamber and the electric field strength. Therefore, we hypothesize that if particle size (as a proxy of surface area) influences the charging and the subsequent separation properties, then particles of different sizes will be separated at different regions on the electrodes.

2. Materials and Methods

2.1. Materials

Whey protein isolate (Davisco Foods International, Le Sueur, MN, USA) with a protein content of 97.6 wt % was ground as that described in Landauer et al. [47]. Barley starch (Altia, Finland) with a starch content of 97.0 wt % was narrowed in particle size distribution in a wheel classifier (ATP 50, Hosokawa Alpine, Augsburg, Germany) under the conditions described in Landauer et al. [47].

2.2. Methods

2.2.1. Separation Setup

The simple experimental setup, originally demonstrated by Landauer et al. [47,48], comprises of an exchangeable charging section and a rectangular separation chamber. The dispersion of powders added to the gas flow is facilitated by a Venturi nozzle. The charging tube (of diameter 10 mm and length 230 mm) was composed of polytetrafluoroethylene (PTFE). An electrical field strength of 109 V/m was applied to the parallel-plate capacitor in a rectangular separation chamber (46 mm × 52 mm × 400 mm). The protein contents of the binary protein–starch mixtures were varied as 15, 35, and 45 wt %, and were determined as described in [48]. Briefly, the powder was dispersed in NaCl buffer (pH 7, 0.15 M) and the protein concentration was photometrically determined at 280 nm. To measure the protein content along the electrodes, the powder was sampled in three colored areas (see Figure 1a).

The amount of particles separated along the electrodes was determined by measuring the height of the separated powder. The measuring points are shown in Figure 1a and marked with gray circles. The powder deposition height was measured homogenously along each electrode in order to get a topography. Note that the powder height was determined using a micrometer screw (according to DIN 863-1:2017-02). The mean was calculated from the results of three independent separation experiments ($n = 3$). Error bars indicate the confidence intervals of the Student's *t*-test with an $\alpha = 0.05$ significance level.

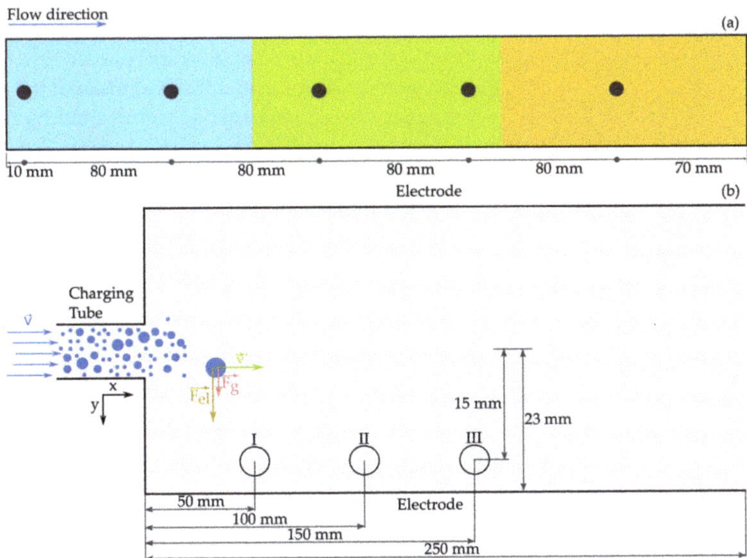

Figure 1. (**a**) Schematic of the sampling points along the electrodes. The powder deposition height on the electrode was measured at the points marked by gray circles. Colored areas mark the areas of powder sampling along the electrode. (**b**) Schematic of the separation chamber and the charging tube. The in situ particle size near the electrodes was measured at positions I, II, and III. Measurements near the anode and cathode were enabled by switching the polarity of the electrical field. The electric force F_{el} and weight force F_g acting on each particle in the separation chamber are visualized.

2.2.2. In Situ Particle Size Analysis

To investigate whether the particles agglomerate along the charging tube, we analyzed the in situ particle size distributions along the charging tube. For this purpose, parts of the charging and separation setup were installed in the measuring gap of a HELOS laser diffraction system (Sympatec, Clausthal-Zellerfeld, Germany). The charging and dispersing setup has been described in previous studies [47,48]. In the charging section, the particle size distributions were determined at the outlet of the Venturi nozzle (inlet of the charging section) and at the outlet of the charging tube. In the separation chamber, the particle size distribution was measured as a function of length. The schematic in Figure 1b shows the measuring positions I, II, and III along the electrodes in the separation chamber. The measuring points were chosen to be close to the electrodes and in the first half of the separation chamber. Due to the triboelectric separation, the particle concentration is decreasing along the separation chamber. Thus, the particle concentration, which is required to determine the particle size distribution, was not accessible. The particle size distributions on the anode and the cathode were obtained by switching the polarity of the electrical field with an electrical field strength of 217.4 kV/m. The mean of six independent separation experiments ($n = 6$) was calculated and the error is indicated by the confidence intervals of the Student's *t*-test with an $\alpha = 0.05$ significance level using error bars.

2.2.3. Flow Simulation and Estimation of the Particle Charge

The change in cross section between the charging tube and the separation chamber is very rigorous. The flow profile in the separation chamber, which might affect the separation characteristics and the particle size distribution along the electrodes (Figure 1b), was investigated in a computational fluid dynamics (CFD) simulation (ANSYS Fluent, version: 17.0, supplier: Ansys, Inc., Canonsburg, PA,

USA) of a realizable k–ε model. The particle motion in the separation chamber was visualized by tracking particles in the flow simulation. The inserted spherical particles followed a Weibull size distribution with a minimum, mean, and maximum (measured in the initial particle size distribution) of 1, 16, and 40 µm, respectively. The powder density was considered as the mean of the true density (1465 kg/m^3), which was measured by a gas pycnometer (Accupyc 1330, Micromeritics Instrument Corp., Norcross, GA, USA). The minimum charge at which the particle will be deflected in the measuring region was determined by simulating the in situ particle size distribution at different gravity levels (emulating different particle charges). The Coulomb force aligns with the weight force, as shown in Figure 1b. The absolute value of the charge q of the particles is estimated as follows:

$$q = \frac{x_3}{6E}\pi\rho_s g(n-1)$$ (1)

where x is the mean particle size, ρ_s is the true density, $\vec{E}$ is the electrical field, $\vec{g}$ is gravity, and the scaling factor n is the increase in particle charge. The scaling factor in the simulation was varied between 1 and 44. Note that the polarity of the charge depends on the electrical field's polarity.

3. Results

3.1. Particle Size Distribution

3.1.1. Agglomeration within the Charging Tube

Figure 2 shows the volume–weight density distributions at the Venturi nozzle outlet (panel a) and at the outlet of the charging tube (panel b) in the 15 and 30 wt % powders. In both particle size distributions, the particle size decreased with increasing initial protein content. The particle size distributions were similar at the outlets of the nozzle and the charging tube. The mean particle size (peak position) and the maximum particle size are the same at the tube inlet and the tube outlet. By comparing the distribution of finer particles, an increase in finer particles is visible. Thus, a dispersion along the charging tube is measured. The reason for this dispersion could be the high particle–particle collision number within the charging tube, due to the high turbulence [47]. The results in Figure 2 indicate breaking up particle agglomerates of fine particles during the charging step that could promote electrostatic separation. Contrarily, no electrostatic agglomeration that could impair electrostatic separation is observed.

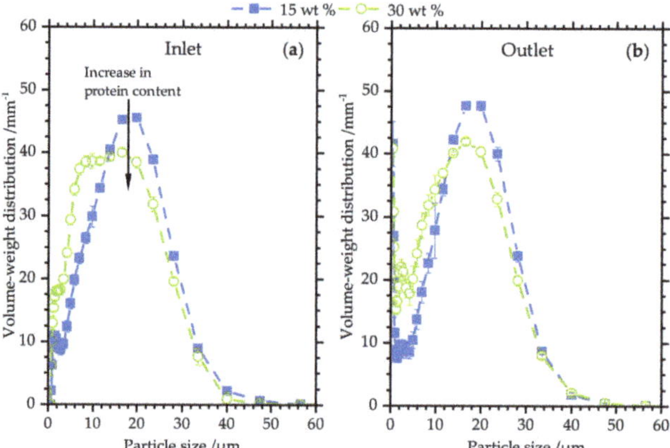

Figure 2. Volume–weight density distribution at (**a**) the Venturi nozzle outlet and (**b**) the outlet of the charging tube. Increasing the initial protein content (from 15 to 30 wt %) refined the particle size distribution. The distributions at the nozzle and tube ends are not obviously different.

3.1.2. Particle Size Distribution along the Electrodes

Figure 3 shows the volume–weight density distributions of the powder close to the cathode (a) and the anode (b) in measuring regions I, II, and III (Figure 1b). Increasing the initial protein content refined the particle size distributions at both the cathode and anode, as evidenced by the higher peak at ~6 µm in the 30 wt % compared to the 15 wt % distribution. This higher peak indicates a higher amount of finer protein particles (cf. Figure 2). In the sample with an initial protein content of 15 wt %, the particle size decreased along the investigated regions I, II, and III (note the lower peak height at 16 µm than that at 6 µm). This stepwise decrease in particle size was observed along both the cathode and the anode, as well as in the sample with higher initial protein content (30 wt %). The peak increases from 15 to 30 wt % are more clearly observed at 6 µm compared to 16 µm because increasing the protein content increases the amount of smaller particles. Comparing the particle size distributions at the cathode and the anode, the particles were finer on the cathode regardless of the initial protein content. These results suggest a higher protein content on the cathode (cf. Figure 2). The protein content of the separated powder on the cathode and the anode is approximately 80 and 2.5 wt %, respectively. Thus, protein is enriched on the cathode and starch is enriched on the anode [47,48]. However, the enhancement of finer protein particles near the cathode cannot be correlated with the protein content because the used protein powder is finer than the starch powder. Nevertheless, the particle size distributions at each measuring position in the separation chamber depended on the initial protein content. Thus, the particle size is influenced by the polarity of the electric field, the distance from the charging tube outlet, and (most strongly) the initial protein content. The region in which the particles separate plays a subordinate role on the particle size distribution. Furthermore, the particle size distributions on the anode and cathode resembled the initial distributions determined at the outlets of the Venturi nozzle and the tube.

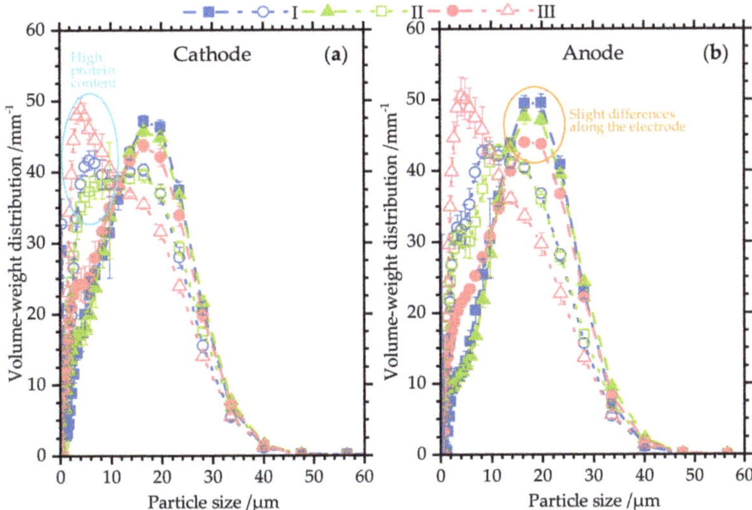

Figure 3. Volume–weight density distributions recorded near the cathode (**a**) and the anode (**b**) in regions I, II, and III. Closed and open symbols denote initial protein contents of 15 and 30 wt %, respectively. Increasing the initial protein content reduces the particle size at both anode and cathode. The particle size distributions differ in the three measuring regions.

3.2. Powder on the Electrodes

3.2.1. Powder Height

Figure 4 shows the powder height along the cathode and the anode. On the cathode, the distribution of powder was approximately homogeneous along the electrode. The powder height varied most extensively at the second measuring point, and was least variable at the first and fourth measuring points. This significant but extremely low variation should not be overinterpreted; however, the powder height severely decreased along the anode. The powder height was constant at the first two measuring points, and then dropped to zero over the next two measurement points. Thus, the powder heights on the cathode and the anode exhibited very different profiles, suggesting different charges of the particles separated on the two electrodes. In particular, the negatively charged particles exhibited a higher net charge than the positively charged particles.

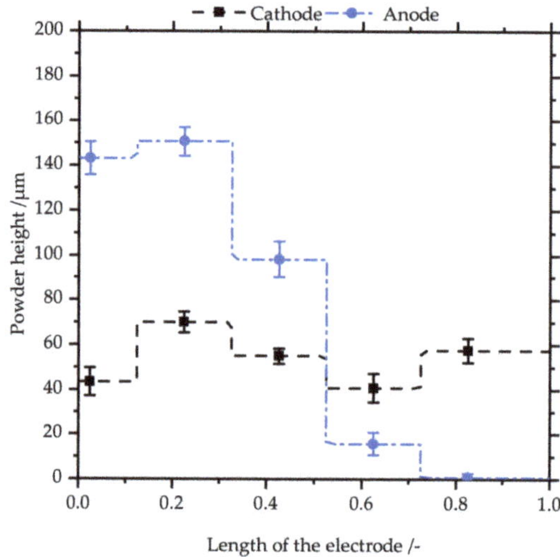

Figure 4. Powder height along the cathode and the anode. Powder height is approximately constant on the cathode, but mostly separates over the first half of the anode.

3.2.2. Protein Content

Figure 5 shows the protein content on the cathode in the three measurement areas at initial protein contents of 15 and 30 wt %. The protein content was consistently higher for the sample with the higher initial protein content. Independently of the initial protein content, the protein content increased in the second area (relative to the first area). In the third area, the protein content decreased at the initial protein content of 15 wt %, but remained high at the higher initial protein content. If we compare the protein content with the powder height, the two quantities are apparently independent because the powder height was approximately constant along the electrode, whereas the protein content extensively varied.

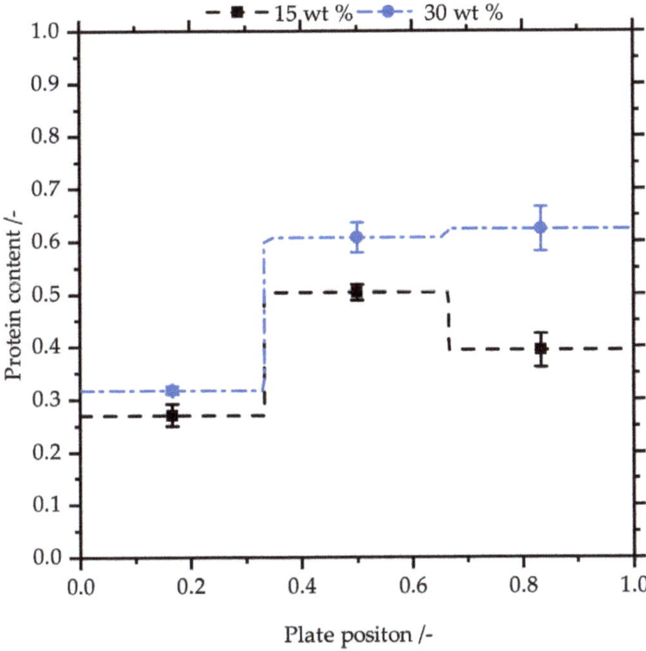

Figure 5. Protein content on the cathode in three different measurement areas for initial protein contents of 15 and 30 wt %. The protein content increases form the first to the second area regardless of initial protein content. In the third area, the protein content decreases (15 wt %) or remains the same (30 wt %).

3.3. Estimation of the Charge Correlated with the In Situ Particle Size Distribution

To estimate the minimum charge at which particles will separate in the separation chamber (enabling an in situ particle size analysis), the particles were tracked in a CFD study. Figure 6 shows the trajectories of spherical particles with different accelerations (varied by changing the electrical force in Equation (1)). In a homogenous electrical field, the net charge of the particles is a multiple of the elementary charge. Uncharged particles might be undetectable in every measuring region. Particles with a net charge of 1.45×10^3 q_e can be detected in regions II and III, whereas those with charges of 7.26×10^3 q_e and 1.16×10^4 q_e might be measurable only in region II. Particles with a net charge of 3.77×10^4 q_e and higher are visible in region I. When generating Figure 6 and calculating the associated particle charge, we assumed spherical particles with a mean diameter of 16 μm corresponding to the mean particle size of the powders used for the experiments. According to Equation (1), the particle size affects both the charge on a single particle and the particle trajectories. In all cases, varying the particle size only slightly affected the trajectories.

The background of Figure 6 shows the velocity profile in the separation chamber. The profile shows a jet at the charging tube outlet followed by homogeneity. The jet formed at the outlet of the tube affected the particle trajectories considerably. Regardless of their net charge, the particles remained within the jet to ~100 mm from the outlet. Then, they lost speed and were deflected toward the electrode by the Coulomb force. Thus, the simulation visualized the influence of the particle net charge on the particle trajectories within a complex velocity profile. Observing these particle trajectories, we can understand how particles might be charged to ensure their separation in the measuring regions of in situ particle size analyses.

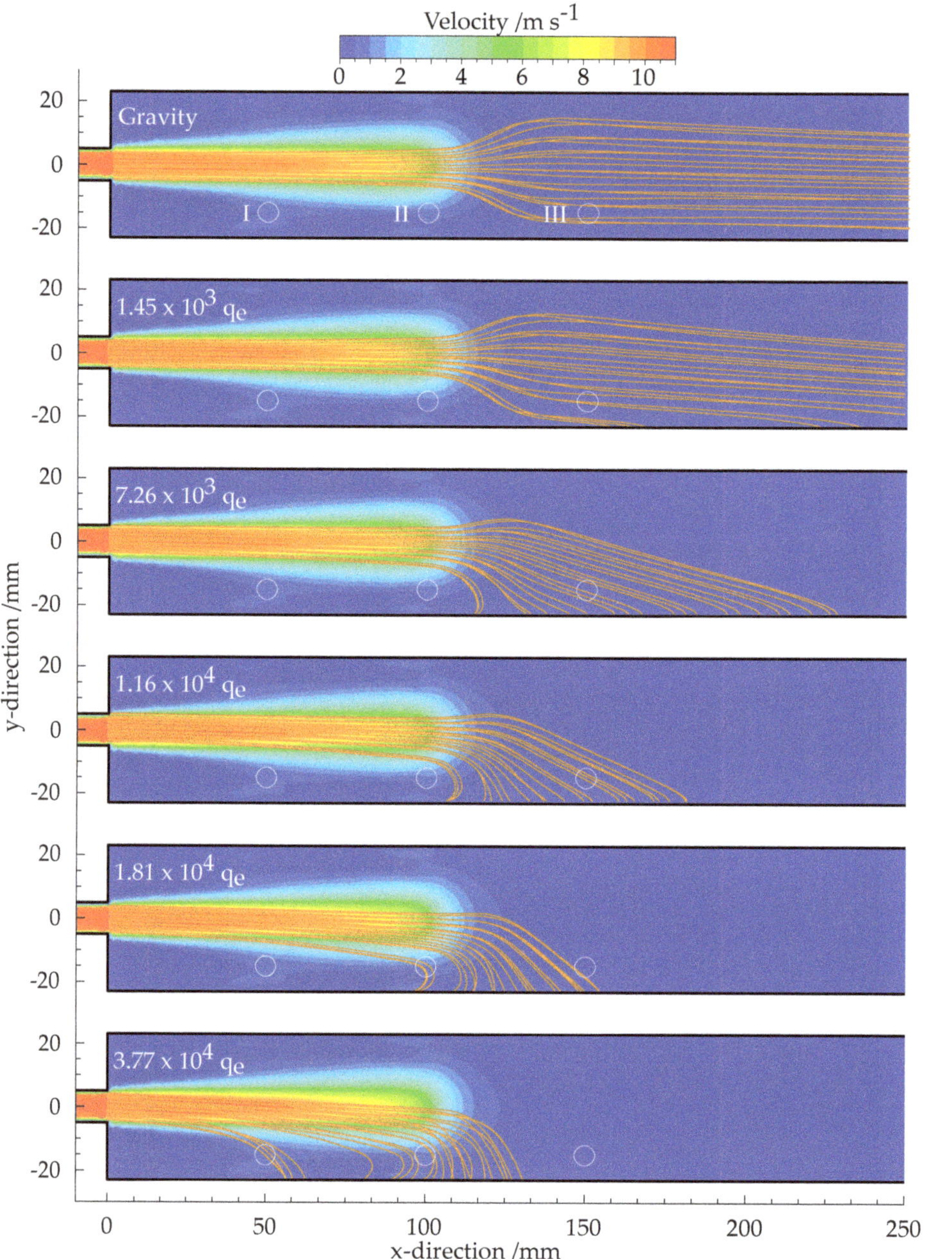

Figure 6. Trajectories of 16 μm diameter spherical particles with different particle charges (multiples of the elementary charge calculated by Equation (1)). The measuring regions I, II, and III of the in situ particle size distribution are indicated by the white open circles. Depending on their net charges, certain particles are not detectable in every measuring region. The background visualizes the velocity profile. The jet formed at the outlet of the charging tube is clearly visible.

4. Discussion

To use triboelectric separation as a tailor-made particle separation tool, one must separate the particles by their specific chargeabilities. Accordingly, it is necessary to disperse the particles before the charging step and avoid their agglomeration during the charging step. The selected setup enables the appropriate conditions for dispersal and aggregation prevention (Figure 2). Hence, detailed investigations of the separation step are required to establish triboelectric separation as an industrial separation technique.

Assuming that the charge distribution of fine particles is sourced from the triboelectric charging of the particles and that the charge distribution also possibly depends on the particle size and the chemical composition [34], the particle size distributions were determined at different locations close to the electrodes (Figure 3). As expected, the particle size distribution was coarser on the anode than on the cathode, because increasing the protein content refined the particle size distribution (Figure 2) and the protein was enriched on the cathode [47,48]. Moreover, along the measuring regions close to the electrodes, the decreased particle size accompanying the refined particles was demonstrated for different initial protein contents. These results were identical on the cathode and the anode, suggesting (as a first hint) that the net charges of the particles after triboelectric charging are independently distributed of the particle sizes.

The local distribution of the separated powder on the electrodes indicates the strength of the particle charges because particles with a higher and lower net charge are separated at the inlet and the near-outlet of the electrode, respectively. The powder height profiles on the cathode and the anode exhibited very different characteristics (Figure 4). The powder was dispersed almost homogeneously on the cathode but was separated close to the inlet on the anode. The absence of powder at the anode outlet indicates that the negative particles were more highly charged than the positive ones. The same results were reported in single-particle charge measurements [49]. This result further indicates independent distributions of the particle charges and sizes because the particle size distributions were similar on the cathode and anode (Figure 3). Furthermore, the homogeneously distributed powder exhibited a distributed protein content with a peak in the middle of the electrode at 15 wt % initial protein content, and level peaks in the second and third parts of the electrode at 30 wt % initial protein content (Figure 5). This suggests a lower net charge of protein particles than of starch particles (Figure 6). These results are underpinned by the decreased particle size (higher protein content) along the cathode than along the anode (Figure 3). The particle trajectories were affected by the inhomogeneous flow profile in the separation chamber; however, in the flow simulation, they were predominantly influenced by the charge. Moreover, they showed a charge-dependent separation region. To summarize, the binary powder mixture with a polydispersed particle size distribution showed no clear relationship between particle size and particle charge in the separation region. These results contradict previous studies, which reported that smaller particles are predominantly negatively charged [25–30]. The results support an effect of particle size on triboelectric charging, but no clear tendency was found regarding the fine and coarse particles. Thus, the hypothesis of this study, i.e., that particle size distribution (as a measure of surface area) plays a major role in triboelectric charging and subsequent separation, is questionable. Indeed, there is a dependence of particle size along the electrodes, but the results show a more complex connection between the particle material and particle size.

5. Conclusions

The dispersing and agglomeration characteristics of powders with different initial protein contents were consistent along the charging tube. The in situ particle size measurements were consistent at different regions in the separation camber. After estimating the minimum charge for particle separation, it was found that large charge differences were required for separation in every measuring region of the chamber. This wide charge distribution might lead to different separation regions of the particles, as indicated by the roughly homogeneous powder height on the cathode and the steep decrease in powder height on the anode. These results show a complex dependency of triboelectric charging and subsequent separation on the size and material of the particles. As the mechanism of triboelectric or contact charging has not been accurately determined, determining the primary influencing factors is very challenging. The present results indicate the high complexity of triboelectric charging and indicate that particle size is not a highly important factor in triboelectric separation but affects the triboelectric charging through surface-area differences.

Author Contributions: Conceptualization, J.L.; methodology, J.L.; data curation, J.L.; writing—original draft preparation, J.L.; writing—review and editing, J.L. and P.F.; visualization, J.L.; supervision, P.F.

Funding: This research received no external funding.

Acknowledgments: The authors would like to thank Lukas Hans for help with performing in situ particle size measurement and Heiko Briesen for the possibility to carry out this study.

Conflicts of Interest: The authors declare no conflict of interest.

References

1. Iversen, P.; Lacks, D.J. A life of its own: The tenuous connection between Thales of Miletus and the study of electrostatic charging. *J. Electrost.* **2012**, *70*, 309–311. [CrossRef]
2. O'Grady, P.F. *Thales of Miletus. The Beginnings of Western Science and Philosophy*; Ashgate: Aldershot, UK, 2002; ISBN 0754605337.
3. Harper, W.R. The Volta Effect as a Cause of Static Electrification. *Proc. R. Soc. A Math. Phys. Eng. Sci.* **1951**, *205*, 83–103. [CrossRef]
4. Harper, W.R. *Contact and Frictional Electrification. Monographs on the Physics and Chemistry of Materials*; Clarendon Press: Oxford, UK, 1967.
5. Davies, D.K. Charge generation on dielectric surfaces. *J. Phys. D Appl. Phys.* **1969**, *2*, 1533. [CrossRef]
6. Lowell, J.; Rose-Innes, A.C. Contact electrification. *Adv. Phys.* **1980**, *29*, 947–1023. [CrossRef]
7. Liu, C.; Bard, A.J. Electrostatic electrochemistry at insulators. *Nat. Mater.* **2008**, *7*, 505–509. [CrossRef]
8. McCarty, L.S.; Whitesides, G.M. Electrostatic charging due to separation of ions at interfaces: Contact electrification of ionic electrets. *Angew. Chem. Int. Ed. Engl.* **2008**, *47*, 2188–2207. [CrossRef]
9. Ducati, T.R.D.; Simoes, L.H.; Galembeck, F. Charge partitioning at gas-solid interfaces: Humidity causes electricity buildup on metals. *Langmuir* **2010**, *26*, 13763–13766. [CrossRef]
10. Baytekin, H.T.; Patashinski, A.Z.; Branicki, M.; Baytekin, B.; Soh, S.; Grzybowski, B.A. The mosaic of surface charge in contact electrification. *Science* **2011**, *333*, 308–312. [CrossRef]
11. Salaneck, W.R. Double mass transfer during polymer-polymer contacts. *J. Appl. Phys.* **1976**, *47*, 144–147. [CrossRef]
12. Schella, A.; Herminghaus, S.; Schröter, M. Influence of humidity on tribo-electric charging and segregation in shaken granular media. *Soft Matter* **2017**, *13*, 394–401. [CrossRef]
13. Németh, E.; Albrecht, V.; Schubert, G.; Simon, F. Polymer tribo-electric charging: Dependence on thermodynamic surface properties and relative humidity. *J. Electrost.* **2003**, *58*, 3–16. [CrossRef]
14. Baytekin, H.T.; Baytekin, B.; Soh, S.; Grzybowski, B.A. Is water necessary for contact electrification? *Angew. Chem. Int. Ed. Engl.* **2011**, *50*, 6766–6770. [CrossRef]
15. Matsusyama, T.; Yamamoto, H. Impact charging of particulate materials. *Chem. Eng. Sci.* **2006**, *61*, 2230–2238. [CrossRef]
16. Matsuyama, T.; Yamamoto, H. Electrification of single polymer particles by successive impacts with metal targets. *IEEE Trans. Ind. Appl.* **1995**, *31*, 1441–1445. [CrossRef]
17. Watanabe, H.; Ghadiri, M.; Matsuyama, T.; Maruyama, H.; Matsusaka, S.; Ghadiri, M.; Matsuyama, T.; Ding, Y.L.; Pitt, K.G.; Maruyama, H.; et al. Triboelectrification of pharmaceutical powders by particle impact. *Int. J. Pharm.* **2007**, *334*, 149–155. [CrossRef] [PubMed]
18. Grosshans, H.; Papalexandris, M.V. Large Eddy simulation of triboelectric charging in pneumatic powder transport. *Powder Technol.* **2016**, *301*, 1008–1015. [CrossRef]
19. Ireland, P.M. Triboelectrification of particulate flows on surfaces: Part I—Experiments. *Powder Technol.* **2010**, *198*, 189–198. [CrossRef]
20. Fotovat, F.; Bi, X.T.; Grace, J.R. A perspective on electrostatics in gas-solid fluidized beds: Challenges and future research needs. *Powder Technol.* **2018**, *329*, 65–75. [CrossRef]

21. Mehrani, P.; Murtomaa, M.; Lacks, D.J. An overview of advances in understanding electrostatic charge buildup in gas-solid fluidized beds. *J. Electrost.* **2017**, *87*, 64–78. [CrossRef]

22. Engers, D.A.; Fricke, M.N.; Storey, R.P.; Newman, A.W.; Morris, K.R. Triboelectrification of pharmaceutically relevant powders during low-shear tumble blending. *J. Electrost.* **2006**, *64*, 826–835. [CrossRef]

23. Karner, S.; Urbanetz, N.A. Arising of electrostatic charge in the mixing process and its influencing factors. *Powder Technol.* **2012**, *226*, 261–268. [CrossRef]

24. Ghori, M.U.; Supuk, E.; Conway, B.R. Tribo-electric charging and adhesion of cellulose ethers and their mixtures with flurbiprofen. *Eur. J. Pharm. Sci.* **2014**, *65*, 1–8. [CrossRef] [PubMed]

25. Forward, K.M.; Lacks, D.J.; Sankaran, R.M. Particle-size dependent bipolar charging of Martian regolith simulant. *Geophys. Res. Lett.* **2009**, *36*, 139. [CrossRef]

26. Waitukaitis, S.R.; Lee, V.; Pierson, J.M.; Forman, S.L.; Jaeger, H.M. Size-Dependent Same-Material Tribocharging in Insulating Grains. *Phys. Rev. Lett.* **2014**, *112*. [CrossRef]

27. Lacks, D.J.; Levandovsky, A. Effect of particle size distribution on the polarity of triboelectric charging in granular insulator systems. *J. Electrost.* **2007**, *65*, 107–112. [CrossRef]

28. Zhao, H.; Castle, G.S.P.; Inculet, I.I.; Bailey, A.G. Bipolar charging of poly-disperse polymer powders in fluidized beds. *IEEE Trans. Ind. Appl.* **2003**, *39*, 612–618. [CrossRef]

29. Mukherjee, R.; Gupta, V.; Naik, S.; Sarkar, S.; Sharma, V.; Peri, P.; Chaudhuri, B. Effects of particle size on the triboelectrification phenomenon in pharmaceutical excipients: Experiments and multi-scale modeling. *Asian J. Pharm. Sci.* **2016**, *11*, 603–617. [CrossRef]

30. Trigwell, S.; Grable, N.; Yurteri, C.U.; Sharma, R.; Mazumder, M.K. Effects of surface properties on the tribocharging characteristics of polymer powder as applied to industrial processes. *IEEE Trans. Ind. Appl.* **2003**, *39*, 79–86. [CrossRef]

31. Karner, S.; Maier, M.; Littringer, E.; Urbanetz, N.A. Surface roughness effects on the tribo-charging and mixing homogeneity of adhesive mixtures used in dry powder inhalers. *Powder Technol.* **2014**, *264*, 544–549. [CrossRef]

32. Lacks, D.J.; Duff, N.; Kumar, S.K. Nonequilibrium accumulation of surface species and triboelectric charging in single component particulate systems. *Phys. Rev. Lett.* **2008**, *100*, 188305. [CrossRef]

33. Burgo, T.A.L.; Silva, C.A.; Balestrin, L.B.S.; Galembeck, F. Friction coefficient dependence on electrostatic tribocharging. *Sci. Rep.* **2013**, *3*. [CrossRef] [PubMed]

34. Diaz, A.F.; Felix-Navarro, R.M. A semi-quantitative tribo-electric series for polymeric materials: The influence of chemical structure and properties. *J. Electrost.* **2004**, *62*, 277–290. [CrossRef]

35. Grosshans, H.; Papalexandris, M.V. A model for the non-uniform contact charging of particles. *Powder Technol.* **2017**, *305*, 518–527. [CrossRef]

36. Haeberle, J.; Schella, A.; Sperl, M.; Schröter, M.; Born, P. Double origin of stochastic granular tribocharging. *Soft Matter* **2018**, *14*, 4987–4995. [CrossRef] [PubMed]

37. Hu, J.; Zhou, Q.; Liang, C.; Chen, X.; Liu, D.; Zhao, C. Experimental investigation on electrostatic characteristics of a single grain in the sliding process. *Powder Technol.* **2018**, *334*, 132–142. [CrossRef]

38. Ireland, P.M. Impact tribocharging of soft elastic spheres. *Powder Technol.* **2019**, *348*, 70–79. [CrossRef]

39. Wang, J.; Zhao, J.; de Wit, M.; Boom, R.M.; Schutyser, M.A.I. Lupine protein enrichment by milling and electrostatic separation. *Innov. Food Sci. Emerg. Technol.* **2016**, *33*, 596–602. [CrossRef]

40. Wang, J.; de Wit, M.; Boom, R.M.; Schutyser, M.A.I. Charging and separation behavior of gluten–starch mixtures assessed with a custom-built electrostatic separator. *Sep. Purif. Technol.* **2015**, *152*, 164–171. [CrossRef]

41. Xing, Q.; de Wit, M.; Kyriakopoulou, K.; Boom, R.M.; Schutyser, M.A.I. Protein enrichment of defatted soybean flour by fine milling and electrostatic separation. *Innov. Food Sci. Emerg. Technol.* **2018**, *50*, 42–49. [CrossRef]

42. Wang, J.; de Wit, M.; Schutyser, M.A.I.; Boom, R.M. Analysis of electrostatic powder charging for fractionation of foods. *Innov. Food Sci. Emerg. Technol.* **2014**, *26*, 360–365. [CrossRef]

43. Wang, J.; Suo, G.; de Wit, M.; Boom, R.M.; Schutyser, M.A.I. Dietary fibre enrichment from defatted rice bran by dry fractionation. *J. Food Eng.* **2016**, *186*, 50–57. [CrossRef]

44. Tabtabaei, S.; Jafari, M.; Rajabzadeh, A.R.; Legge, R.L. Development and optimization of a triboelectrification bioseparation process for dry fractionation of legume flours. *Sep. Purif. Technol.* **2016**, *163*, 48–58. [CrossRef]

45. Tabtabaei, S.; Jafari, M.; Rajabzadeh, A.R.; Legge, R.L. Solvent-free production of protein-enriched fractions from navy bean flour using a triboelectrification-based approach. *J. Food Eng.* **2016**, *174*, 21–28. [CrossRef]
46. Chen, Z.; Liu, F.; Wang, L.; Li, Y.; Wang, R.; Chen, Z. Tribocharging properties of wheat bran fragments in air–solid pipe flow. *Food Res. Int.* **2014**, *62*, 262–271. [CrossRef]
47. Landauer, J.; Aigner, F.; Kuhn, M.; Foerst, P. Effect of particle-wall interaction on triboelectric separation of fine particles in a turbulent flow. *Adv. Powder Technol.* **2019**, *30*, 1099–1107. [CrossRef]
48. Landauer, J.; Foerst, P. Triboelectric separation of a starch-protein mixture—Impact of electric field strength and flow rate. *Adv. Powder Technol.* **2018**, *29*, 117–123. [CrossRef]
49. Landauer, J.; Tauwald, S.M.; Foerst, P. A Simple μ-PTV Setup to Estimate Single-Particle Charge of Triboelectrically Charged Particles. *Front. Chem.* **2019**, *7*, 1. [CrossRef]

Article

Research on the Pressure Dropin Horizontal Pneumatic Conveying for Large Coal Particles

Daolong Yang [1],*, Yanxiang Wang [1] and Zhengwei Hu [2]

[1]   School of Mechatronic Engineering, Jiangsu Normal University, Xuzhou 221116, China; 6020180134@jsnu.edu.cn
[2]   Research and Development Centre, Jiangsu Kampf Machinery Technology Co. LTD, Xuzhou 221116, China; zhengwei.hu1988@hotmail.com
*   Correspondence: yangdl@jsnu.edu.cn; Tel.: +86-1595-525-2518

Received: 17 April 2020; Accepted: 28 May 2020; Published: 30 May 2020

Abstract: As a type of airtight conveying mode, pneumatic conveying has the advantages of environmental friendliness and conveying without dust overflow. The application of the pneumatic conveying system in the field of coal particle conveying can avoid direct contact between coal particles and the atmosphere, which helps to reduce the concentration of air dust and improve environmental quality in coal production and coal consumption enterprises. In order to predict pressure drop in the pipe during the horizontal pneumatic conveying of large coal particles, the Lagrangian coupling method and DPM (discrete particle model) simulation model was used in this paper. Based on the comparison of the experimental results, the feasibility of the simulation was verified and the pressure drop in the pipe was simulated. The simulation results show that when the flow velocity is small, the simulation results of the DPM model are quite different from that of the experiment. When the flow velocity is large, the large particle horizontal pneumatic conveying behavior predicted by the model is feasible, which can provide a simulation reference for the design of the coal pneumatic conveying system.

Keywords: pneumatic conveying; large coal particles; Euler–Lagrange approach; DPM; pressure drop

---

1. Introduction

Coal is an important fossil energy and many countries have strong dependence on coal resources [1]. The open-conveying way is the most common conveying way for coal particles. The main conveying equipment is a belt conveyor. In the process of mining, unloading, separation and conveying, the pulverized coal and dust escape into the air, causing environmental pollution and serious dust explosion accidents [2]. The coal particles come in direct contact with the atmosphere, which leads to coal dust diffusion, pollution of the surrounding environment, waste of resources and coal dust explosion accidents [3].

How to convey coal cleanly, efficiently and safely has become an urgent environmental problem, one that needs to be solved. It also puts forward higher requirements for environmental protection, economy and reliability of coal conveying equipment. As a kind of airtight conveying mode, pneumatic conveying has the advantages of environmental friendliness and no-dust overflow [4]. Applying the pneumatic conveying system to the field of coal particle conveying can avoid the direct contact between coal particles and the atmosphere [5]. It is of great significance to improve the surrounding environment of coal enterprises, avoid safety production accidents and realize the green use of coal resources.

Many scholars have made contributions in the field of particle pneumatic conveying. Rabinovich [6] proposed the generalized flow pattern of vertical pneumatic conveying and a fluidized bed system that considers the effects of particle and gas properties, pipe diameter and particle concentration. Pahk [7] studied the friction between the plug and the pipe wall in the dense phase pneumatic conveying system. The results showed that the friction between the plug and the pipe wall increased with the contact area, and the friction between the spherical particles was greater than that of the cylindrical particles. Njobuenwu [8] used the particle trajectory model and wear model to predict the wear of elbows with different square sections in dilute particle flow. The results show that the maximum wear position is in the range of 20°–35° of elbows. Watson [9] carried out the vertical pneumatic conveying test of alumina particles with a particle size of 2.7 mm, measured the solid-phase mass flow rate, gas-phase mass flow rate and inlet and outlet pressure, as well as the pressure distribution and other parameters, and proved that the dense plug flow pneumatic conveying system has advantages in conveying coarse particles. Ebrahimi [10] established a horizontal pneumatic conveying test-bed based on a laser Doppler velocimeter, and carried out the conveying experiments of spherical glass powder with particle size of 0.81 mm, 1.50 mm and 2.00 mm. Ogata [11] studied the influence of different glass particle properties on the fluidization dense-phase pneumatic conveying system in a horizontal rectangular pipeline. Makwana [12] studied the causes of the fluctuation of the pneumatic conveying in the horizontal pipeline and showed that the pressure loss in the pipeline increased sharply with the formation of sand dunes, and the pressure drop value was related to the axial position of sand dunes. Anantharaman [13] studied the relationship between particle size, density, sphericity and minimum pickup speed of particles in the pneumatic conveying system and showed that the influence of particle size on pickup speed is greater than that of particle density. Akira [14,15] compared the pneumatic conveying system with a soft wing and sand dune model with the traditional pneumatic conveying system and showed that the pressure loss of the pneumatic conveying system with a soft wing and sand dune model was less than that of the traditional pneumatic conveying system at low air speed. Yang [16] carried out simulation and experimental research on the pneumatic suspension behavior of large irregular coal particles and obtained the suspension speed of coal particles under different particle sizes. Yang [17,18] studied the influence of structural parameters of a coal particle gas–solid injection feeder on the pure flow field injection performance and particle injection performance through multi-factor orthogonal experiments.

Previous work in the field has looked at fine particle [19,20], powder [21,22] and seed [23,24], all of which are less than 5 mm in size and are within the range of Geldart A to Geldart C. However, there are a few studies on large sizes (larger than Geldart D) [25,26]. When using the common computational fluid dynamics and discrete element method (CFD-DEM) simulation, due to the coupling method between the discrete element method and the finite element method, the simulation time is long and there is a lot to calculate, so it is difficult to get the results quickly. In order to quickly predict the pressure drop in horizontal pneumatic conveying for large (5–25 mm) coal particles, this paper uses the coupling method based on the Euler–Lagrange approach, DPM (discrete particle model) and the particle trajectory equations. The experiment and the simulation of horizontal pneumatic conveying for large coal particles was carried out to verify the feasibility of the simulation method. The multi-factor simulations were carried out to analyze the effects of particle size, flow field velocity, solid-gas rate and pipe diameter on pressure drop.

2. Theory

The flow field provides the energy required by the particles' motion, and the exchange of momentum and energy between the flow field and particles occurs in the pneumatic conveying flow field. In the Euler–Lagrange approach, the fluid phase is treated as a continuum by solving the Navier–Stokes equations, while the dispersed phase is solved by tracking a large number of particles through the calculated flow field. The dispersed phase can exchange momentum, mass and energy with the fluid phase.

2.1. Gas Phase Equations

The gas phase is a continuous medium, considering the influence of solid phase to flow field, and the continuity equation adds the volume fraction term ξ to exclude the gas volume occupied by the solid phase. It is assumed that the temperature of both gas and solid phases in pneumatic conveying is the same as that of the atmosphere, and no exothermic reaction occurs between the two phases. Therefore, the energy equation of gas and solid phases can be ignored.

(1) Gas phase continuity equation

According to the law of mass conservation, the gas phase continuity equation can be obtained and is shown in Equations (1) and (2).

$$\frac{\partial \xi \rho}{\partial t} + \nabla \cdot \rho \varepsilon v = 0;\tag{1}$$

$$\nabla \equiv \frac{\partial}{\partial x}\vec{i} + \frac{\partial}{\partial y}\vec{j} + \frac{\partial}{\partial z}\vec{k},\tag{2}$$

where ρ is the density of gas phase, v is the velocity of gas phase and ξ is the volume fraction term.

(2) Gas phase momentum equation

The momentum equation of the gas phase can be obtained from the law of momentum conservation, which is similar to the continuity equation.

$$\frac{\partial \xi \rho v}{\partial t} + \nabla \cdot \rho \xi \mu v = -\nabla p + \nabla \cdot (\xi \mu \nabla v) + \rho \xi g - S_m.\tag{3}$$

The momentum transfer S_m refers to the sum of the fluid drag in the fluid unit.

$$S_m = \frac{\Sigma F_d}{V}.\tag{4}$$

(3) Turbulence transmission equations

The Realizable k-ε model [27] has the advantage of a more accurate prediction for the divergence ratio of flat and cylindrical jets, and its transmission equations are shown in Equations (5)–(7).

$$\frac{\partial}{\partial t}(\rho k) + \frac{\partial}{\partial x_j}(\rho k v_j) = \frac{\partial}{\partial x_j}\left[\left(\mu + \frac{\mu_t}{\sigma_k}\right)\frac{\partial k}{\partial x_j}\right] + G_k + G_b - \rho \varepsilon - Y_M + S_k;\tag{5}$$

$$\frac{\partial}{\partial t}(\rho \varepsilon) + \frac{\partial}{\partial x_j}(\rho \varepsilon v_j) = \frac{\partial}{\partial x_j}\left[\left(\mu + \frac{\mu_t}{\sigma_\varepsilon}\right)\frac{\partial \varepsilon}{\partial x_j}\right] + \rho C_1 S \varepsilon - \frac{\rho C_2 \varepsilon^2}{k + \sqrt{v \varepsilon}} + \frac{C_{1\varepsilon} C_{3\varepsilon} G_b \varepsilon}{k} + S_\varepsilon;\tag{6}$$

where

$$\begin{cases} C_1 = \max\left[0.43, \frac{\eta}{\eta+5}\right] \\ \eta = \sqrt{2S_{ij}S_{ij}}\frac{k}{\varepsilon} \end{cases}.\tag{7}$$

2.2. Motion Equations of Solid Phase

The solid phase is a discrete phase as the motion law of solid particles obeys Newton's second law.

(1) Particle Force Balance

The trajectory of the solid phase is predicted by integrating the force balance on the particle which is written in a Lagrangian reference frame. This force balance equates the particle inertia with the forces acting on the particle and can be written as

$$\frac{d\vec{u}_p}{dt} = \frac{\vec{u} - \vec{u}_p}{\tau_r} + \frac{\vec{g}(\rho_p - \rho)}{\rho_p} + \vec{F},$$
(8)

where $\vec{F}$ is an additional acceleration term, $\vec{g}$ is the force of gravity on the particle, $\frac{\vec{u} - \vec{u}_p}{\tau_r}$ is the drag force per unit particle mass and τ_r is the particle relaxation time [28], which is defined as

$$\tau_r = \frac{\rho_p d_p^2}{18\mu} \frac{24}{C_d \mathrm{Re}},$$
(9)

where $\vec{u}$ is the fluid phase velocity, $\vec{u}_p$ is the particle velocity, μ is the molecular viscosity of the fluid, ρ is the fluid density, ρ_p is the density of the particle and d_p is the particle diameter. Re is the relative Reynolds number, which is defined as

$$\mathrm{Re} \equiv \frac{\rho_p d_p \left| \vec{u} - \vec{u}_p \right|}{\mu}.$$
(10)

For non-spherical particles, Haider and Levenspiel [29] developed the correlation

$$C_D = \frac{24}{\mathrm{Re}_{sph}}(1 + b_1 \mathrm{Re}_{sph}{}^{b_2}) + \frac{b_3 \mathrm{Re}_{sph}}{b_4 + \mathrm{Re}_{sph}},$$
(11)

where

$$\begin{aligned}
b_1 &= \exp(2.3288 - 6.4581\phi + 2.4486\phi^2); \\
b_2 &= 0.0964 + 0.5565\phi; \\
b_3 &= \exp(4.9050 - 13.8944\phi + 18.4222\phi^2 - 10.2599\phi^3); \\
b_4 &= \exp(1.4681 + 12.2584\phi - 20.7322\phi^2 + 15.8855\phi^3).
\end{aligned}$$
(12)

The shape factor ϕ is defined as

$$\phi = \frac{s}{S},$$
(13)

where s is the surface area of a sphere having the same volume as the particle and S is the actual surface area of the particle.

The additional forces $\vec{F}$ in the particle force are virtual mass and pressure gradient forces which are not important when the density of the fluid is much lower than the density of the particles. For this study, the virtual mass and pressure gradient forces are ignored.

(2)   Particle Torque Balance

Particle rotation is a natural part of particle motion and can have a significant influence on the trajectory of a particle moving in a fluid. The impact is even more pronounced for large particles with high moments of inertia. In this case, if particle rotation is disregarded in simulation studies, the resulting particle trajectories can significantly differ from the actual particle paths. The torque $\vec{T}$ results from equilibrium between the particle inertia and the drag.

$$\vec{T} = I_p \frac{d\vec{\omega}_p}{dt} = \frac{\rho_f}{2} \left(\frac{d_p}{2} \right)^5 C_\omega \vec{\Omega},$$
(14)

where I_p is the moment of inertia, $\vec{\omega}_p$ is the particle angular velocity, ρ_f is the fluid density, d_p is the particle diameter, C_ω is the rotational drag coefficient, $\vec{T}$ is the torque applied to a particle in a fluid domain and $\vec{\Omega}$ is the relative particle–fluid angular velocity calculated by:

$$\vec{\Omega} = \frac{1}{2}\nabla \times \vec{u}_f - \vec{\omega}_p. \tag{15}$$

The particles will have impact with the wall and other particles in the pipe. The collision recovery factor is obtained by Forder's recovery factor equations [30].

$$
\begin{aligned}
e_n &= 0.988 - 0.78\theta + 0.19\theta^2 - 0.024\theta^3 + 0.027\theta^4; \\
e_r &= 1 - 0.78\theta + 0.84\theta^2 - 0.21\theta^3 + 0.028\theta^4 - 0.022\theta^5,
\end{aligned}
\tag{16}
$$

where e_n is the normal recovery factor, e_r is the tangential recovery factor and θ is the impact angle.

3. Simulations

3.1. Simulation Model

The Euler–Lagrange approach, DPM and the two-way coupling method were used in the simulations. The boundary conditions and injection parameters are shown in Figure 1. The simulation pipe diameters were 70, 100 and 150 mm, and the length was 6 m. The wall roughness of seamless steel pipes used in the experiments was 0.05 mm. The particle density was 2100 kg/m^3. The solid-gas rate was the ratio of the particle's mass flow ratio between the air mass flow ratio during pneumatic conveying. The particles were injected into the pipe inlet uniformly. The transmission medium was air, which was considered an incompressible gas. The gas density was 1.225 kg/m^3 and the dynamic viscosity was 1.8×10^{-5} kg/m^3 (20 °C, 1 atm).

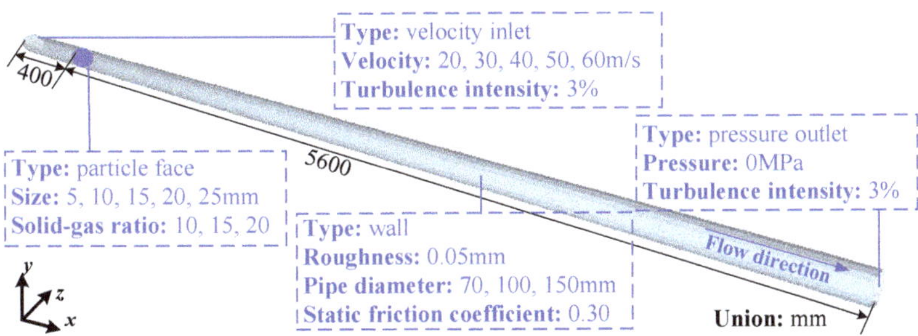

Figure 1. Simulation model.

The DPM model is based on ANSYS FLUENT and the meshes of the simulation models are the hexahedral orthogonal meshes. The simulation considered the interaction between the gas and particle phase; however, the shape characteristics of particles was ignored.

3.2. Effects of Particle Size and Flow Field Velocity on Pressure Drop

The different particle sizes and flow field velocities were used in the simulations to obtain the effects of particle size and flow field velocity on pressure drop. The simulation parameters are shown in Table 1.

Table 1. Simulations parameters.

Feeding Rate	Pipe Diameter	Roughness	Particle Density	Particle Size	Flow Field Velocity
0.5 kg/s	70 mm	0.05 mm	2100 kg/m^3	5, 10, 15, 20, 25 mm	20, 30, 40, 50, 60 m/s

As an example, the variation trend of pressure drop was analyzed by taking the simulation results under the condition of 10 mm particle size and 50 m/s flow field velocity. The variation trend in the stable conveying part in the horizontal conveying pipe is shown in Figure 2. The coupled static pressure, the coupled dynamic pressure and the coupled total pressure refer to the coupling simulation process of the coal particles and the flow field, while the pure flow static pressure, the pure flow dynamic pressure and the pure flow total pressure refer to the simulation process of pure flow fluid conveying.

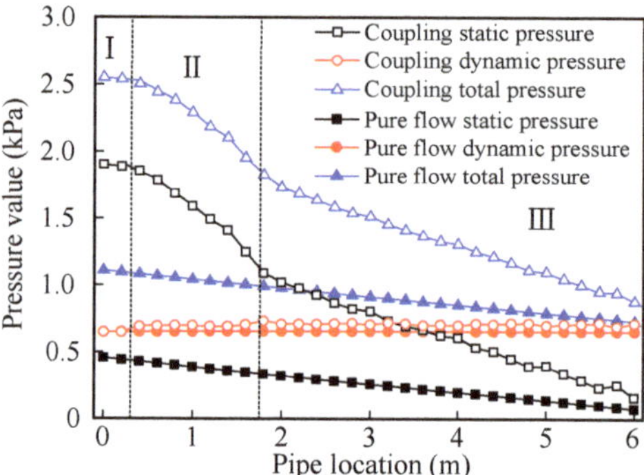

Figure 2. Pressure variation in a horizontal pipe.

Both in coupling and pure flow simulation results, the changes of dynamic pressure were small, which means there was a small change of flow field velocity. The change of the flow field total pressure was mainly due to the change of static pressure. The coupling static pressure and coupling total pressure curves can be divided into three regions. Region I is the pure flow region, region II is the particle dropping and rebounding region and region III is the stable conveying region. Region I is behind the particle factory. Since there is no particle formation, it is still the pure fluid field, meaning the slope of static pressure drop is the same as that of pure flow static pressure. At region II, the particles start to form and exchange momentum and energy with the flow field. Since particles are randomly generated in the particle factory, the interaction between the particles and the flow field is sufficient when the particles enter the flow field. There is an initial collision between particles and wall, so the static pressure of this region declines faster than other regions and the energy conversion efficiency between the flow fluid and the particles is also the highest. Region III is in a stable conveying stage-the particles have sufficient velocity and the particle–particle collisions and the particle–wall collisions are basically in a stable and eased off state. This means the static pressure drop trend of the flow field is to slow down. The unit distance static pressure drop (hereinafter referred to as the pressure drop) of region III under different particle size and flow field velocity is shown in Figure 3.

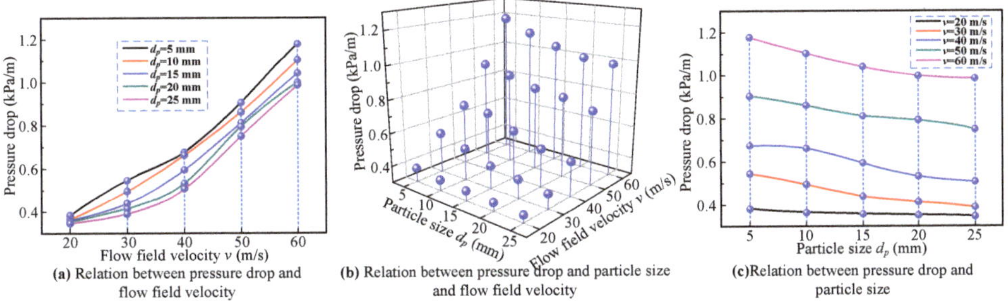

(a) Relation between pressure drop and flow field velocity

(b) Relation between pressure drop and particle size and flow field velocity

(c) Relation between pressure drop and particle size

Figure 3. Relationship between pressure drop, particle size and flow field velocity.

The pressure drop increases with the flow field velocity and decreases with the particle size. At a lower flow field velocity (v = 20 m/s), the difference in pressure drop under different sizes is small. This is because the flow field is stratified due to the small flow field velocity. Only some particles participate in the interaction with flow field, so the particle size has a small effect on the pressure drop. The stratification state of flow field disappears gradually with the flow velocity, and more particles participate in the interaction with the flow field. However, under the same feeding rate, the smaller the particle size, the higher the number of particles and the easier the interaction with the flow field. More particle collisions probability result in more energy consumption of particles. Therefore, the pressure drop is also higher. However, when the particle size is larger, the number of particles is smaller and the probability of particle collisions is less. This means the energy consumption of particles is also less, which leads the pressure to continue dropping. When the flow velocity increases further, the particle collision probability is more severe. The more energy consumption that is needed leads to a greater pressure drop.

3.3. Effects of Pipe Diameter and Solid-Gas Ratio on Pressure Drop

The different solid-gas rates and pipe diameters were used in the simulations to obtain the effects of pipe diameter and solid-gas ratio on pressure drop. The simulation parameters are shown in Table 2. The relationship between pressure drop, pipe diameter and solid-gas rate is shown in Figure 4.

Table 2. Simulation parameters.

Particle Size	Flow Field Velocity	Roughness	Solid-Gas Rate	Pipe Diameter
15 mm	60 m/s	0.05 mm	10, 15, 20	70, 100, 150 mm

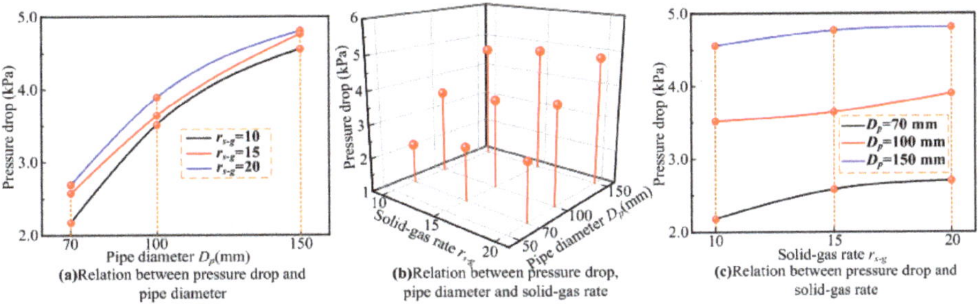

(a) Relation between pressure drop and pipe diameter

(b) Relation between pressure drop, pipe diameter and solid-gas rate

(c) Relation between pressure drop and solid-gas rate

Figure 4. Relationship between pressure drop, pipe diameter and solid-gas rate.

The pressure drop increases with the pipe diameter and the solid-gas rate. Under the same solid-gas ratio, the pressure drop increases greatly with the pipe diameter. This is because the mass flow rate of the flow field increases with the pipe diameter under the fixed flow field velocity. The mass flow rate of particles increases with the flow field due to the fixed solid-gas rate, which leads to a significant increase in pressure drop. The influence of the solid-gas rate on pressure drop is less than that of pipe diameter, but the increase of the solid-gas rate makes the mass flow rate of particles and the pressure drop of the flow field increase. The pressure drop increases slightly with the solid-gas rate under the same pipe diameter D_p = 100 and 150 mm. It is because there is enough space for the particles (size = 15 mm) to move in the large diameter pipe when the solid-gas rate increases, so the increment of collision probability is less. The pressure drop increases greatly with the solid-gas rate under the pipe diameter D_p = 70 mm. This is because the small pipe diameter and the increase of solid-gas rate both cause an increase in collision probability. The increase of the solid-gas rate leads to an increase in collision probability and also leads to high energy consumption and a drop in pressure.

4. Experimental Verification

The experimental system of horizontal pneumatic conveying is shown in Figure 1. The total length of the conveying pipelines was 10 m and the diameter was 70 mm. The test system included two pressure transducers, a signal amplifier, a data acquisition instrument and a computer. The position of two pressure transducers and one transparent pipe is shown in Figure 5.

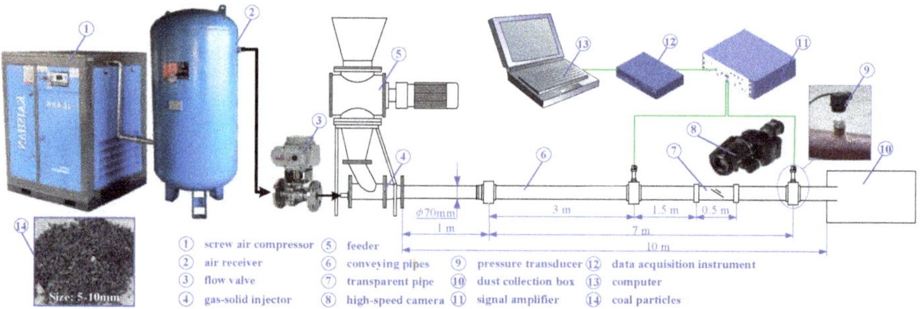

Figure 5. Experimental system of horizontal pneumatic conveying.

The air flow is pressurized by the screw air compressor and stabilized by the air receiver which enters into the gas-solid injector through the flow valve. The coal particles gain kinetic energy from air flow and are conveyed into the dust collection box through conveying pipes. There is some back pressure in the dust collection box, but far less than the pressure of air flow. Therefore, the dust collector can be considered as atmospheric pressure. The first pressure transducer is installed 3 m from the outlet of the gas-solid injector and the second is 4 m downstream of the first pressure transducer. The transparent pipe is installed 1.5 m downstream of the first pressure transducer to monitor the particle motion state.

When the output pressure of the air compressor is certain, the opening of the flow valve determines the air flow rate in the pipe. It is called a pure flow field when no particles enter. However, when the particles do enter the flow field, some of flow field dynamic pressure turns to static pressure to transfer momentum and energy to the particles. At this point, it amounts to add back pressure into the pipe. As result, the air flow rate will be reduced by this back pressure. Therefore, the air flow rate of the pure flow field is regarded as the reference standard in the experiments. When the particles enter into the flow field, the opening of the flow valve will be appropriately increased to complement the reduction.

The size of experimental coal particles was 5–10 mm and the feeding rate was controlled by the frequency converter. The mass density of the experimental coal particle was 2100 kg/m^3. The experimental scheme is shown in Table 3.

Table 3. The experimental scheme of horizontal pneumatic conveying.

No.	Output Pressure of Air Compressor	Flow Field Velocity	Feeding Rate	Solid-Gas Rate
1	0.3 MPa	21 m/s	1.12 kg/s	11.37
2	0.4 MPa	29 m/s	1.12 kg/s	8.23
3	0.5 MPa	37 m/s	1.12 kg/s	6.45
4	0.6 MPa	44 m/s	1.12 kg/s	5.43
5	0.3 MPa	21 m/s	2.25 kg/s	22.74
6	0.4 MPa	29 m/s	2.25 kg/s	16.49
7	0.5 MPa	37 m/s	2.25 kg/s	12.91
8	0.6 MPa	44 m/s	2.25 kg/s	10.85

4.1. Experiment Results

In all the experimental results, the pressure signal curves obtained by the two pressure transducers were basically the same trend, but the output pressure of the air compressor and feeding rate had obvious influence on the pressure signal curves. Due to the highest output pressure of the air compressor and the largest feeding rate in the No.8 experiment, the pressure signal curves obtained by the two pressure sensors are best visualized. Therefore, the No.8 experiment is taken as an example to analyze the trend of pressure drop in the pipe during the pneumatic conveying experiment. The static pressure signals of first and second pressure transducers are shown in Figure 6.

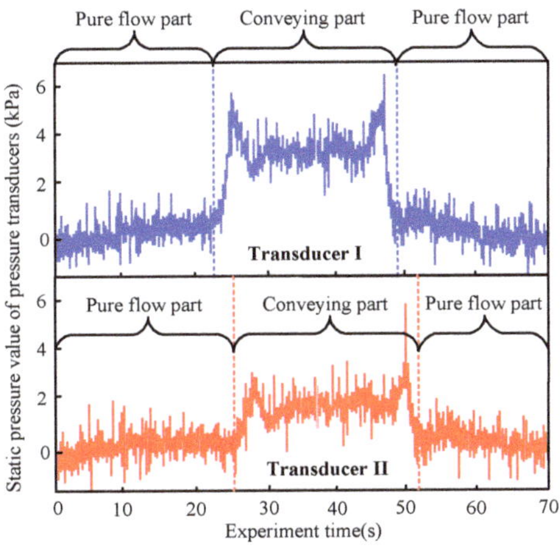

Figure 6. Pressure signal in horizontal pneumatic conveying pipe.

The pressure signal curves were clearly divided into pure flow part and conveying part. At 0–5 s, the pressure value gradually increased, which was the initial stage of single-phase gas flow in the pipe. After that, the pressure value was relatively stable, although there was noise fluctuation. Basically, the pressure value floated up and down at zero, which indicates that the air flow was fully developed and the stable flow field was formed, which is called the pure flow part. The pressure value rose rapidly and maintained its high value when the particles entered the flow field. The pressure curve has a large fluctuation with the particle conveying which eliminates the noise fluctuation. It is because the coal particles in the pipeline hinder the flow field that the pressure value of this part increased.

This is called the conveying part. Next, the pressure value gradually decreased with the decline of coal particles in the pipe, and then the pure flow part was restored.

The unit-distance pressure drop ($\Delta p/l$) of the horizontal pneumatic conveying experiment was the ratio of the pressure difference obtained by the two pressure transducers of the distance between the two measuring points, which is shown in Figure 7. According to the zero temperature drift of pressure transducers (±0.15%FS/°C), sensitivity temperature drift of pressure transducers (±0.15%FS/°C) and signal fluctuation (±1.0%) in the measurement process, the pressure error measured in the experiment (20 °C) was estimated to be ±7%. Pressure drop curves have the same trend, where the pressure drop is first reduced and then increased. The pressure drop increased with the feeding rate under the same flow field velocity.

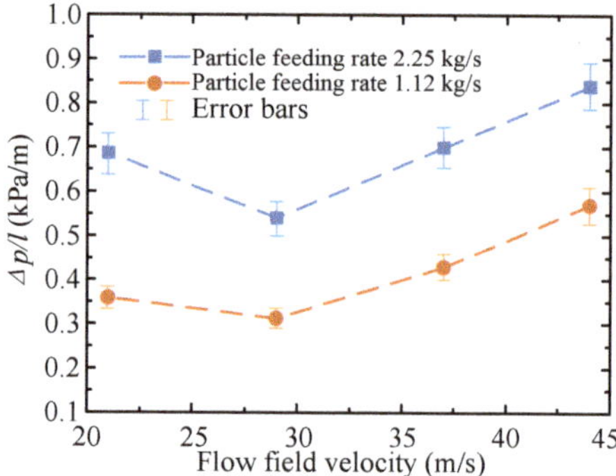

Figure 7. Relationship between pressure drop and flow field velocity.

4.2. Comparison Pressure Drop

The comparative simulations were carried out using the simulation particles with the diameter of 5–10 mm and the R-R particle size distribution [31]. The simulation scheme is the same as Table 3. The pressure drop comparison between the simulation and the experiment results is shown in Figure 8. The experiment variation trend of the pressure drop shows the parabola shape with the flow field velocity, and the simulation variation trend shows a linear growth pattern. From the numerical analysis, the pressure drop of simulation results were very different from the experimental results when the flow velocity was small. However, the simulation and experimental results changed gradually when the flow velocity increased. This indicates that the simulation results of DPM are similar to experiment results when flow velocity is large. Therefore, larger flow field velocities are used in the subsequent simulation to ensure the accuracy of the simulation results.

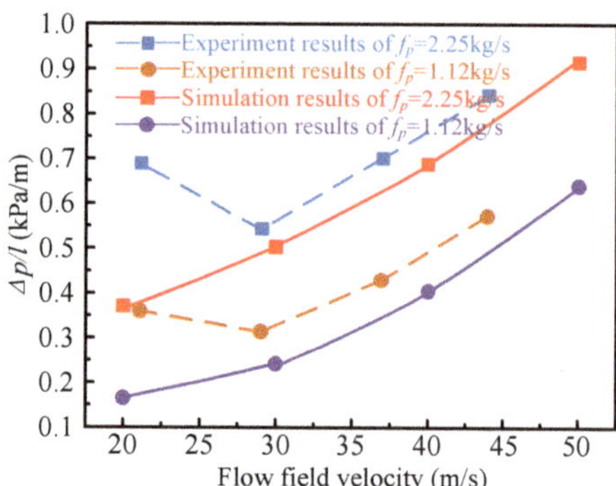

Figure 8. Comparison of pressure drop between experiment and simulation.

5. Discussion and Limitation

5.1. Discussion of Simulation Results Accuracy

The pressure drop obtained by simulation and experiments varied greatly with respect to flow field velocity. This is because the DPM used in this paper does not consider particle shape characteristics. Particle shape characteristics play a key role in particle–particle and particle–wall collisions. When the flow velocity is low, the stratification of flow field leads to more frequent particle–particle and particle–wall interactions, and some of the particles even accumulate at the pipe bottom. Therefore, the accuracy of DPM simulation results is very low at low flow field velocity. However, when the flow velocity increases, the particles get more energy from the flow field and some particles are suspended in the flow field, which leads to a decrease in collision probability. Therefore, the DPM simulation results are similar to the experimental results at the high flow field velocity.

5.2. Limitation of Experiments

Coal is an inflammable and explosive material. Coal dust will inevitably be produced in the pneumatic conveying process and there is a chance of sparks when large sized particles make an impact with the pipe wall at high speed. This in turn could cause accidents such as pipeline explosions. Therefore, under the premise of ensuring a safe experiment, the pneumatic conveying experiment of large sized particles cannot be carried out, only simulation research can be carried out.

5.3. Future Research

The goal of this paper is to find a simple and effective numerical simulation method to predict the pneumatic conveying of large coal particles. According to the results of the experimental and simulation comparison, this is only the first step in finding a suitable model. The simulation model used in this paper is only suitable for high-speed flow field, and we will continue to study and modify the simulation model to obtain an effective and fast prediction model for all conditions in the pneumatic conveying of large particles.

6. Conclusions

This paper uses the coupling method based on the Euler–Lagrange approach, DPM and particle trajectory equations to quickly predict the pressure drop in horizontal pneumatic conveying for large coal particles. The multi-factor simulations are carried out to analyze the effects of particle size, flow field velocity, solid-gas rate and pipe diameter on pressure drop. In the simulation results, the change of the flow field total pressure was mainly due to the change of static pressure. The pressure curves can be divided into three regions: the pure flow region, the particle dropping and rebounding region and the stable conveying region.

In the stable conveying region, the pressure drop increases with the flow field velocity and decreases with the particle size. At lower flow field velocity, the difference of pressure drop under different particle sizes is small. In the stable conveying region, the pressure drop increases with the pipe diameter and the solid-gas rate. The influence of the solid-gas rate on the pressure drop is less than that of the pipe diameter, but the increase of solid-gas rate makes the mass flow rate of particles and the pressure drop of flow field increase.

Comparing simulation results with verified experimental results of pressure drop, the pressure drop of simulation results greatly differs from experimental results when the flow velocity is small. However, the simulation and experimental results gradually become more similar when the flow velocity increases. This indicates that the DPM model is feasible in predicting horizontal pneumatic conveying for large coal particles at large flow field velocity.

Author Contributions: Methodology, D.Y.; software, D.Y.; validation, Y.W.; writing—original draft preparation, D.Y.; writing—review and editing, Y.W. and Z.H. All authors have read and agreed to the published version of the manuscript.

Funding: This research was funded by the National Natural Science Foundation of China (51705222), the National Natural Science Foundation of Jiangsu Province (BK20170241) and the National Natural Science Foundation of the Jiangsu Normal University (17XLR028).

Conflicts of Interest: The authors declare no conflict of interest.

References

1. Yang, D.L.; Li, J.P.; Zheng, K.H.; Jiang, H.X.; Xu, H.D.; Liu, S.Y. High-hardness alloy substituted by low hardness during drilling and cutting experiments of conical pick. *Int. J. Rock Mech. Min. Sci.* **2017**, *95*, 73–78. [CrossRef]
2. Yang, D.L.; Wang, Y.X.; Xing, B.S.; Yu, Y.T.; Wang, Y.T.; Xia, Y.T. Recent patents on roll crushing mills for selective crushing of coal and gangue. *Recent Patents Mech. Eng.* **2020**, *13*, 2–12. [CrossRef]
3. Yang, D.L.; Li, J.P.; Zheng, K.H.; Du, C.L.; Liu, S.Y. Impact-crush separation characteristics of coal and gangue. *Int. J. Coal Prep. Util.* **2018**, *38*, 127–134. [CrossRef]
4. Yang, D.L.; Xing, B.S.; Li, J.P.; Wang, Y.X. Recent patents on pressurization and dedusting for pneumatic conveying. *Recent Patents Mech. Eng.* **2018**, *11*, 180–189. [CrossRef]
5. Yang, D.L.; Li, J.P.; Xing, B.S.; Wang, Y.X. Recent patents on gangue pneumatic filling for coal auger mining method. *Recent Patents Mech. Eng.* **2018**, *11*, 31–40. [CrossRef]
6. Rabinovich, E.; Kalman, H. Flow regime diagram for vertical pneumatic conveying and fluidized bed systems. *Powder Technol.* **2011**, *207*, 119–133. [CrossRef]
7. Pahk, J.B.; Klinzing, G.E. Frictional force measurement between a single plug and the pipe wall in dense phase pneumatic conveying. *Powder Technol.* **2012**, *222*, 58–64. [CrossRef]
8. Njobuenwu, D.O.; Fairweather, M. Modelling of pipe bend erosion by dilute particle suspensions. *Comput. Chem. Eng.* **2012**, *42*, 235–247. [CrossRef]
9. Watson, R.J.; Thorpe, R.B.; Davidson, J.F. Vertical plug-flow pneumatic conveying from a fluidised bed. *Powder Technol.* **2012**, *224*, 155–161. [CrossRef]
10. Ebrahimi, M.; Crapper, M.; Ooi, J.Y. Experimental and simulation studies of dilute horizontal pneumatic conveying. *Part. Sci. Technol.* **2014**, *32*, 206–213. [CrossRef]
11. Ogata, K.; Hirose, T.; Yamashita, S. Effect of particle properties on fluidized powder conveying in a horizontal channel. *Procedia Eng.* **2015**, *102*, 968–975. [CrossRef]
12. Makwana, A.B.; Patankar, A.; Bose, M. Effect of dune formation on pressure drop in horizontal pneumatic conveying system. *Part. Sci. Technol.* **2015**, *33*, 59–66. [CrossRef]
13. Anantharaman, A.; Cahyadi, A.; Hadinoto, K.; Chew, J.W. Impact of particle diameter, density and sphericity on minimum pickup velocity of binary mixtures in gas-solid pneumatic conveying. *Powder Technol.* **2016**, *297*, 311–319. [CrossRef]
14. Rinoshika, A. Effect of oscillating soft fins on particle motion in a horizontal pneumatic conveying. *Int. J. Multiph. Flow* **2013**, *52*, 13–21. [CrossRef]
15. Rinoshika, A. Dilute pneumatic conveying of a horizontal curved 90° bend with soft fins or dune model. *Powder Technol.* **2014**, *254*, 291–298. [CrossRef]
16. Yang, D.L.; Xing, B.S.; Li, J.P.; Wang, Y.X.; Hu, N.N.; Jiang, S.B. Experiment and simulation analysis of the suspension behavior of large (5–30 mm) nonspherical particles in vertical pneumatic conveying. *Powder Technol.* **2019**, *354*, 442–455. [CrossRef]
17. Yang, D.L.; Li, J.P.; Du, C.L.; Jiang, H.; Zheng, K. Injection performance of a gas-solid injector based on the particle trajectory model. *Adv. Mater. Sci. Eng.* **2015**, *2015*, 871067. [CrossRef]
18. Yang, D.L.; Xing, B.S.; Li, J.P.; Wang, Y.X.; Gao, K.D.; Zhou, F.; Xia, Y.T.; Wang, C. Experimental study on the injection performance of the gas-solid injector for large coal particles. *Powder Technol.* **2020**, *364*, 879–888. [CrossRef]
19. Sanchez, L.; Vasquez, N.; Klinzing, G.E.; Dhodapkar, S. Characterization of bulk solids to assess dense phase pneumatic conveying. *Powder Technol.* **2003**, *138*, 93–117. [CrossRef]

20. Zhou, F.B.; Hu, S.Y.; Liu, Y.K.; Liu, C.; Xia, T.Q. CFD-DEM simulation of the pneumatic conveying of fine particles through a horizontal slit. *Particuology* **2014**, *16*, 196–205. [CrossRef]
21. Watano, S. Mechanism and control of electrification in pneumatic conveying of powders. *Chem. Eng. Sci.* **2006**, *61*, 2271–2278. [CrossRef]
22. Choi, K.; Endo, Y.; Suzuki, T. Experimental study on electrostatic charges and discharges inside storage silo during loading of polypropylene powders. *Powder Technol.* **2018**, *331*, 68–73. [CrossRef]
23. Güner, M. Pneumatic conveying characteristics of some agricultural seeds. *J. Food Eng.* **2007**, *80*, 904–913. [CrossRef]
24. Ghafori, H.; Sharifi, M. Numerical and experimental study of an innovative design of elbow in the pipe line of a pneumatic conveying system. *Powder Technol.* **2018**, *331*, 171–178. [CrossRef]
25. Zhou, J.W.; Liu, Y.; Du, C.L.; Liu, S.Y.; Li, J.P. Numerical study of coarse coal particle breakage in pneumatic conveying. *Particuology* **2018**, *38*, 204–214. [CrossRef]
26. Zhou, J.W.; Liu, Y.; Du, C.L.; Liu, S.Y. Effect of the particle shape and swirling intensity on the breakage of lump coal particle in pneumatic conveying. *Powder Technol.* **2017**, *317*, 438–448. [CrossRef]
27. Shih, T.H.; Liou, W.W.; Shabbir, A.; Yang, Z.G.; Zhu, J. A new k-εeddy-viscosity model for high reynolds number turbulent flows. *Comput. Fluids* **1995**, *24*, 227–238. [CrossRef]
28. Gosman, A.D.; Ioannides, E. Aspects of computer simulation of liquid-fuelled combustors. *J. Energy* **1983**, *7*, 482–490. [CrossRef]
29. Haider, A.; Levenspiel, O. Drag coefficient and terminal velocity of spherical and nonspherical particles. *Powder Technol.* **1989**, *58*, 63–70. [CrossRef]
30. Dritselis, C.D.; Vlachos, N.S. Large eddy simulation of gas-particle turbulent channel flow with momentum exchange between the phases. *Int. J. Multiph. Flow* **2011**, *37*, 706–721. [CrossRef]
31. Yang, D.L.; Li, J.P.; Du, C.L.; Zheng, K.H.; Liu, S.Y. Particle size distribution of coal and gangue after impact-crush separation. *J. Cent. South Univ.* **2017**, *24*, 1252–1262. [CrossRef]

Article

CFD Optimization Process of a Lateral Inlet/Outlet Diffusion Part of a Pumped Hydroelectric Storage Based on Optimal Surrogate Models

Xueping Gao [1], Hongtao Zhu [1], Han Zhang [1], Bowen Sun [1,*], Zixue Qin [1] and Ye Tian [2]

[1] State Key Laboratory of Hydraulic Engineering Simulation and Safety, Tianjin University, Tianjin 300072, China; xpgao@tju.edu.cn (X.G.); htzhutju@163.com (H.Z.); tjuzhang@tju.edu.cn (H.Z.); zxqintju@163.com (Z.Q.)

[2] China Water Resources Beifang Investigation, Design and Research Co. Ltd. (BIDR), Tianjin 300072, China; tjtianye@sina.com

* Correspondence: bwsun@tju.edu.cn; Tel.: +86-022-27890287

Received: 5 March 2019; Accepted: 4 April 2019; Published: 10 April 2019

Abstract: The lateral inlet/outlet plays a critical role in the connecting tunnels of a water delivery system in a pumped hydroelectric storage (PHES). Therefore, the shape of the inlet/outlet was improved through computational fluid dynamics (CFD) optimization based on optimal surrogate models. The CFD method applied in this paper was validated by a physical experiment that was carefully designed to meet bidirectional flow requirements. To determine a good compromise between the generation and pump mode, reasonable weights were defined to better evaluate the overall performance. In order to find suitable surrogate models to improve the optimization process, the best suited surrogate models were identified by an optimal model selection method. The optimal configurations of the surrogate model for the head loss and the velocity distribution coefficient were the Kriging model with a Gaussian kernel and the Kriging model with an Exponential kernel, respectively. Finally, a multi-objective surrogate-based optimization method was used to determine the optimum design. The overall head loss coefficient and velocity distribution coefficients were 0.248 and 1.416. Compared with the original shape, the coefficients decrease by 6.42% and 40.28%, respectively. The methods and findings of this work may provide practical guidelines for designers and researchers.

Keywords: pumped hydroelectric storage; inlet/outlet; surrogate model selection; multi-objective optimization process

---

1. Introduction

In recent decades, many electricity generation problems have been caused by their high production cost and environmental impact [1]. It is generally accepted that renewable energy can provide promising solutions for the problems, and much research has been focused on numerous techniques, such as hydropower [2–4], fuel cells [5], biofuels [6], wind power [7–9], and solar power generation [10]. Pumped hydroelectric storage (PHES) has attracted widespread interest because of its great flexibility and storage capacity in improving grid stability of other renewable energy sources [11,12]. The lateral inlet/outlet plays a critical role in the connecting tunnels of the water delivery system in a PHES, and its flow conditions can affect the economic benefit and operation stability of the PHES to a great extent. The PHES operates under pumped or generation conditions according to the command, and the water flows between the upper and lower reservoir through the inlet/outlet. Compared with ordinary hydropower stations, the hydraulic requirements for the inlet/outlet structure of the PHES require more attention.

Much research in recent years has focused on the hydraulic performances of the inlet/outlet experimentally and numerically. Müller et al. [13] carried out prototype measurements by Acoustic Doppler Current Profiler (ADCP) to study flow velocities in the reservoir; subsequently a numerical model was built to further understand the flow development near the inlet/outlet. In order to ensure satisfactory hydraulic performances, Bermúdez et al. [14] conducted numerical and physical model studies on the inlet/outlet of the Belesar-III power station in Northwest Spain. The initial design was optimized by numerical method, and the new design presented a more homogenous flow distribution and a reduced head loss. Cai et al. [15] performed an experimental study on the effects of the shape and position of separation piers on the flow characteristics with two operations. Gao et al. [16] tried to improve the velocity distribution at the intake-outlet orifices through 20 types of shape optimization experiments. Sun et al. [17] found the incorrect arrangement of separation piers probably contributes to disadvantageous velocity distribution for the trash rack. Ye et al. [18] simulated turbulent flow in a lateral inlet/outlet to explore the diffusion segment shape that can obtain better velocity and discharge distributions. Based on the validation of the experiment, flows in the lateral inlet/outlet were numerically simulated focusing on the variations in discharge distributions over different orifices; the results showed that the flow non-uniformity was improved successfully by adjusting the diffusion segment and width between the separation piers [19]. The investigations indicate that computational fluid dynamics (CFD) has been used frequently to optimize the structure parameters of the inlet/outlet that can easily influence the hydraulic objectives. However, the process mentioned above is a trial-and-error approach requiring intensive work and a lot of time [20], and the results may be subjective to some extent.

CFD coupled with optimization techniques has been used to determine the best design scheme for wind turbine devices [21–23], and the adopted methods provide a reference for inlet/outlet optimization. For complex optimization problems, two methods have been generally applied to achieve optimization. When there are many individual objective functions, the functions can be combined into a single composite one; nonetheless it can be difficult to determine the proper and typical selection of the weights. In addition, the combination method would provide only a single result that cannot be chosen to make a trade-off. Therefore, multi-objective evolutionary algorithms (MOEAs) are more desired by decision makers. For MOEAs, it is common to use the concept of Pareto solutions. Pareto solutions consist of many non-dominated solutions where some gain in objectives always causes some sacrifice to others. MOEAs, such as Non-dominated Sorting Genetic Algorithm II (NSGA-II) [24,25], can offer many Pareto results in one single simulation process and have been extensively applied in engineering optimization.

Although MOEAs can obtain many solutions, these algorithms also require an even larger number of simulations and an unacceptable computation time in practice. Thus, multi-objective optimization based on the surrogate methods is receiving increasingly more support and attention. These approaches present objective functions by surrogate models, replacing expensive high-fidelity design simulations. The model is an effective engineering technique that can considerably save computing expense while still providing reasonably accuracy. Therefore, the multi-objective surrogate-based optimization method makes it more feasible to perform exhaustive global searches in the design space for complicated shape optimization. The optimization result will be more satisfactory because the exploration is not subjected to a limited number of simulating data and computing resources. General surrogate models contain Response Surface Methodology (RSM) [26,27], Radial Basis Function (RBF) [28–32], and Kriging [33–35]. However, owing to the diverse functional characteristics of the popular models, it is important to choose the most appropriate models for the responses of hydraulic objectives to diffusion segment parameters. For that reason, a recently surrogated model selection method proposed by Mehmani et al. [36], called the Concurrent Surrogate Model Selection (COSMOS), was adopted to construct suitable models in this paper.

Although a multi-objective CFD optimization process for the inlet/outlet was developed in previous works [37,38], there are some drawbacks in the original process: (1) The CFD method was indirectly validated only by comparing the result with that of a published reference [18], which weakened the reliability of the optimization process. (2) Only a single surrogated model (RSM) was considered to accelerate the optimization procedure. However, there is no universal surrogate model for every practical problem, and suitable model selection is critical for accuracy. (3) The objectives for the pump and generation mode were simply added up to estimate the total performance. Nevertheless, the significance of the pump mode and generation mode is probably not equal in many cases. Therefore, in this study a physical model experiment was well-designed to meet the requirement of bi-directional flow, and a more extensive validation of the CFD model was performed. Based on the validated CFD model, the COSMOS method was applied to select suitable surrogate models for the objectives of the inlet/outlet, to further improve the accuracy of the original optimization process. In addition, in order to better evaluate the overall hydraulic performance, the selection of objective functions was reconsidered and weight coefficients of different operation modes were introduced. In the present article, the shape of an inlet/outlet was improved through CFD optimization based on optimal surrogate models. The aim of this work was to apply the optimal surrogate optimization process to determine the optimum shape of the inlet/outlet, to simultaneously obtain better hydraulic performances under bidirectional flow conditions.

2. Problem Description

The typical shape of the lateral intake is shown in Figure 1. When the PHES is in the generation mode, water flows into the lower reservoir through the tunnel and the inlet/outlet structure. Because water flows out from the inlet/outlet in this mode, the phenomenon is also called "outflow". Along the outflow direction, the structure generally contains a transition part (a structure connecting a circular cross-section tunnel and a diffusion part), a diffusion part, a rectification part, and an anti-swirl part.

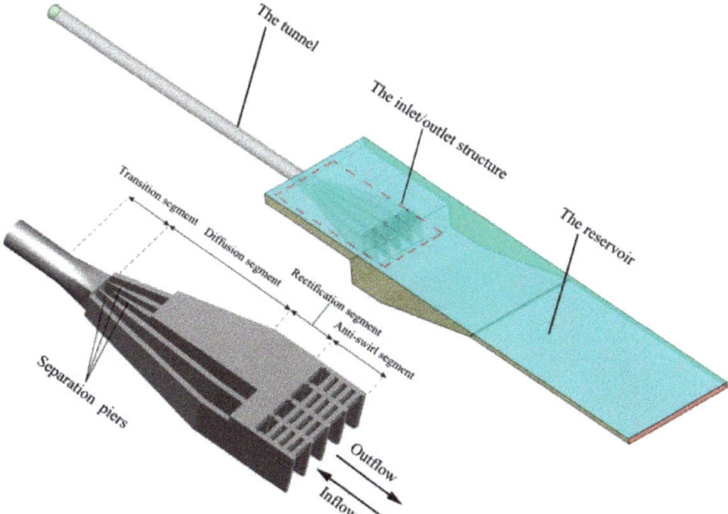

Figure 1. General structure of the lateral intake-outlet.

The diffusion part is very sensitive to the performance of the intake, and it is generally improved for the structure design. Figure 2 present flow passages and basic parameters in the diffusion part. This parameters include the length of the diffusion part (L_{DS}),height of the inlet/outlet orifices ($H_{I/O}$),wide of the separation pier in the middle orifices (W_{DSM}),width of the separation pier in the side orifices (W_{DSS}),horizontal diffusion angle (η), and vertical angle (θ). The left side displays the 3D model for the flow passage. On the right side, plane sections of the model are demonstrated, and the flow passage and separation piers are marked out with light grey and dark grey, respectively. It should be noted that the tunnel diameter D in the PHES is generally decided beforehand by the difference of the water head between the upper and lower reservoir, the installed capacity of turbines, topographic and geological conditions etc. For the inlet/outlet optimization, the value of D is fixed (D = 7.2 m in the current study) because a change may need adjustment of the whole layout of the water conveyance system resulting in excessive costs. Besides, the number of orifices (4) is also fixed in view of it being the most commonly used form of the lateral inlet/outlet in the PHES of China. For more information about the shape parameters, readers can refer to the previous work [38].

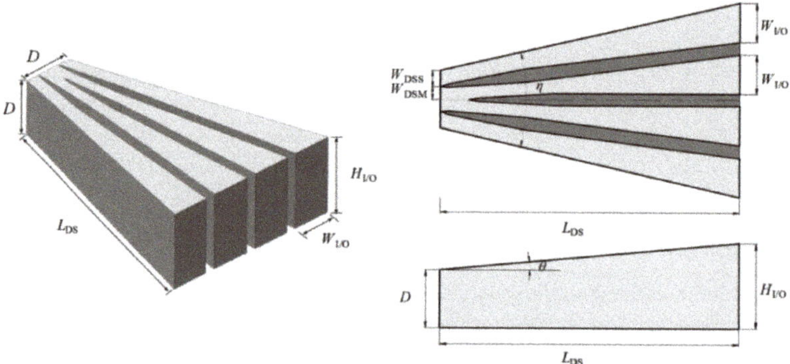

Figure 2. Basic parameters for the diffusion part.

The lateral inlet/outlet hydraulic performance is characterized by three parameters: the head loss, the velocity, and the flowrate distribution [37–39]. According to the Bernoulli equation, the head loss is computed as:

$$h_{0\text{-}1} = \nabla_0 - \nabla_1 - \frac{\alpha v^2}{2g} \tag{1}$$

$$h_{1\text{-}0} = \nabla_1 - \nabla_0 + \frac{\alpha v^2}{2g} \tag{2}$$

$$\xi = \frac{2gh_f}{\alpha v^2} \tag{3}$$

where the head loss $h_f = h_{0\text{-}1}$ and $h_f = h_{1\text{-}0}$ are used for the inflow and outflow mode, respectively; ξ is the coefficient of the head loss; ∇_0 and ∇_1 are the piezometric head in the reservoir and tunnel; v is the bulk velocity at the boundary of the tunnel; α is the kinetic energy correction coefficient.

Numerous vibration-caused trash rack failures have occurred because the trash rack may be exposed to high velocities at the orifice cross-section [40]. Hence C_V is proposed as a typical index evaluating the uniformity of the flow velocity at orifices [39,41], and a lower C_V implies a better velocity distribution in the inlet/outlet. Considering there are four orifices divided by the separation pier, C_V could change with different orifices due to the influence of horizontal diffusion. In order to represent the overall performance of the velocity distribution, the worst case i.e., the maximum value of C_V among the four orifices, is selected. Thus C_V is defined as:

$$C_V = \max\left(\frac{v_{\max,i}}{v_i}\right) \tag{4}$$

where $v_{\max}$ is the maximum value among the velocities at the orifice cross-section; i indicates the index of orifices; v indicates the average velocity at the orifice cross-section.

In order to further improve the discharge distribution in the horizontal direction, the separation piers are designed to divide the flow passage into sub-tunnels. Then C_Q is used to estimate the degree of flux uniformity by comparing the discharge between adjacent orifices, which can be defined as:

$$C_Q = \frac{\max(Q_{i,i+1}) - \min(Q_{i,i+1})}{\min(Q_{i,i+1})} \times 100\% \tag{5}$$

where Q_i indicates the rate of flow at the different orifices ($i = 1, 2, 3$).

For the purpose of determining a good compromise between the two modes of operation, these parameters under inflow and outflow conditions should be carefully combined. In a recent study, Gao et al. [38] simply added up the parameters for the pump mode and the generation mode to estimate the total performance. However, the performance of the generation mode is probably more significant than that of the pump mode in many cases. The pumped water can be stored in the upper reservoir under the pump mode during the low load period of the power grid, while under the generation mode the water will flow into the lower reservoir to generate electricity during the peak load period. Designers tend to pay more attention to the head loss under the generation mode in order to achieve higher economic benefits. Therefore, a more reasonable approach is to use weights to deal with the combinations of the two modes. For that reason, the new combined objectives to evaluate the overall performance of the inlet/outlet are defined as follows:

$$\xi_{\text{total}} = \omega_1 \xi_{\text{in}} + \omega_2 \xi_{\text{out}} \tag{6}$$

$$C_{V,\text{total}} = \omega_1 C_{V,\text{in}} + \omega_2 C_{V,\text{out}} \tag{7}$$

where ξ_{total} and $C_{V,\text{total}}$ are the overall head loss coefficient and velocity distribution coefficient, respectively; ω_1 and ω_2 are the weight coefficients of the objective function under the conditions of inflow and outflow, and $\omega_1 + \omega_2 = 1$. In this article, the values of ω_1 and ω_2 are taken as 0.33 and 0.67 based on the specific engineering designer's suggestion. However, they can be flexibly selected according to different projects.

Another drawback of their study [37,38] is the definition for the overall C_Q, which is expressed as the difference of C_Q between the pump and generation mode. $C_{Q,\text{in}}$ is the discharge uneven distribution coefficient under inflow condition, $C_{Q,\text{out}}$ is the discharge uneven distribution coefficient under outflow condition. When it is regarded as an objective function, it seeks to minimize the difference between $C_{Q,\text{in}}$ and $C_{Q,\text{out}}$, rather than $C_{Q,\text{in}}$ and $C_{Q,\text{out}}$ themselves. If there are some bad points in the design space, on which the values of both $C_{Q,\text{in}}$ and $C_{Q,\text{out}}$ may be large, the difference (i.e., the overall C_Q) may be small. Besides, according to the guideline the discharge distribution is sufficiently uniform when C_Q is less than 10% [39]. As a result it is more appropriate to choose C_Q as the constraint, and the head loss and velocity distribution (ξ_{total}, $C_{V,\text{total}}$) can be optimized more flexibly.

Therefore, the objective functions were set to find the minimum values for ξ and C_V, while C_Q, η, and θ were treated as constraint conditions. Table 1 lists the design parameters and constraints in this study.

Table 1. Design parameters and constraints.

Item	Symbol	Range
Design parameters	L_{DS}	34–46 m
	$H_{I/O}$	8–11 m
	W_{DSM}	1.5–1.7 m
Constraints	η	25–45°
	θ	1–5°
	$C_{Q,\text{in}}, C_{Q,\text{out}}$	<10%

3. Numerical Model and Validation

3.1. Numerical Model

Figure 3 shows a detailed view of the structured grids in the research area. To increase the accuracy of the simulation, boundary layer grids were added to ensure that the dimensionless wall distance $y+$ of the first points near the walls was between 30 and 300. The distribution of $y+$ values for the inlet/outlet is shown in Figure 4.

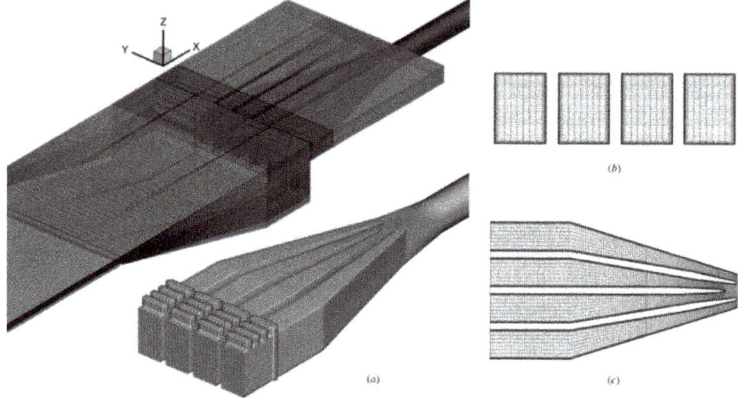

Figure 3. Grids generation for the computational domain: (**a**) grids on the boundary; (**b**) grids at the orifice cross-section, and (**c**) grids on the orifice horizontal section.

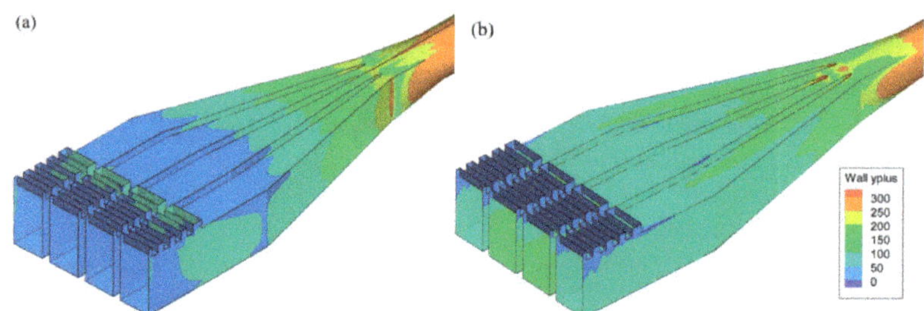

Figure 4. The Distribution of y+ values for the inlet/outlet: (**a**) the inflow condition; (**b**) the outflow condition.

Considering the accuracy and efficiency of the CFD simulations, it is important to work out a reasonable number of mesh elements [42–44]. Based on a previous grid independence study [38], a grid amount of 1.4×10^6 was sufficient for CFD simulation, and similar grid sizes were thus adopted in this article.

As presented in Figure 1, a three-dimensional computational domain far larger than the inlet/outlet structure was developed to accurately simulate the flow pattern in the inlet/outlet. As to the boundary conditions, a uniform velocity was prescribed at the tunnel boundary, and a pressure condition was imposed at the reservoir boundary. The flow discharge changed slightly according to the operating conditions of the reversible turbines. A summary of boundary conditions for the Computational Fluid Dynamics (CFD) case under the generation and pump mode is listed in Table 2.

Table 2. Boundary conditions applied in the inlet/outlet CFD simulation.

Operation Modes	Parameters	Notes	Values
Outflow	Inlet	The inlet of the pipe was defined as the velocity inlet.	$v_d = 3.738$ m/s
	Outlet	The reservoir boundary was defined as the pressure outlet.	Relative pressure: 0 Pa
Inflow	Inlet	The reservoir boundary was defined as the pressure inlet.	Relative pressure: 0 Pa
	Outlet	The inlet of the pipe was defined as the velocity outlet.	$v_d = 4.376$ m/s

In the present study, the Reynolds Averaged Navier Stokes (RANS) equations were solved using ANSYS Fluent 17.0 [45]. The simulation was performed in a segregated manner adopting the realizable k-ε model combined with the wall function. The Navier–Stokes equations were solved by means of pressure implicit with splitting of the operator's algorithm, and a second-order upwind method was employed to realize the discretization. The converged solution was obtained after the residual levels became less than 1×10^{-4}. For detailed information about the governing equation and numerical settings, the reader can refer to references [38,45,46].

3.2. Experimental Setup

The CFD method in the original study [38] was validated by comparing the result with that of a published reference [18]. Although the results were found to be in good agreement, there were obvious differences in the velocity distribution between the case in the reference [18] and the case in this study, which weakened the reliability of the validation in the original study. Therefore, the CFD method still needs to be validated further, and a physical model experiment adopting a scale of 1/60 was conducted in this paper. As shown in Figure 5, an experimental setup was established which was 5.2 m long, 1.2 m wide, and 0.6 m high.

Figure 5. The experimental setup of the inlet/outlet.

In order to show the experiment more clearly, the side view of the layout plan is illustrated in Figure 6. Control valves and electromagnetic flowmeters with an accuracy of 0.5% for the measured ranges were used to control the flow discharge. By switching the different control valves, bidirectional flow conditions could be realized in the experiment. In the inflow condition, the inflow control valves were open, while the outflow control valves were closed. Water from the high constant-head tower flowed into the reservoir channel through the pipe. Before the water entered the channel, a device was set up to align the flow pattern. Finally water flowed into a lower collecting tank, from which it was pumped back to the constant-head tank using a centrifugal pump. While in the outflow condition, the inflow control valves were closed, and the outflow control valves were open. Water was conveyed to the reservoir tank through a horizontal pipe with a length 50D that was long enough to guarantee full development of the flow in the pipe. The discharge in the inflow condition was 4.359 L/s, corresponding to a Reynolds number Re = 46,277 in the tunnel; while in the outflow condition, the discharge was 5.795 L/s and the Reynolds number was 61,519.

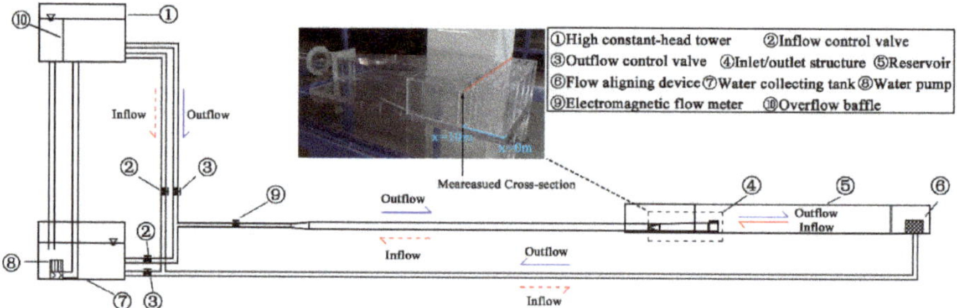

Figure 6. The side layout plan for the experimental system of the inlet/outlet.

A Vectrino ADV instrument, developed by Nortek AS, was used to instantaneously measure stream-wise velocity of the diversion orifices at various depths. Based on the Doppler shift principle, the ADV instrument can measure velocities with ±1% accuracy in a measurement range of 1 mm/s. Each measurement point with a sampling volume of 7 mm worked at a sampling frequency of 50 Hz for 60 s of sampling time.

3.3. Validation of the CFD Model

The head loss coefficient in the inflow condition is obviously smaller than that in the outflow condition. In the inflow condition, the coefficients obtained from numerical simulation and experiment were 0.164 and 0.171 respectively, and their relative error was 4.09%. The coefficients of the simulation and experiment in the outflow condition were 0.315 and 0.338, and the relative error was 6.81%. Comparing the simulation results with the experimental data, although the numerical model slightly underestimates the coefficients, it is still acceptable to predict the results of the head losses.

For the purpose of comprehensively comparing the velocity distribution at the orifice cross-section (x = 10 m), there are three measured lines with equal distance at each orifice. As shown in Figure 7, the velocity data are obtained along the measured lines (A—A, B—B and C—C) at the cross-section.

Figure 7. Measured lines at the cross-section (x = 10 m).

Figures 8 and 9 compare the numerical solutions with the experimental ones for the velocity distribution in both conditions. In the legend, M and S indicate the middle and side orifice, respectively; A, B, and C indicate the measured lines A—A, B—B, and C—C, respectively. For example, M-A represents the measured line A—A at the middle orifice. The *y*-axis indicates the relative height of the measured point normalized by the orifice height H, and the *x*-axis indicates the relative velocity also normalized by the average velocity of the measured line.

Figure 8a demonstrates that a good matching between the two approaches is obtained in the middle orifice. The results of both methods show that velocities among the three lines are very close to each other, and the distribution is also uniform along the height. In the side orifice, the numerical results are also in good agreement with the experimental ones. There are only slight differences in the velocity distribution among the measured lines. Compared with the flow pattern in the middle orifice, the more obvious change is that the mainstream appears at the lower part of the orifice, and the range of *y*/H is about 0.13–0.52.

While in the outflow condition, differences of the velocity distribution among the measured lines are also small in the side orifice. However, in the middle orifice the experimental results show that there are obvious differences among the velocities of three measured lines, which is also predicted by the simulation. In addition, the velocity distribution along the height is also not uniform, and negative velocities are generated at the top. For the M-B line, the *y*/H range of the negative region obtained from the experiment is about 0.8–1.0, and the numerical range is about 0.7–1.0. This difference is partly due to errors caused by the limited number of measured points in the experiment, and partly due to the slight overestimation for the strength of flow separation by the simulation. At the bottom of the orifice, the mainstream in the experiment is located a little higher than that in the simulation. Although there are some minor deviations in the outflow velocity distribution, the agreement is quite acceptable considering that the more complex flow separation occurs in the outflow condition.

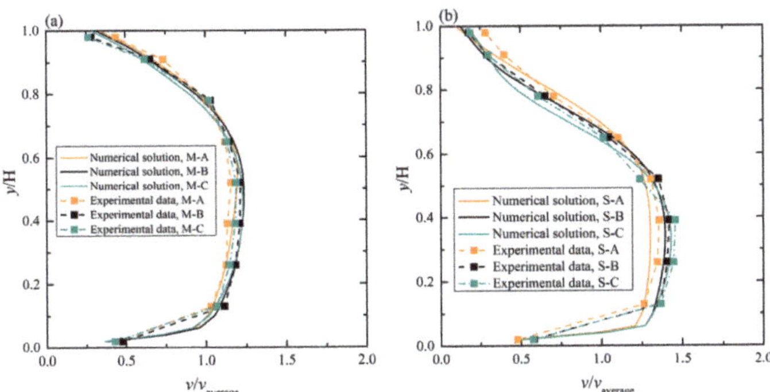

Figure 8. Validation for the velocity distribution in the inflow condition: (**a**) middle orifice; (**b**) side orifice.

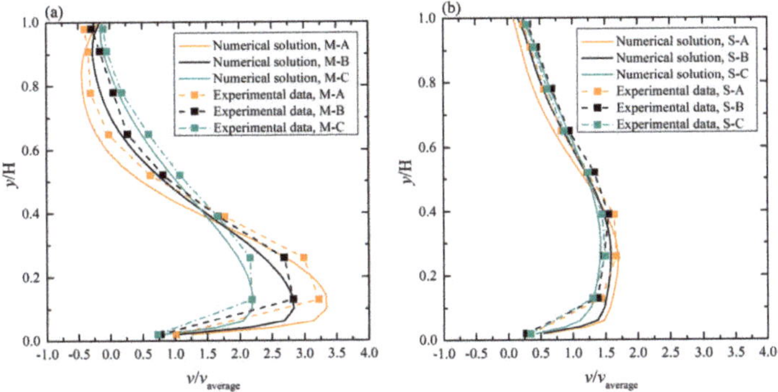

Figure 9. Validation for the velocity distribution in the outflow condition: (**a**) middle orifice; (**b**) side orifice.

4. Optimization Methodology

4.1. Surrogate Models Selection

The optimization algorithm coupled with the computationally intensive simulation requires intensive work and a lot of time. It is necessary to adopt the surrogate model because the CFD-optimization procedure requires too many computing resources, which weakens its application in practice. Therefore, surrogate models are employed to engage the optimization procedure in the paper. There are several surrogate model types as listed in Table 3, such as Response Surface Methodology (RSM), Radial Basis Function (RBF), and Kriging, which have been extensively used in recent years to solve engineering optimization design. Unlike other approaches, RSM applies simple algebraic expressions to generate the function relationship, and it is very suitable for complex engineering applications [47]. In addition, RBF has the best approach for high dimensionality optimization problems according to a literature survey [48]. For low dimensional optimization problems, the Kriging method can achieve good results in both accuracy and time. According to a survey of popular surrogate models in similar situations, RSM, RBF, and Kriging are considered as surrogate candidates in the researches.

Table 3. A list of shape optimization investigated using surrogate models in recent years.

Author	Design Parameter	Objective Function	Sampling Points	Surrogate Model	Optimization Algorithm
Zhou et al. [49]	3	2	30	RSM	NSGA-II
Jiang et al. [24]	10	2	150	RSM	NSGA-II
Yang et al. [26]	3	1	20	RSM	GA
Brar et al. [50]	3	3	30	RBF	NSGA-II
Li et al. [51]	5	1	60	RBF	GA
Zhang et al. [52]	16	2	262	Kriging	NSGA-II
Liu et al. [33]	4	2	50	Kriging	NSGA-II
Singh et al. [53]	7	2	71	RBF, Kriging	NSGA-II
Wang et al. [54]	7	2	80	RSM, RBF, Kriging	NSGA-II
Halder et al. [55]	2	1	16	RSM, RBF, Kriging	GA

When employing the surrogate model in the hydraulic optimization procedure, the challenge is how to create a model which is as accurate as possible. As mentioned above, RSM, RBF, and Kriging can be used to generate surrogate models, and their parameters are listed in Table 4. The kernel function and hyper-parameter also need to be chosen carefully, which requires an effective surrogate model selection method.

As we know, existing surrogate model selection methods generally contain three levels: selecting a model type, selecting a kernel function, and optimizing the hyper-parameters. Some researchers have used different error measures to separately select the model type, kernel function, or hyper-parameter [56–58]. With the intention of performing thorough selection, the Predictive Estimation of Model Fidelity (PEMF) was suggested by Mehmani et al. [59]. Furthermore, Mehmani et al. [36] advanced the Concurrent Surrogate Model Selection (COSMOS) on the foundation of the PEMF, and applied it to select suitable models by minimizing the model error. The error selection criterion is based on different situations and preferences. For the exploration of design space and parameter analysis, the median error can be used. On the other hand, it is more appropriate to choose the conservative maximum error for the structural safety optimization (e.g., to model vehicle crash simulation). Therefore, these criteria might be mutually conflicting or mutually promoting [36]. Considering the optimization for the hydraulic performance of the inlet/outlet may have certain requirements for the above two situations, the median and maximum errors of the surrogate model are treated as two selection criteria estimated by the PEMF method. Moreover, the framework will solve the multi-objective problem (minimizing the errors) to find the optimal models through the One-Step technique. In the technique, a single mixed integer nonlinear programming problem

(MINLP) can be solved by the NSGA-II. Considering different models have different numbers of kernels and hyper-parameters, the candidates are classed as three smaller categories according to the hyper-parameters [36]. The computation of the Kriging model is performed with the MATLAB toolbox DACE [60], as it has been widely used in many applications concerning the Kriging model. The problem is usually anisotropic, which is accounted for in the construction of the surrogate models.

Table 4. Parameters of surrogate model candidates.

Model Type	Kernel Function	Hyper-Parameter	Lower/Upper Bounds
Response surface methodology	-	-	-
Radial Basis Function	Linear	-	-
	Cubic	-	-
	Gaussian	Shape parameter, α	$0.1 < \alpha < 3$
	Multiquadric	Shape parameter, α	$0.1 < \alpha < 3$
Kriging	Linear	Correlation parameter, β	$0.1 < \beta < 20$
	Exponential	Correlation parameter, β	$0.1 < \beta < 20$
	Gaussian	Correlation parameter, β	$0.1 < \beta < 20$
	Spherical	Correlation parameter, β	$0.1 < \beta < 20$

4.2. Optimization Algorithm

Extensive literature based on the NSGA-II [61] has been reported in engineering applications (listed in Table 3), which is also chosen for this investigation. In NSGA-II, a random parent population is initially produced, and individuals are sorted according to rank and crowding distance. Then the optimum resolutions are chosen to generate offspring populations using genetic operators that are the same in the standard GA. Specifically, a binary tournament selection, Simulated Binary Crossover (SBX) and polynomial mutation are employed in NSGA-II. SBX simulates the binary crossover observed in nature, and the crossover distribution index determines how far away the children solutions are from their parents [62]. Deb et al. [63] suggested a polynomial mutation operator with a mutation distribution index that can control the amount of perturbation in a variable. Table 5 lists the parameters for the optimization process.

Table 5. The parameters for the optimization process.

Parameter	Values
Population size (even value)	100
Number of Generations	100
Crossover Probability	0.9
Crossover Distribution Index	10.0
Mutation Distribution Index	20.0

To pick the optimum point from the Pareto frontier, Technique for Order Preference by Similarity to Ideal Solution (TOPSIS) is implemented in this article. TOPSIS, proposed by Hwang et al. [64], is a ranking method that can rank possible solutions on the Pareto frontier and select a trade-off optimum by measuring Euclidean distances.

4.3. Optimization Process Scheme

The multi-objective optimization process is built by combining the CFD numerical simulation, the design of experiment (DOE), the surrogate model selection based on the COSMOS, and the optimization method NSGA-II. The flow diagram of the process is presented in Figure 10.

First, the Optimized Latin Hypercube Sampling (OLHS) algorithm is applied in the Design of Experiment (DOE) to choose representative points. Then sample points are simulated using the CFD method, and the results are exacted to evaluate the objectives of the hydraulic performance. Based on the database of the DOE, a suitable surrogate model is constructed using COSMOS to improve the optimization efficiency and precision. The different surrogate models, such as RSM, RBF, and Kriging, are generated until the PEMF is deemed acceptable. Finally, NSGA-II is adopted to conduct the optimization. If the objective function does not change, or the maximum iteration number is reached, the optimization is terminated. Otherwise, the process will continue to loop with a new set of proposed design parameters.

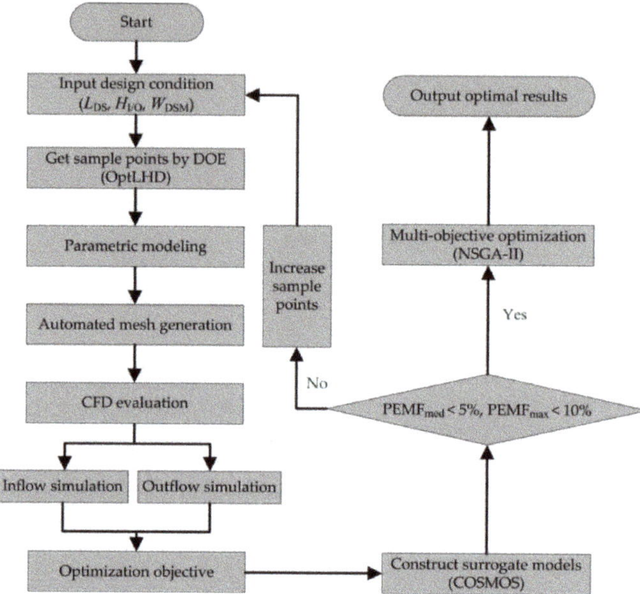

Figure 10. Flow diagram of the multi-objective optimization process.

5. Results and Discussion

5.1. Optimal Surrogate Models

In the present study, COSMOS is applied to choose the appropriate surrogate models. First of all, OLHS as an efficient sampling strategy is applied in DOE sampling, in which 60 points are randomly selected to construct these surrogate models. The Predictive Estimation of Model Fidelity (PEMF) method is chosen to check the accuracies of these surrogate models.

By solving the MINLP problem with NSGA-II, the final results of the models for ξ, C_V, $C_{Q,in}$ and $C_{Q,out}$ are presented in Figure 11. Φ_i indicates the surrogate models containing i hyper-parameter(s). β_H, β_L, and β_W indicate the correlation parameter of the design variable $H_{I/O}$, L_{DS}, and W_{DSM}, respectively. A Pareto frontier is provided in each figure, and a trade-off solution is selected under comprehensive consideration of median and maximum errors. In Figure 11a, optimal parameters of the surrogate model of ξ are: the Kriging with a Gaussian kernel and $\beta_H = 2.20$, $\beta_L = 9.41$, $\beta_W = 0.11$. As shown in Figure 11b, the optimal configuration of C_V is: the Kriging model with an Exponential kernel and $\beta_H = 1.05$, $\beta_L = 0.14$, $\beta_W = 0.13$. Figure 11c shows the optimal configuration of $C_{Q,in}$ is: the Kriging model with a Gaussian kernel and correlation parameters, $\beta_H = 0.11$, $\beta_L = 0.13$, $\beta_W = 0.17$. Figure 11d shows the optimal configuration for the $C_{Q,out}$ is: the Kriging model with a Gaussian kernel and correlation parameters, $\beta_H = 5.52$, $\beta_L = 0.10$, $\beta_W = 2.50$. The details are also listed in Table 6. For all hydraulic indices, the performance of the Kriging model is better than that of the other models. The median errors of ξ and C_V are only 0.98% and 0.57% respectively, and the maximum errors are around 5%. Although the maximum error of the $C_{Q,out}$ is slightly higher than the others, the value is acceptable as the complex flow separation phenomenon in the outflow condition makes the index more volatile.

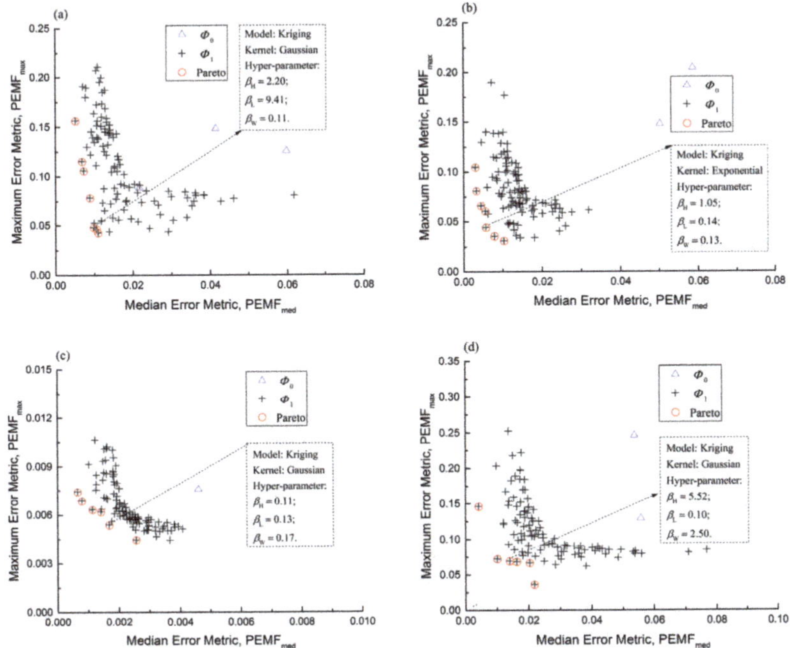

Figure 11. Surrogate model selection of hydraulic performance indices: (**a**) ξ; (**b**) C_V; (**c**) $C_{Q,in}$; (**d**) $C_{Q,out}$.

Table 6. Optimal surrogate model configurations and their Predictive Estimation of Model Fidelity (PEMF) errors.

Index	Optimal Surrogate Model Configuration			PEMF Error	
	Model Type	**Kernel Function**	**Hyper-Parameter**	**Median**	**Max**
ξ_{total}	Kriging	Gaussian	$\beta_H = 2.20, \beta_L = 9.41, \beta_W = 0.11$	0.0098	0.0481
$C_{V,total}$	Kriging	Exponential	$\beta_H = 1.05, \beta_L = 0.14, \beta_W = 0.13$	0.0057	0.0443
$C_{Q,in}$	Kriging	Gaussian	$\beta_H = 0.11, \beta_L = 0.13, \beta_W = 0.17$	0.0017	0.0054
$C_{Q,out}$	Kriging	Gaussian	$\beta_H = 5.52, \beta_L = 0.10, \beta_W = 2.50$	0.0141	0.0695

5.2. Discussion of Optimized Results of the Inlet/Outlet

Based on the improved process, objective functions are set to find the minimum values for ξ and C_V. Figure 12 demonstrates the optimization process for finding the optimal shape, in which the red star point indicates the final selection. For the purpose of presenting the optimal result better, the comparisons between the optimal and the original case are listed in Table 7. In the multi-objective optimum situation, ξ_{total} is reduced by 6.42%, $C_{V,total}$ is reduced by 40.28% compared with the original shape.

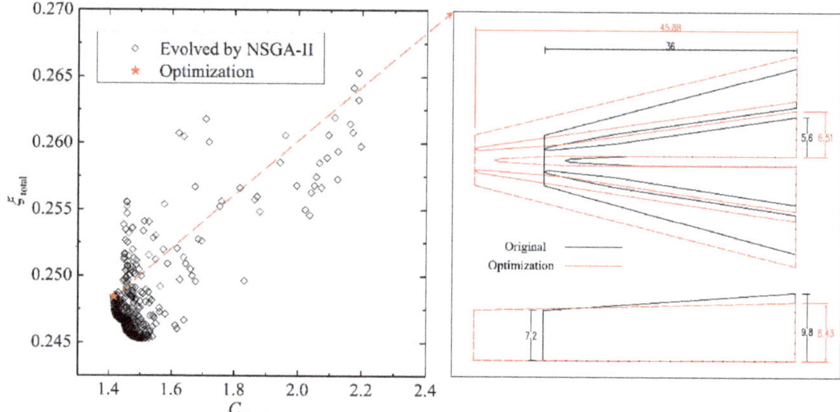

Figure 12. The process of objectives evolved through NSGA-II (dimensions in m).

Table 7. The comparison of parameters between baseline case and optimum case.

Case	Design Variables			Objectives		Variation (%)	
	$H_{I/O}$ (m)	L_{DS} (m)	W_{DSM} (m)	ξ_{total}	$C_{V,total}$	$\Delta\xi_{total}$	$\Delta C_{V,total}$
Baseline	9.800	36.000	1.580	0.265	2.371	−6.42	−40.28
Optimum	8.432	45.875	1.513	0.248	1.416		

Figure 13 compares velocity contours in the orifices under the inflow condition between the baseline and the optimum design. The inflow velocity contours in the optimum case are found to change little, and the flow velocities in the anti-swirl segment are slightly increased due to the decrease of the optimized inlet height.

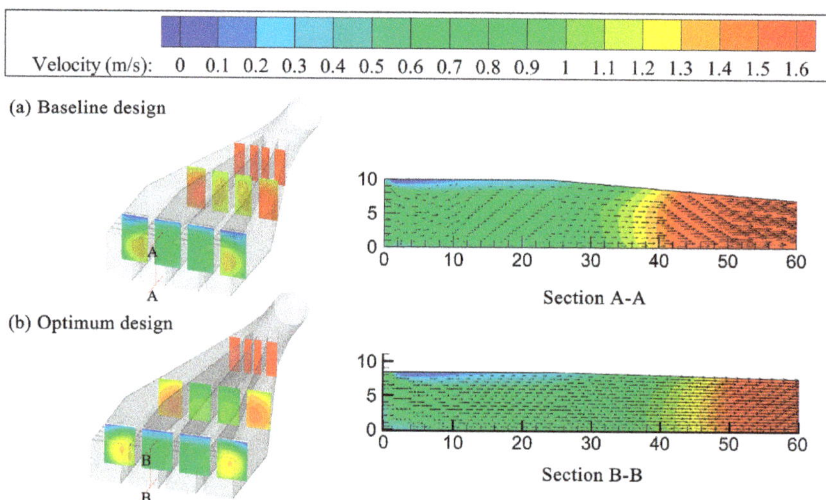

Figure 13. The comparison of inflow velocity distribution between the baseline and optimum design.

As shown in Figure 14, the change between the baseline and optimum case is more obvious when the water flows out from the tunnel. For the baseline case, the flow separation phenomenon occurs at the upper part of the cross sections in the diffusion segment. Due to insufficient flow diffusion, the separation phenomenon deteriorates into a backflow at the top of the trash rack cross sections, which should be prevented as much as possible in the design. Figure 14a shows the flow separation arises early at $x = 40$ m because of the bad original design. As the flow continues to develop, the flow separation deteriorates and induces reverse velocity at the top of the whole rectification segment and part of the anti-vortex segment.

For the design optimized by NSGA-II, the flow separation region is very close to the walls in the diffusion segment, and the reasonable shape brings in a more desired flow phenomenon. Figure 14b shows the desired flow pattern after optimization. Because of the optimization for the diffusion segment, more uniform velocity distributions are observed in the section plane. The flow separation appears to be slight in the small areas at the junction of the diffusion and rectification segment. Consequently, there is hardly any backflow at the top of the orifices.

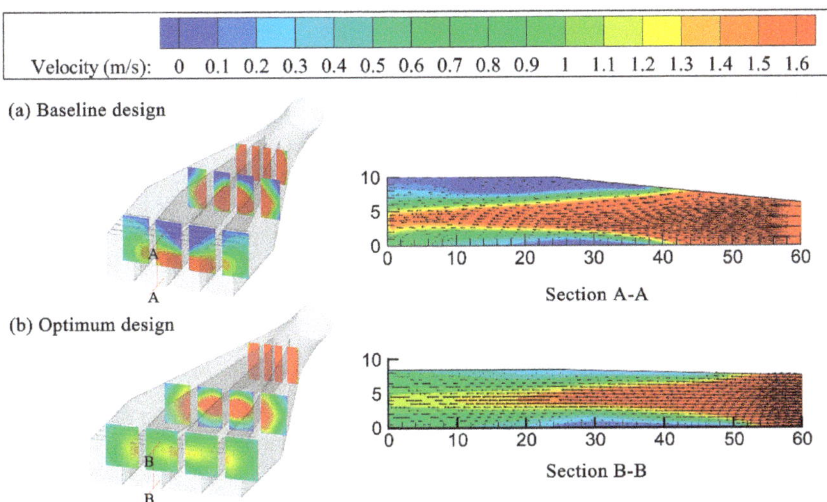

Figure 14. The comparison of outflow velocity distribution between the baseline and optimum design.

In order to further compare the velocity distribution at the orifice cross-section ($x = 10$), the velocity data from the central line of the section is analyzed. As we can see from Figure 15a, there is little change for the normalized velocity profiles between the baseline and optimum design. Figure 15b demonstrates the velocity profiles along the height direction under the outflow condition. In the baseline case, the main stream of the velocity locates at the lower part, and even negative values appear at the top of the middle orifices. For the optimum design, the velocity distributes more uniformly along the height direction, and the mainstream is in the central section of the height. The maximum value of the normalized velocity is 1.44, which is located at the middle orifice with the corresponding y/H value of 0.47. Compared with the original shape, the maximum value is reduced by 49.36%.

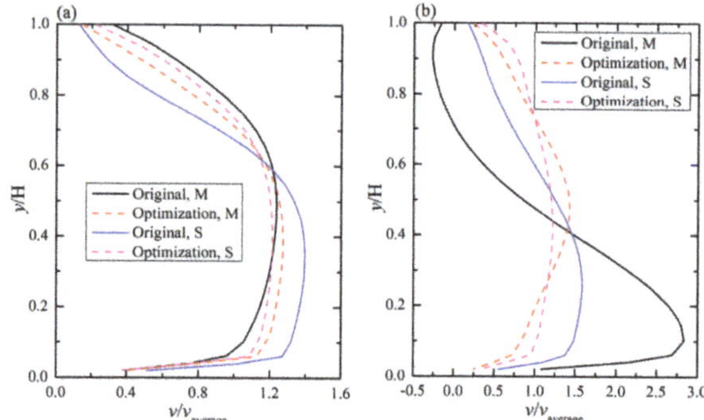

Figure 15. Velocity profiles at the orifice cross-section: (**a**) inflow; (**b**) outflow.

6. Conclusions

In this article, the CFD numerical simulation and multi-objective surrogate-based optimization strategy method (COSMOS and NSGA-II) were combined to optimize intake shape effectively. The CFD method applied in this paper was validated by a physical experiment carefully designed to meet bidirectional flow requirements for the inlet/outlet. For the purpose of determining a good compromise between the two modes of operation, reasonable weights were defined to better evaluate the overall performance. Then, suitable surrogate models based on COSMOS were utilized to build functions, and NSGA-II was chosen to complete the optimization.

The optimum shape of the diffusion part in a PHES can be achieved automatically through the whole process, including numerical model building, CFD simulation, optimal surrogate model selection, and multi-objective optimization strategy. There were 60 sampling points generated by the OLHS employed to establish suitable surrogate models based on COSMOS, while the PEMF method was applied to estimate errors. The results indicate that the Kriging with a Gaussian kernel is best for the overall head loss coefficient and the shape parameter $\beta_H = 2.20$, $\beta_L = 9.41$, $\beta_W = 0.11$; the best configuration of the overall velocity distribution coefficient is the Kriging model with an Exponential kernel and $\beta_H = 1.05$, $\beta_L = 0.14$, $\beta_W = 0.13$.

Finally, the NSGA-II algorithm was applied to generate Pareto optimal results, and the final optimal point was selected from the Pareto set through the TOPSIS decision making method. The optimization results show that compared with the original shape the overall head loss coefficient decreases by 6.42% and the overall velocity distribution coefficient decreases by 40.28%. This study demonstrates this new optimization process is a good choice for inlet/outlet engineering designers.

Author Contributions: Conceptualization, X.G. and B.S.; methodology, H.Z. (Hongtao Zhu) and Y.T.; software, H.Z. (Han Zhang) and Z.Q.; validation, Z.Q.; formal analysis, H.Z. (Han Zhang); investigation, H.Z. (Hongtao Zhu) and Y.T.; data curation, Z.Q.; writing—original draft preparation, H.Z. (Hongtao Zhu) and Y.T.; writing—review and editing, H.Z. (Hongtao Zhu) and B.S.; visualization, Z.Q.; supervision, X.G. and B.S.

Funding: This research was funded by the Science Fund for Creative Research Group of the National Natural Science Foundation of China (51621092), the National Natural Science Foundation of China (51279125) and Tianjin Municipal Natural Science Foundation (15JCYBJC22600).

Conflicts of Interest: The authors declare no conflict of interest.

References

1. Frosina, E.; Buono, D.; Senatore, A. A Performance Prediction Method for Pumps as Turbines (PAT) Using a Computational Fluid Dynamics (CFD) Modeling Approach. *Energies* **2017**, *10*, 103. [CrossRef]
2. Itsukushima, R.; Ikematsu, S.; Nakano, M.; Takagi, M.; Shimatani, Y. Optimal structure of grated bottom intakes designed for small hydroelectric power generation. *J. Renew. Sustain. Energy* **2016**, *8*, 034501. [CrossRef]

3.   Kolekar, N.; Banerjee, A. A coupled hydro-structural design optimization for hydrokinetic turbines. *J. Renew. Sustain. Energy* **2013**, *5*, 053146. [CrossRef]

4.   Li, D.; Gong, R.; Wang, H.; Wei, X.; Liu, Z.; Qin, D. Numerical investigation on transient flow of a high head low specific speed pump-turbine in pump mode. *J. Renew. Sustain. Energy* **2015**, *7*, 063111. [CrossRef]

5.   Dong, W.; Li, Y.; Xiang, J. Optimal Sizing of a Stand-Alone Hybrid Power System Based on Battery/Hydrogen with an Improved Ant Colony Optimization. *Energies* **2016**, *9*, 785. [CrossRef]

6.   Singh, A.; Sevda, S.; Abu Reesh, I.; Vanbroekhoven, K.; Rathore, D.; Pant, D. Biohydrogen Production from Lignocellulosic Biomass: Technology and Sustainability. *Energies* **2015**, *8*, 13062–13080. [CrossRef]

7.   Belabes, B.; Youcefi, A.; Paraschivoiu, M. Numerical investigation of Savonius wind turbine farms. *J. Renew. Sustain. Energy* **2016**, *8*, 053302. [CrossRef]

8.   Gosselin, R.; Dumas, G.; Boudreau, M. Parametric study of H-Darrieus vertical-axis turbines using CFD simulations. *J. Renew. Sustain. Energy* **2016**, *8*, 053301. [CrossRef]

9.   Shen, X.; Avital, E.; Paul, G.; Rezaienia, M.A.; Wen, P.; Korakianitis, T. Experimental study of surface curvature effects on aerodynamic performance of a low Reynolds number airfoil for use in small wind turbines. *J. Renew. Sustain. Energy* **2016**, *8*, 053303. [CrossRef]

10.   Kumar, A.; Kim, M.-H. Thermal Hydraulic Performance in a Solar Air Heater Channel with Multi V-Type Perforated Baffles. *Energies* **2016**, *9*, 564. [CrossRef]

11.   Katsaprakakis, D.A.; Christakis, D.G.; Pavlopoylos, K.; Stamataki, S.; Dimitrelou, I.; Stefanakis, I.; Spanos, P. Introduction of a wind powered pumped storage system in the isolated insular power system of Karpathos–Kasos. *Appl. Energy* **2012**, *97*, 38–48. [CrossRef]

12.   Kocaman, A.S.; Modi, V. Value of pumped hydro storage in a hybrid energy generation and allocation system. *Appl. Energy* **2017**, *205*, 1202–1215. [CrossRef]

13.   Müller, M.; De Cesare, G.; Schleiss, A.J. Flow field in a reservoir subject to pumped-storage operation—In situ measurement and numerical modeling. *J. Appl. Water Eng. Res.* **2016**, *6*, 109–124. [CrossRef]

14.   Bermudez, M.; Cea, L.; Puertas, J.; Conde, A.; Martin, A.; Baztan, J. Hydraulic model study of the intake-outlet of a pumped-storage hydropower plant. *Eng. Appl. Comput. Fluid Mech.* **2017**, *11*, 483–495. [CrossRef]

15.   Cai, F.; Hu, M.; Zhang, Z. Study on guide piers in frank inlet-outlet with double flow directions. *J. Hohai Univ.* **2000**, *28*, 74–77.

16.   Gao, X.; Wang, Z. The Hydraulic Characteristics at Vertical Pipe Intake of Xilongchi Pumped Storage Plant. *J. Hydroelectr. Eng.* **2002**, *1*, 52–60.

17.   Sun, S.; Liu, H.; Li, Z.; Sun, C. Study on velocity distribution behind the trashrack in lateral intake/outlet of pumped storage power station. *Shuili Xuebao* **2007**, *38*, 1329–1335.

18.   Ye, F.; Gao, X. Numerical simulations of the hydraulic characteristics of side inlet/outlets. *J. Hydrodyn. Ser. B* **2011**, *23*, 48–54. [CrossRef]

19.   Gao, X.; Li, Y.; Tian, Y.; Sun, B. Flow distribution in side intake/outlet tunnel of pumped storage power stations. *J. Hydroelectr. Eng.* **2016**, *35*, 87–94.

20.   Ansoni, J.L.; Seleghim, P. Optimal industrial reactor design: Development of a multiobjective optimization method based on a posteriori performance parameters calculated from CFD flow solutions. *Adv. Eng. Softw.* **2016**, *91*, 23–35. [CrossRef]

21.   Ding, Y.; Zhang, X. An optimal design method of swept blades for HAWTs. *J. Renew. Sustain. Energy* **2016**, *8*, 043303. [CrossRef]

22.   Gao, Q.; Cai, X.; Zhu, J.; Guo, X. Multi-objective optimization and fuzzy evaluation of a horizontal axis wind turbine composite blade. *J. Renew. Sustain. Energy* **2015**, *7*, 063109. [CrossRef]

23.   Lopez, N.; Mara, B.; Mercado, B.; Mercado, L.; Pascual, M.; Promentilla, M.A. Design of modified Magnus wind rotors using computational fluid dynamics simulation and multi-response optimization. *J. Renew. Sustain. Energy* **2015**, *7*, 063135. [CrossRef]

24.   Jiang, Y.T.; Lin, H.F.; Yue, G.Q.; Zheng, Q.; Xu, X.L. Aero-thermal optimization on multi-rows film cooling of a realistic marine high pressure turbine vane. *Appl. Therm. Eng.* **2017**, *111*, 537–549. [CrossRef]

25.   Elsayed, K.; Lacor, C. CFD modeling and multi-objective optimization of cyclone geometry using desirability function, artificial neural networks and genetic algorithms. *Appl. Math. Model.* **2013**, *37*, 5680–5704. [CrossRef]

26. Yang, Y.T.; Tang, H.W.; Ding, W.P. Optimization design of micro-channel heat sink using nanofluid by numerical simulation coupled with genetic algorithm. *Int. Commun. Heat Mass Transf.* **2016**, *72*, 29–38. [CrossRef]

27. Sun, W.Q.; Cheng, W. Finite element model updating of honeycomb sandwich plates using a response surface model and global optimization technique. *Struct. Multidiscip. Optim.* **2016**, *55*, 121–139. [CrossRef]

28. Gan, W.B.; Zhang, X.C. Design optimization of a three-dimensional diffusing S-duct using a modified SST turbulent model. *Aerosp. Sci. Technol.* **2017**, *63*, 63–72. [CrossRef]

29. Naseri, S.; Tatar, A.; Shokrollahi, A. Development of an accurate method to prognosticate choke flow coefficients for natural gas flow through nozzle and orifice type chokes. *Flow Meas. Instrum.* **2016**, *48*, 1–7. [CrossRef]

30. An, Z.; Zhounian, L.; Peng, W.; Linlin, C.; Dazhuan, W. Multi-objective optimization of a low specific speed centrifugal pump using an evolutionary algorithm. *Eng. Optim.* **2015**, *48*, 1251–1274. [CrossRef]

31. Yin, H.; Fang, H.; Wen, G.; Wang, Q.; Xiao, Y. An adaptive RBF-based multi-objective optimization method for crashworthiness design of functionally graded multi-cell tube. *Struct. Multidiscip. Optim.* **2015**, *53*, 129–144. [CrossRef]

32. Lotfan, S.; Ghiasi, R.A.; Fallah, M.; Sadeghi, M.H. ANN-based modeling and reducing dual-fuel engine's challenging emissions by multi-objective evolutionary algorithm NSGA-II. *Appl. Energy* **2016**, *175*, 91–99. [CrossRef]

33. Liu, C.; Bu, W.; Xu, D. Multi-objective shape optimization of a plate-fin heat exchanger using CFD and multi-objective genetic algorithm. *Int. J. Heat Mass Transf.* **2017**, *111*, 65–82. [CrossRef]

34. Jakumeit, J.; Herdy, M.; Nitsche, M. Parameter optimization of the sheet metal forming process using an iterative parallel Kriging algorithm. *Struct. Multidiscip. Optim.* **2005**, *29*, 498–507. [CrossRef]

35. Zhang, Y.; Hu, S.; Wu, J.; Zhang, Y.; Chen, L. Multi-objective optimization of double suction centrifugal pump using Kriging metamodels. *Adv. Eng. Softw.* **2014**, *74*, 16–26. [CrossRef]

36. Mehmani, A.; Chowdhury, S.; Meinrenken, C.; Messac, A. Concurrent surrogate model selection (COSMOS): Optimizing model type, kernel function, and hyper-parameters. *Struct. Multidiscip. Optim.* **2017**, *57*, 1093–1114. [CrossRef]

37. Gao, X.; Tian, Y.; Sun, B. Shape optimization of bi-directional flow passage components based on a genetic algorithm and computational fluid dynamics. *Eng. Optim.* **2017**, *50*, 1287–1303. [CrossRef]

38. Gao, X.; Tian, Y.; Sun, B. Multi-objective optimization design of bidirectional flow passage components using RSM and NSGA II: A case study of inlet/outlet diffusion segment in pumped storage power station. *Renew. Energy* **2018**, *115*, 999–1013. [CrossRef]

39. *Design Guide for Pumped Storage Power Station (DL/T 5208-2005)*; China Electric Power Press: Beijing, China, 2005.

40. Johnson, P.L. HydroPower Intake Design Considerations. *J. Hydraul. Eng.* **1988**, *114*, 651–661. [CrossRef]

41. Sell, L.E. Hydroelectric Power Plant Trashrack Design. *J. Power Div.* **1971**, *97*, 115–121.

42. Lei, T.; Shan, Z.B.; Liang, C.S.; Chuan, W.Y.; Bin, W.B. Numerical simulation of unsteady cavitation flow in a centrifugal pump at off-design conditions. *Proc. Inst. Mech. Eng. Part C J. Mech. Eng. Sci.* **2013**, *228*, 1994–2006. [CrossRef]

43. Tan, L.; Zhu, B.; Wang, Y.; Cao, S.; Gui, S. Numerical study on characteristics of unsteady flow in a centrifugal pump volute at partial load condition. *Eng. Comput.* **2015**, *32*, 1549–1566. [CrossRef]

44. Liu, Y.; Tan, L.; Hao, Y.; Xu, Y. Energy Performance and Flow Patterns of a Mixed-Flow Pump with Different Tip Clearance Sizes. *Energies* **2017**, *10*, 191. [CrossRef]

45. Fluent. *User's and Theory Guide*; ANSYS, Inc.: Canonsburg, PA, USA, 2015.

46. Shih, T.-H.; Liou, W.W.; Shabbir, A.; Yang, Z.; Zhu, J. A new k-ε eddy viscosity model for high reynolds number turbulent flows. *Comput. Fluids* **1995**, *24*, 227–238. [CrossRef]

47. Sun, H.; Shin, H.; Lee, S. Analysis and optimization of aerodynamic noise in a centrifugal compressor. *J. Sound Vib.* **2006**, *289*, 999–1018. [CrossRef]

48. Khalfallah, S.; Ghenaiet, A. Shape optimization of a centrifugal compressor impeller. In Proceedings of the Institution of Mechanical Engineers, 8th International Conference on Compressors and their Systems, London, UK, 9–10 September 2013; Woodhead Publishing Limited: Cambridge, UK, 2013.

49. Zhou, C.; Yang, C.; Zhang, X.; Wang, C. A simulation-based two-step method for optimal thermal design of multiple compartments. *Appl. Therm. Eng.* **2017**, *122*, 535–545. [CrossRef]

50. Brar, L.S.; Elsayed, K. Analysis and optimization of multi-inlet gas cyclones using large eddy simulation and artificial neural network. *Powder Technol.* **2017**, *311*, 465–483. [CrossRef]

51. Li, Z.; Liu, Y. Blade-end treatment for axial compressors based on optimization method. *Energy* **2017**, *126*, 217–230. [CrossRef]

52. Zhang, M.; Gou, W.; Li, L.; Yang, F.; Yue, Z. Multidisciplinary design and multi-objective optimization on guide fins of twin-web disk using Kriging surrogate model. *Struct. Multidiscip. Optim.* **2017**, *55*, 361–373. [CrossRef]

53. Singh, P.; Couckuyt, I.; Elsayed, K.; Deschrijver, D.; Dhaene, T. Shape optimization of a cyclone separator using multi-objective surrogate-based optimization. *Appl. Math. Model.* **2016**, *40*, 4248–4259. [CrossRef]

54. Wang, B.; Li, T.; Ge, L.; Ogawa, H. Optimization of combustion chamber geometry for natural gas engines with diesel micro-pilot-induced ignition. *Energy Convers. Manag.* **2016**, *122*, 552–563. [CrossRef]

55. Halder, P.; Samad, A.; Thevenin, D. Improved design of a Wells turbine for higher operating range. *Renew. Energy* **2017**, *106*, 122–134. [CrossRef]

56. Martin, J.D.; Simpson, T.W. Use of Kriging Models to Approximate Deterministic Computer Models. *Aiaa J.* **2005**, *43*, 853–863. [CrossRef]

57. Viana, F.A.C.; Haftka, R.T.; Steffen, V. Multiple surrogates: How cross-validation errors can help us to obtain the best predictor. *Struct. Multidiscip. Optim.* **2009**, *39*, 439–457. [CrossRef]

58. Gorissen, D.; Couckuyt, I.; Demeester, P.; Dhaene, T.; Crombecq, K. A Surrogate Modeling and Adaptive Sampling Toolbox for Computer Based Design. *J. Mach. Learn. Res.* **2010**, *11*, 2051–2055.

59. Mehmani, A.; Chowdhury, S.; Messac, A. Predictive quantification of surrogate model fidelity based on modal variations with sample density. *Struct. Multidiscip. Optim.* **2015**, *52*, 353–373. [CrossRef]

60. Lophaven, S.N.; Nielsen, H.B.; Sondergaard, J. *DACE—A MATLAB Kriging Toolbox—Version 2.0, Informatics and Mathematical Modeling*; Technical University of Denmark: Lyngby, Denmark, 2002.

61. Deb, K.; Pratap, A.; Agarwal, S.; Meyarivan, T. A fast and elitist multiobjective genetic algorithm: NSGA-II. *IEEE Trans. Evol. Comput.* **2002**, *6*, 182–197. [CrossRef]

62. Deb, K.; Agrawal, R.B. Simulated Binary Crossover for Continuous Search Space. *Complex Syst.* **1994**, *9*, 115–148.

63. Deb, K.; Goyal, M. A Combined Genetic Adaptive Search (GeneAS) for Engineering Design. *Comput. Sci. Inform.* **1996**, *26*, 30–45.

64. Hwang, C.L.; Yoon, K.P. *Multiple Attribute Decision Making—Methods and Applications: A State-of-the-Art Survey*; Springer: Berlin/Heidelberg, Germany, 1981; pp. 11, 269.

Article

In Situ Measurements and CFD Numerical Simulations of Thermal Environment in Blind Headings of Underground Mines

Wenhao Wang [1], Chengfa Zhang [1,2], Wenyu Yang [1,2,3,*], Hong Xu [1], Sasa Li [1], Chen Li [1], Hui Ma [1] and Guansheng Qi [1]

[1] College of Mining and Safety Engineering, Shandong University of Science and Technology, Prefecture of Shandong, Qingdao 266590, China; wenhaowang@aliyun.com (W.W.); chengfazhang@aliyun.com (C.Z.); xuhong316@aliyun.com (H.X.); lisasa121919@aliyun.com (S.L.); lc17660990724@163.com (C.L.); Hughie_Ma@163.com (H.M.); qiguansheng@126.com (G.Q.)

[2] Zhaojin Mining Industry Co., Ltd., Zhaoyuan 265418, China

[3] School of Resources and Civil Engineering, Northeastern University, Prefecture of Liaoning, Shenyang 110819, China

* Correspondence: wenyu@sdust.edu.cn; Tel.: +86-159-6425-9897

Received: 9 May 2019; Accepted: 22 May 2019; Published: 24 May 2019

Abstract: In order to gain a knowledge of the heat emitted from a variety of sources at the blind heading of an underground gold mine, the present study conducts an in situ measurement study in a blind heading within the load haul dumps (LHDs) that are operating. The measurements can provide a reliable data basis for the setting of numerical simulations. The results demonstrate that the distances between the forcing outlet and the mining face (denoted as Z_m), as well as the heat generation from LHDs (denoted as Q_L), has brought significant impacts on the airflow velocity, relative humidity, and temperature distributions in the blind heading. Setting Z_m to 5 m could achieve a relative optimal cooling performance, also indicating that when the LHD is fully operating in the mining face, employing the pure forcing system has a limited effect on the temperature decrease of the blind heading. According to the numerical simulations, a better cooling performance can be achieved based on the near-forcing-far-exhausting (NFFE) ventilation system.

Keywords: thermal environment; numerical simulations; ventilation cooling; duct position; the heat dissipation of LHD; auxiliary ventilation

---

1. Introduction

With the development of the global economy, the demands for mineral resources are persistently increased by human beings. In recent years, shallow resources have displayed a declining trend annually and the underground mining depth will become deeper and deeper in the future [1–3]. In addition, the higher-powered machinery that increases production levels impose an increased burden on ventilation systems to maintain an acceptable working environment [4]. As a result, the worsening thermal environments in underground mining resulted from higher virgin rock temperature and heat emitted from mechanical equipment gave rise to a lot of concerns regarding the health and safety of miners [5]. In the blind headings, the heat transferred from fragmented rock, explosives and LHDs (or roadheaders, continuous miners) with the existence of an auxiliary ventilation system could make the aerodynamic much more complex in the face area [6]. Auxiliary ventilation can be classified into three basic types, Pure Forcing (PF), Pure Exhausting (PE), and overlap systems (FFNE and NFFE) as shown in Figure 1. Therefore, to design the optimal ventilation and cooling systems, it is of great significance to carry out in-depth research on the thermal environment for high-speed ramp and tunnel developments [7].

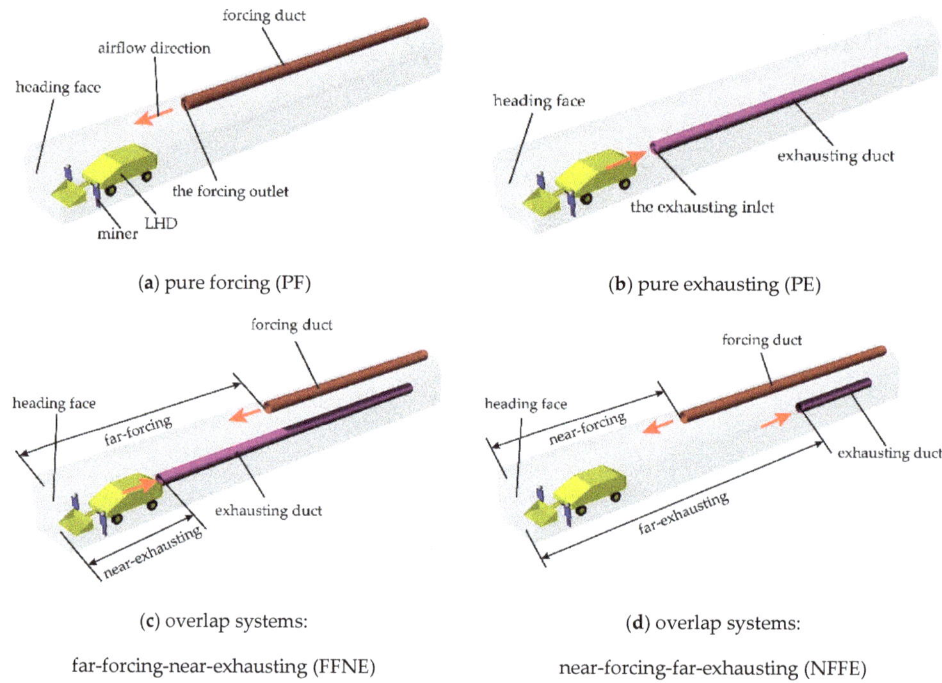

(**a**) pure forcing (PF)

(**b**) pure exhausting (PE)

(**c**) overlap systems:

far-forcing-near-exhausting (FFNE)

(**d**) overlap systems:

near-forcing-far-exhausting (NFFE)

Figure 1. Geometry model of the PF (**a**), PE (**b**), FFNE (**c**), and NFFE (**d**) ventilation systems.

Traditionally, researchers mainly adopt the following three methods for examining the thermal environment in blind headings: in situ measurement, laboratory experiment and numerical simulation [8]. The in situ measurement and experimental research has always been a challenge and only a few related studies are available due to the safety issue. For example, Bluhm et al. quantified the different heat sources by field measurements [7]. The heat flow from the rock was found to be the principal source of heat. Lowndes et al. measured the thermal environmental conditions in a representative development drivage in a coal mine during the non-production period and production period. The input values for the tunnel model were modified based on the measured data [9]. Shimada utilized a one-tenth scale model experimental apparatus to investigate the cooling effectiveness of the auxiliary ventilation systems and heat transfer at the heading face [10]. Jia et al. designed a thermal exchange simulation test platform for a blind heading that can be employed to investigate the cooling factor for the ventilation and cooling system [11]. However, most of the previous studies are mainly focused on the heat load transferred from rock.

With the emergence of computational fluid dynamics (CFD) in the 1960s, numerical simulations aiming at the thermal management in heading faces have become efficient and low cost [12]. Gao et al. employed the finite volume method for simulating the airflow velocity and temperature fields in a heading face. They obtained the distribution of heat transfer coefficient in the heading face [13]. Sasmito et al. developed a three-dimensional blind heading model for studying the effects of the ventilation flow rate, cooling load, virgin rock temperature, and the heat rejection from mining machinery [12]. To date, much more researchers have conducted a large number of numerical simulations to investigate the airflow ventilation-methane behavior and airflow ventilation-dust control [14–18] in PF and PE, respectively. Based on these research findings, the constraint conditions for Z_m and Z_e (the distances between the exhausting outlet and the heading face) in PF and PE can be written as Equations (1) and (2), respectively [19]:

$$Z_m \leq (4 \sim 5) \sqrt{S} \tag{1}$$

$$Z_e \leq 1.5 \sqrt{S} \tag{2}$$

where S is the section area of a blind heading (m^2).

As stated above, in all numerical studies performed so far, much more attention was concentrated on the methane and dust distributions at the face area. The numerical models targeting thermal environment research were over-simplified. The heat sources such as the heat generation from diesel engines were mainly calculated by a simple equation proposed by McPherson [20]. The key parameters are regulated mostly based on field experience. There is still no ventilation scheme that has been proposed from the view point of thermal environment. In practical ventilation and cooling engineering, doubts still exist when determining Z_m or Z_e based on the Equations (1) and (2).

This study was organized as follows: firstly, an in situ measurement was performed in a selected blind heading. The heat emitted from surrounding rock and LHDs was calculated based on the measured airflow velocity, temperature and relative humidity. Secondly, FLUENT software was employed to study the airflow velocity, temperature and relative humidity distributions when the forcing outlet was set at different positions. An optimal position that can achieve a favorable cooling performance was determined. Then, effect of the heat emitted from LHDs is also discussed with respect to the temperature distributions. Finally, NFFE ventilation was proposed in the numerical simulation. Moreover, this study could help extend the existing thermal environment research and provide theoretical support for the design and operation of the auxiliary ventilation systems in blind headings that are mined using LHDs.

2. Field Measurements

2.1. Measuring Setup

This case study relates to a 40 m multi-blast development heading about 230 m below surface in an underground gold mine located in Japan. The drill and blast methods with the use of diesel equipment (such as LHD) were applied in the mine. There was no wet source in the heading and the surface of the walls was dry. To verify the heat load from the surrounding rock and diesel equipment in an actual blind heading, the dry bulb temperatures were continuously measured and digitally recorded within the mining stope during production periods. Due to the existence of hot spring water in the deep strata, the virgin rock temperature reaches above 80 °C and the rock stratum is andesite. The mining stope was 40 m in length with an arch cross-sectional area of 21.6 m^2 and a perimeter of 15.5 m. Hence, the hydraulic diameter of the airway was 5.6 m. The heading face was ventilated by PF. The air duct (80 cm diameter) was being installed in the roof of the airway which was producing an airflow velocity of 12.0 m/s with a volume airflow of 7.0 m^3/s. The average airflow velocity can be calculated as 0.3 m/s in the airway. The heat capacity of the airflow per unit time (W/°C) in this airway can be determined as 8442 W/°C by Equation (3):

$$C_f = F \varrho_a C_a \tag{3}$$

where F is volume flow of air (m^3/s), ϱ_a is air density (kg/m^3), C_a is specific heat capacity of air (J/(kg·°C)). In this paper, ϱ_a is 1.2 kg/m^3 and C_a is 1005 J/(kg·°C).

Figure 2 illustrates the location of the air duct and the measuring points near the heading from the side view. Z_m was maintained about 20 m. The blind heading would advance about three meters in each blasting. Data loggers (midi LOGGER GL200, Graphtec Corporation, Yokohama City, Prefecture of Kanagawa, Japan) and thermocouples of Type K (diameter is 0.5 mm, Graphtec Corporation, Yokohama City, Prefecture of Kanagawa, Japan) were used for these measurements. The measuring interval of the data loggers was set at 30 s. The thermocouples were installed in nine locations of the cross-section (Figure 3). One of them was employed to measure the air temperature in the air duct. A blast may cause damage to the data loggers and sensors. Therefore, the sensors had to be installed about one meter from the surface of the airway and the measuring points were 30 m from the face and thus the heat exchange area was 424.6 m^2.

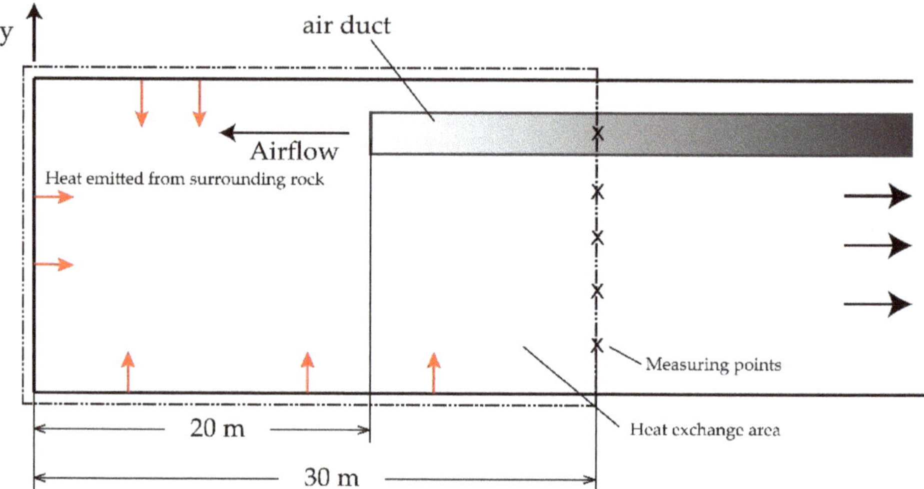

Figure 2. Side view of the blind heading, the measuring points and the heat exchange area.

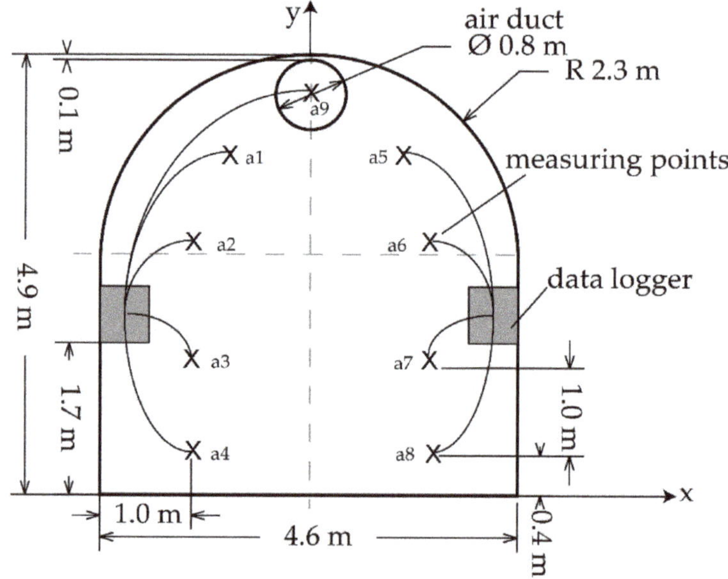

Figure 3. Cross-sectional view of the measuring points.

2.2. Results and Discussion

Figure 4 presents the measured results of air temperature variation with time. The average air temperature of the airway represented the average measured results of the eight thermocouples. The air temperature of the outlet of the air duct represented the measure results at the outlet of the air duct. Table 1 introduces the mining operation process. The brief discussion of the measured results is shown below:

a. No mining operation before 20:40;

b. Air temperature increased due to the blasting at 20:40; and

c. Air temperature increased due to the moving of LHD between 21:50 and 23:30.

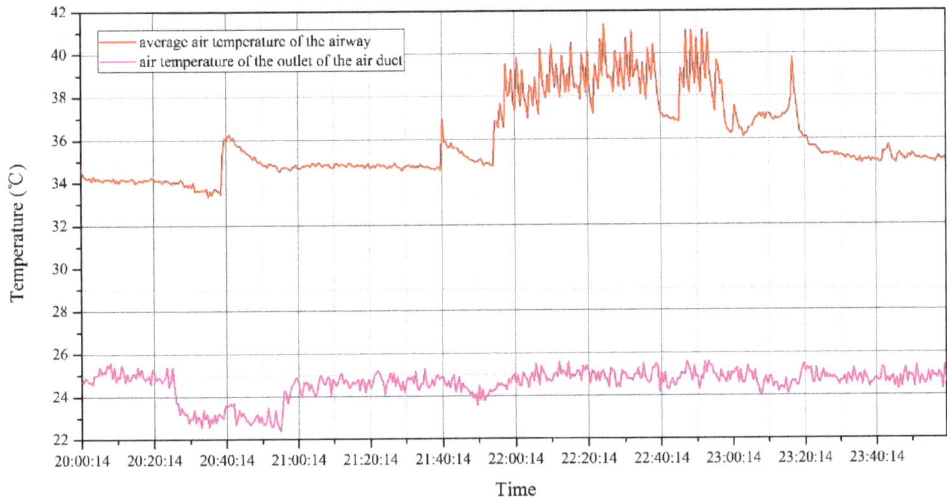

Figure 4. Measuring result of the variation of air temperature with time.

Table 1. Mining process record.

Starting Time	End Time	Operation	Heavy-Duty Machinery	Charge Quantity (kg)	Ore Amount (ton)
20:40	—	Blasting	—	41.8	49.6
21:50	23:30	Loading and hauling the Ore	LHD	—	—

2.2.1. Heat Emitted from Surrounding Rock

Before 20:40, there were no working activities that took place. Hence, it can be considered that the surrounding rock would be the major heat source to the air stream. Figure 5 which was magnified from Figure 4 shows the strata heat resulted in airflow temperatures increased by about 9.7 °C within the heat exchange area. The strata heat before a blasting in the vicinity of the head end could be determined as 81.9 kW by Equation (4), which can serve as the boundary conditions for subsequent numerical simulation:

$$Q_r = C_f\left(\Theta_{f0} - \Theta_i\right) \tag{4}$$

where Q_r is strata heat before the blasting (W), C_f is heat capacity of air in per unit time (W/°C) as shown in Equation (3), Θ_{f0} is the stable air temperature before blasting (°C), Θ_i is air temperature at the outlet of the air duct (°C), here $\Theta_{f0} - \Theta_i = 9.7$ °C.

The Reynolds number could be determined as 1.1×10^5 by Equation (5):

$$R_{e_f} = \frac{d_e u_m}{v_f} \tag{5}$$

where d_e is hydraulic diameter of the airway (m), u_m is average airflow velocity (m/s), v_f is kinematic viscosity (N·m s/kg). In this paper d_e is 5.6 m, u_m is 0.3 m/s and v_f is 15.3×10^{-6} N·m s/kg.

The local heat-transfer coefficient mostly depends on the distribution of air velocity which is influenced by the shape of the airway, the size and position of the duct outlet and the airflow rate.

When the geometrical conditions are identical, it only depends on the airflow rate. Numerous studies have been carried out on the correlation between heat-transfer coefficient and the Reynolds number in a straight pipe with through airflow. Gao et al. recommended the following Equation (6) to calculate the heat transfer coefficient [13]:

$$h_0 = 0.023 Re^{0.8} Pr^{1/3} \lambda / d_e \tag{6}$$

where λ is heat conductivity of air (W/(m·°C)). Pr is Prandtl number given by:

$$Pr = \frac{\mu C_p}{\lambda} \tag{7}$$

where μ is viscosity of air (Pa·s). C_p is isobaric specific heat capacity (J/(kg·°C)). In this paper λ is 0.02 W/(m·°C), C_p is 1005 J/(kg·°C), Pr can be calculated as 0.77, h_0 can be calculated as 0.8 W/m²·°C.

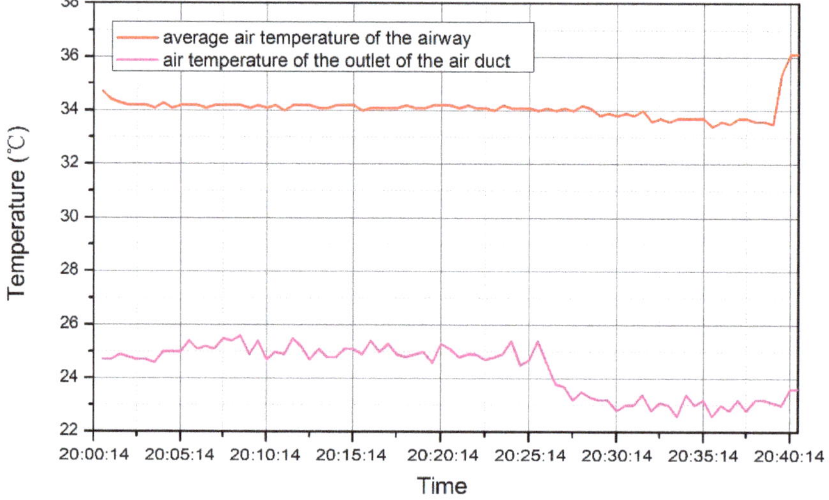

Figure 5. Measuring result of the variation of air temperature with time before 20:40.

As airflow in the heading face is the typical forced turbulence in the semi-closed space, the airflow velocity distribution is uneven near the mining face. The local heat transfer coefficient varies with location in a wide range. Gao et al. suggested that the dimensionless heat transfer coefficients (h/h_0) of roof, side wall, floor and mining face were 9.7, 5.1, 3.5, and 6.8, respectively, with the average of 5.6 [13]. Hence, the heat transfer coefficient can be determined at 4.5 W/(m²·K).

2.2.2. Heat Transferred from LHD

The cyclical nature of mechanized loading and hauling operations can produce periodic fluctuation in the thermal environmental conditions created within these workings [9]. Table 2 shows the specifications of the LHD. Figure 6 presents the variation of the air temperature from 21:50 to 23:30 which was magnified from Figure 4. There are two kinds of heat source, respectively, heat produced by LHD and strata heat. The average temperature difference increases to 12.9 °C. The total heat that is produced by LHD and strata can be calculated as 108.9 kW. Hence, the LHD can produce approximately 27 kW heat during the working period.

Table 2. Specifications of the LHD.

Combustion engine	Displacement	6.7 L
	Output Power	144 kW
Physical dimension	Length	8.0 m
	Width	2.5 m
	Height	2.0 m

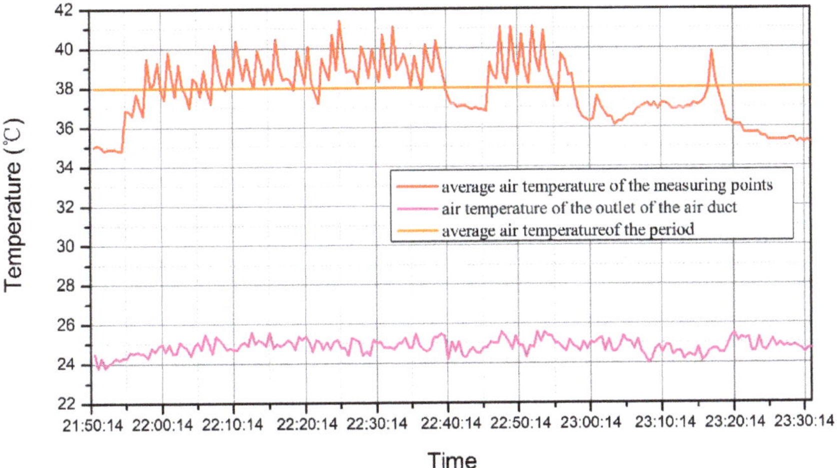

Figure 6. Variation of air temperature with time from 21:50 to 23:30.

3. Numerical Study

3.1. Geometry Model

Two three-dimensional blind headings without and with the use of an LHD employing PF auxiliary ventilation system were built by Rhino 6 (Robert McNeel & Assoc, Seattle City, State of Washington, the United States) as shown in Figure 7a,b, respectively. The positive X direction points to the mining face from the entrance of the blind heading, the positive Y direction points to the left side-walls from the right side-walls and the positive Z direction points to the airway roof from the airway floor. The geometry model was an arch tunnel, 40 m long, 4.6 m wide, and 4.9 m high. A forcing duct with a diameter of 0.8 m was hung at 4 m from the floor. Its setback distance from the mining face was 20 m. An LHD was 8.0 m long, 2.5 m wide and 2 m high. The front part of the LHD was 2 m from the mining face and the miner's height was 1.75 m, 3 m away from the mining face.

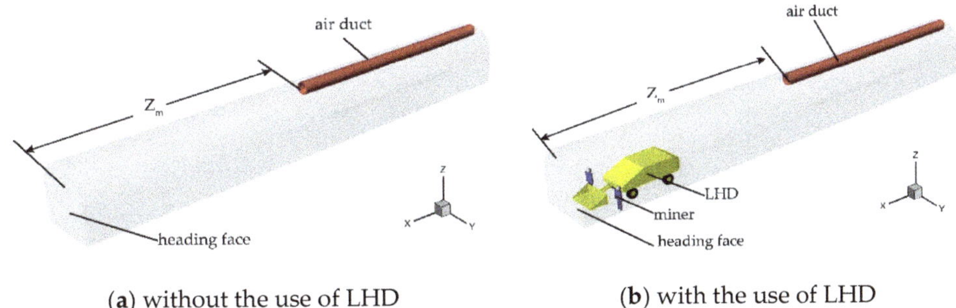

(**a**) without the use of LHD   (**b**) with the use of LHD

Figure 7. Geometric model (**a**,**b**).

3.2. Mesh Generation

The mesh quality determines the computational results and accuracy [21]. The mesh generation was completed with ANSYS Meshing (ANSYS, Inc., Canonsburg City, State of Pennsylvania, the United States) and the globe mesh controls were applied. The element size was set to 1 m. Both Capture Curvature and Capture Proximity were set to Yes. With these settings, the combined effect of both the proximity and curvature sizing can be obtained. Finally, 0.83×10^6 elements were generated. Skewness and orthogonal quality are two of the primary quality measures for a mesh. The statistics of the skewness and orthogonal quality is shown in Table 3. The maximal skewness was lower than 0.95 and the minimal orthogonal quality was higher than 0.1, suggesting that the high-quality mesh had been generated as shown in Figure 8.

Table 3. The mesh metric information.

	Min	Max	Average	Standard Deviation
Skewness	1.72×10^{-4}	0.84	0.30	0.17
Orthogonal quality	0.16	0.99	0.78	0.15

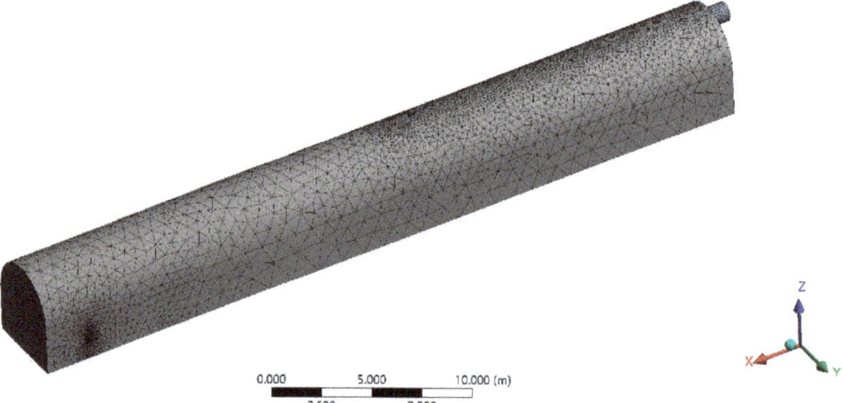

Figure 8. The original computational mesh.

3.3. Mesh Independence Test

The 3D inflation capability provides high quality mesh generation close to wall boundaries to resolve changes in physical properties. In order to ensure a mesh independent solution, another two different amount of elements 1.28×10^6 and 1.75×10^6 were generated by setting different element size and local inflation mesh controls. The comparison with the element amount of 0.83×10^6 and 1.28×10^6 is shown in Figure 9. It can be observed that a finer mesh and the inflation layers were generated in the boundary layers between the airflow and LHD. Figure 10 shows the distribution curves of the airflow velocities and temperatures at the position of Y = 4.1 m, Z = 1.7 m along the airway with three different amount of elements. It can be found that the element amount of 1.28×10^6 gives about 2% deviation compared to the element size of 1.75×10^6. Whereas, the results from the mesh size of 0.83×10^6 deviate up to 15% as compared to those from the finest one. Therefore, a mesh of 1.28×10^6 elements was sufficient for the numerical simulation purposes.

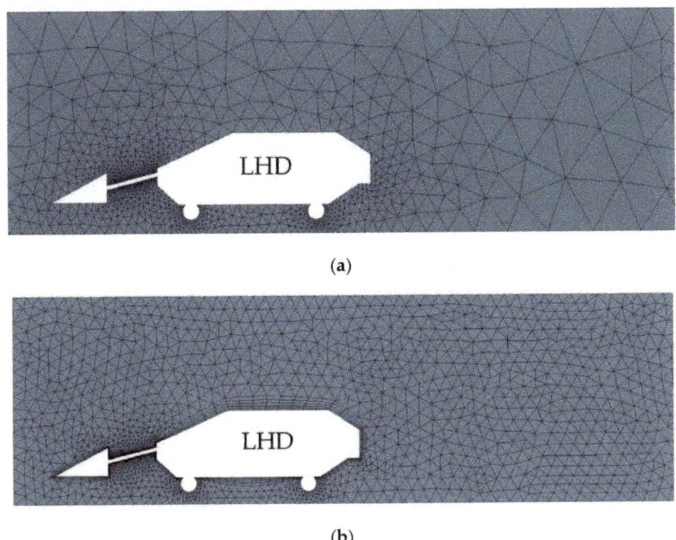

(a)

(b)

Figure 9. Comparison with the original mesh (**a**) and first adaptive mesh (**b**).

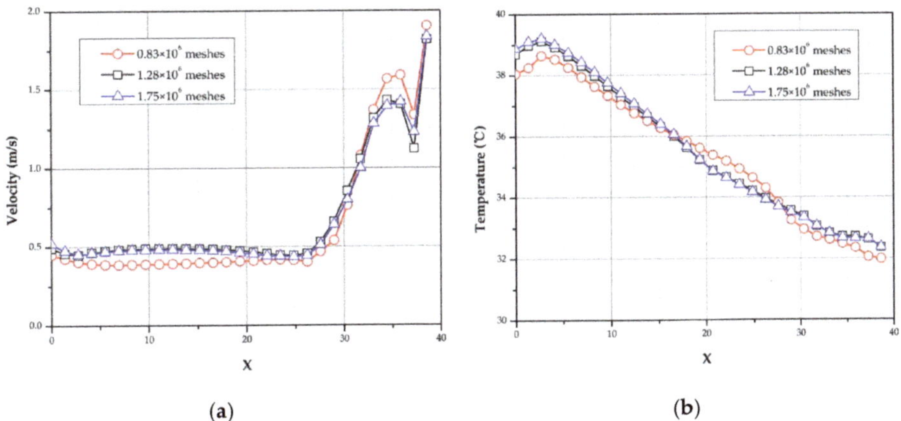

(a)   (b)

Figure 10. Airflow velocities and temperatures distribution of calculating three times along the airway. (**a**) Airflow velocities distribution; and (**b**) airflow temperatures distribution.

3.4. Boundary Conditions

The boundary conditions for the model are summarized in Table 4.

Table 4. Boundary conditions of the CFD method.

Boundary	Conditions
Air duct outlet	Supply airflow temperature 25 °C supply airflow velocity 12 m/s, supply airflow relative humidity 70%
Airway outlet	Pressure outlet
Wall of airway	The heat thermal conductivity: 2.8 W/m·K [22]; Outside heat flux 192.9 W/m^2
Wall of LHD	Fixed heat: Total power: 27 kW
Wall of miners	Metabolic rate: 180 W/m^2 [23]

3.5. Turbulence Model

The turbulence model is the key component in representing flow behavior in underground thermal environment [24]. A commercial CFD code (Fluent, ANSYS, Inc., Canonsburg City, PA, USA) was applied to solve the mass conversion equation (continuous equation), momentum equation (Navier–Stokes equation), energy conversion equation and the turbulence model equation. The flow behavior of various turbulence models, i.e., Spalart-Allmaras, Standard K-Epsilon, Standard K-Omega and Reynolds stress model (RSM) had been compared with the experimental data while the Standard K-Epsilon model provided better results [25]. Table 5 shows the measured results of velocity value as compared to simulation results for various models.

Table 5. Comparison between the measurement data of airflow velocity and the simulation data for various models.

Measurement Point	Measured Results		Spalart–Allmaras		Standard K-Epsilon		Standard K-Omega		Reynolds Stress Model	
	with LHD	without LHD	with LHD	without LHD	with LHD	without LHD	with LHD	without LHD	with LHD	without LHD
a1	Unmeasurable	0.31	0.15	0.33	0.14	0.31	0.15	0.28	0.17	0.35
a2	0.32	Unmeasurable	0.34	0.26	0.33	0.27	0.30	0.25	0.36	0.28
a3	0..50	0.75	0.54	0.77	0.50	0.77	0.47	0.74	0.55	0.79
a4	0.85	1.18	0.86	1.19	0.86	1.17	0.83	1.15	0.87	1.21
a5	Unmeasurable	Unmeasurable	0.16	0.31	0.16	0.29	0.18	0.26	0.19	0.30
a6	Unmeasurable	Unmeasurable	0.29	0.30	0.29	0.30	0.29	0.28	0.31	0.30
a7	0.43	0.78	0.45	0.79	0.44	0.79	0.40	0.74	0.46	0.82
a8	0.65	1.25	0.67	1.28	0.65	1.25	0.61	1.23	0.69	1.29
p	\	\	0.039	0.041	0.083	0.414	0.041	0.042	0.042	0.034

To test whether there were differences between the measurements and the simulation results, a Wilcoxon signed rank test (IBM SPSS Statistics 19, IBM Corporation, Armonk City, State of New York, the United States) was conducted (significance was accepted at $p < 0.05$) as shown in Table 5. Because the minimum threshold airflow velocity measured by the anemometer was 0.3 m/s, "unmeasurable" indicates that the airflow velocities at the measured points were less than 0.3 m/s. Hence, only the measured points of a2, a3, a4, a7, and a8, with LHD and the measured points of a1, a3, a4, a7, and a8 without LHD were analyzed. No significant differences have been found between the measurements and the simulation without and with LHD, respectively ($p = 0.414$ and $p = 0.083$), when K-Epsilon model was employed. Figure 11a,b show the airflow temperature distributions in the cross-section of X = 14 m without and with LHD, respectively, using the K-Epsilon model.

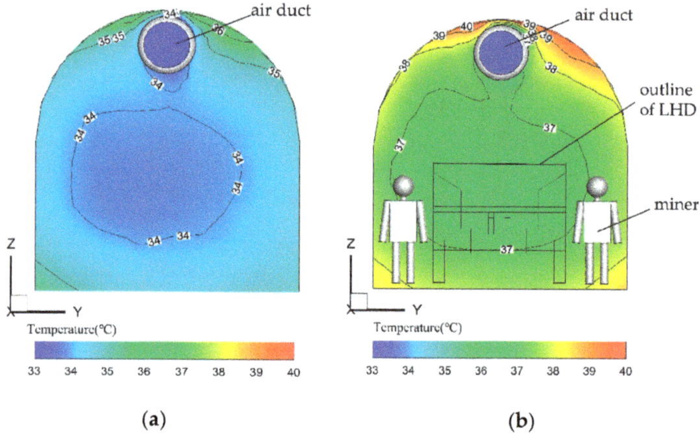

Figure 11. Airflow temperature distributions in the cross section of X = 14 m without and with LHD. (**a**) without LHD; (**b**) with LHD.

The average temperatures were 34.1 °C and 37.4 °C, respectively. Figure 12a,b display the comparison between the measured data of the airflow temperature and the simulation results without and with LHD respectively. The results suggest that K-Epsilon model can give reasonably good prediction, which is in line with the finding of Agus P. Sasmito [26]. The effectiveness of the K-Epsilon model with regard to airflow velocity and temperature simulation was verified. K-Epsilon model considers two-equation model which deals with turbulent kinetic energy, k, and its rate of dissipation, ε, which is coupled with turbulent viscosity. This model is given as:

$$\frac{\partial}{\partial t}(\rho k) + \nabla \cdot (\rho U k) = \nabla \cdot \left[\left(\mu + \frac{\mu_t}{\sigma_k}\right)\nabla k\right] + G_k - \rho\varepsilon \tag{8}$$

$$\frac{\partial}{\partial t}(\rho\varepsilon) + \nabla \cdot (\rho U \varepsilon) = \nabla \cdot \left[\left(\mu + \frac{\mu_t}{\sigma_\varepsilon}\right)\nabla\varepsilon\right] + C_{1\varepsilon}\frac{\varepsilon G_k}{k} + C_{2\varepsilon}\rho\frac{\varepsilon^2}{k} \tag{9}$$

where G_k represents the generation of turbulence kinetic energy due to the mean velocity gradients, $C_{1\varepsilon}$ and $C_{2\varepsilon}$ are model constants, σ_k and σ_ε are the turbulent Prandtl numbers corresponding to the k equation and the ε equation, respectively, and μ_t is turbulent viscosity given by:

$$\mu_t = \rho C_\mu \frac{k^2}{\varepsilon} \tag{10}$$

The values of $C_{1\varepsilon}$, $C_{2\varepsilon}$, C_μ, σ_k, and σ_ε are 1.44, 1.92, 0.09, 1, and 1.3, respectively.

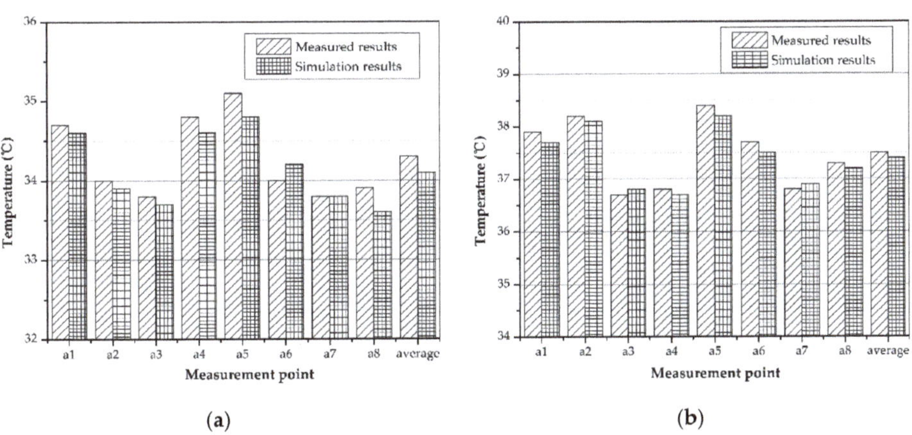

Figure 12. Comparison between the measured data and the simulation results without and with LHD: (a) without LHD; and (b) with LHD.

4. Results and Discussion

The simulation results were mainly focused on the two different planes as shown in Figure 13 including the Y = 2.3 m plane that is vertical to Y-axis (Plane A), as well as Z = 1.7 m (Plane B) which are vertical to Z-axis. The miners are typically located at the area near to the mining face. Hence, the region surrounded by X = 27 m to 37 m and Z = 0 m to 4 m was defined as the occupied zone while the remaining region was defined as the unoccupied zone in accordance with the regional distribution of miners.

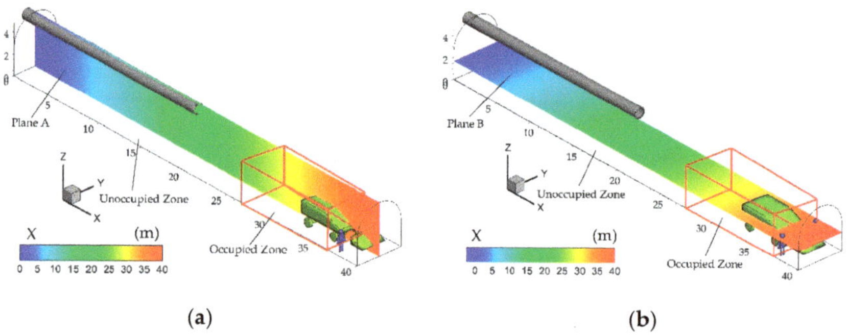

(a)   (b)

Figure 13. Selection of planes. (**a**) Plane A; and (**b**) Plane B.

4.1. Effect of Different Z_m on the Thermal Environment

Figure 14 shows the airflow velocity distributions in Plane A, when Z_m was set at five different values (specifically 5 m, 10 m, 15 m, 20 m, and 25 m). According to the International Tunnelling Association, the minimum airflow velocity should be greater than 0.3 m/s [27]. Obviously, the airflow velocity was maintained below 0.3 m/s in many regions from Figure 14 when Z_m = 5 m, Z_m = 10 m, and Z_m = 15 m, respectively. On the contrary, when Z_m = 20 m, it can be observed that the regions of low airflow velocity can achieve less coverage. With this consideration, when Z_m = 5 m, Z_m = 10 m and Z_m = 15 m, the forcing duct outlet was too close to the mining face, and the airflow could rapidly enter the relatively closed mining region after free expansion. The LHD greatly obstructed the return airflow, resulting in low reflux velocity.

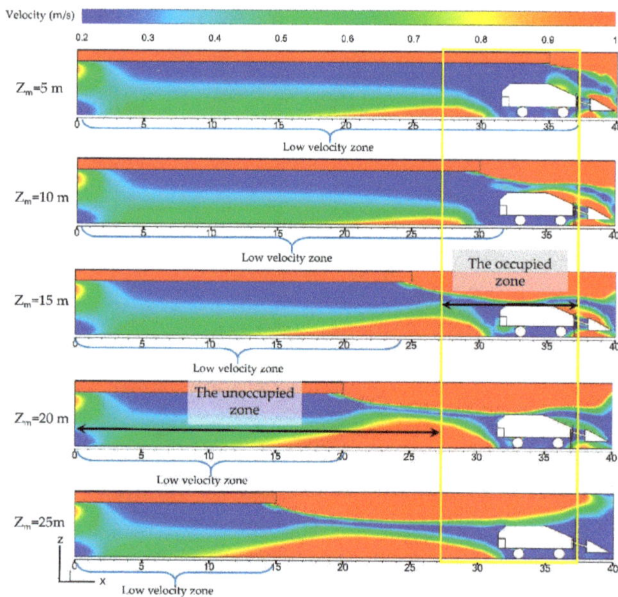

Figure 14. Airflow velocity distributions in plane A when Z_m = 5 m, Z_m = 10 m, Z_m = 15 m, Z_m = 20 m, and Z_m = 25 m.

When Z_m = 20 m, the tail of the LHD restricted the development and diffusion of the jet flow and it reflowed back to the airway due to the blocking effect of the LHD. When Z_m = 25 m, the distance exceeded the effective range of the jet flow. The airflow will form a circulating eddy zone before reaching the mining face and, thus, the effective reflux cannot be formed. Therefore, it can be concluded that the blocking effect of ventilation barriers such as LHDs or roadheaders on the airflow velocity distributions should be considered when repositioning the forcing duct.

Figure 15 depicts the airflow temperature distributions in Plane A when Z_m was set at five different values as mentioned above. It can be observed that, there are no obvious differences in the temperature distributions in the unoccupied zone under different values of Z_m. It can be considered that the airflow in unoccupied zone is mostly constituted by reflux and similar heat transfer has taken place between the airflow and the LHD as well as the surrounding rock. On the contrary, obvious differences in the temperature distributions are observed in the occupied zone. Obviously, the smaller Z_m has better cooling performance in the occupied zone from Figure 15 considering that the smaller Z_m can encourage more fresh airflow to involve the cooling process.

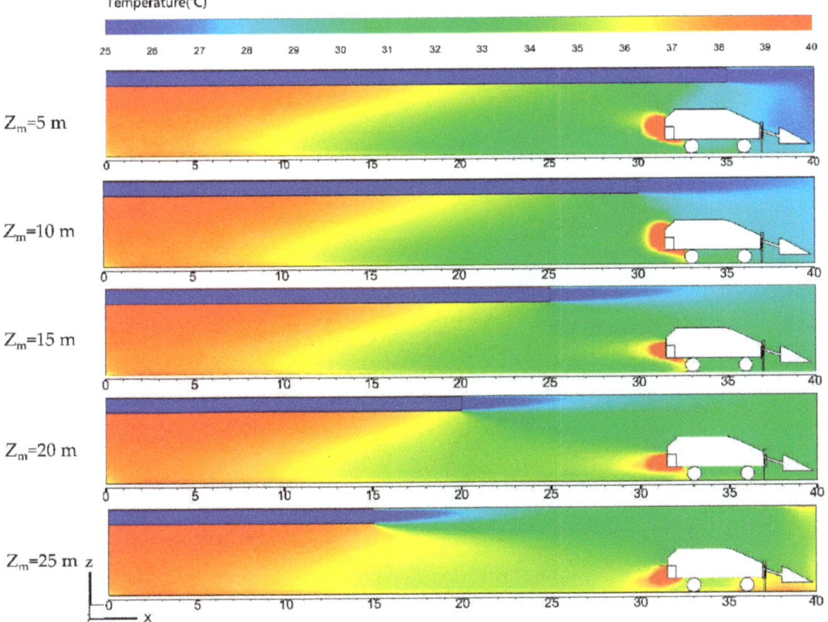

Figure 15. Airflow temperature distributions in plane A when Z_m = 5 m, Z_m = 10 m, Z_m = 15 m, Z_m = 20 m, and Z_m = 25 m.

Figure 16 shows the distribution curves of the airflow temperature at the position of Y = 4.1 m, Z = 1.7 m along the airway. It can be concluded that the different positions of the forcing duct will remarkably affect the airflow temperature distributions in the blind heading, especially in the occupied zone. It can be observed that the airflow temperature fluctuated to various degrees when Z_m is set at five different values at X = 30 m of the airway due to the heat emitted from the LHD. More obvious fluctuations can be seen at Z_m = 5 m, Z_m = 10 m, and Z_m = 25 m.

Figure 17 depicts the airflow relative humidity distributions in Plane A when Z_m was set at five different values. It can be observed that, the smaller Z_m, the higher relative humidity in the occupied zone. The relative humidity gradually decreases along the length of the airway (from the heading face to the outlet of the airway). In the case of the same moisture content of the airflow (the moisture content of the airflow was 13.7 g/m^3), the relative humidity decreases with the increasing dry bulb temperature. If Z_m is set to a higher value, the cooling airflow cannot fully involve in the cooling process in the occupied Zone due to the blocking effect of the LHD. To sum up, the temperature in the occupied zone is about 29 °C and the relative humidity is about 55% when $Z_m = 5$ m, which is more comfortable for miners. Hence, $Z_m = 5$ m should be adopted in subsequent analysis. Much care should be taken that the position of the LHD and the heat generated from the LHD are the important factors when optimizing the ventilation and cooling system.

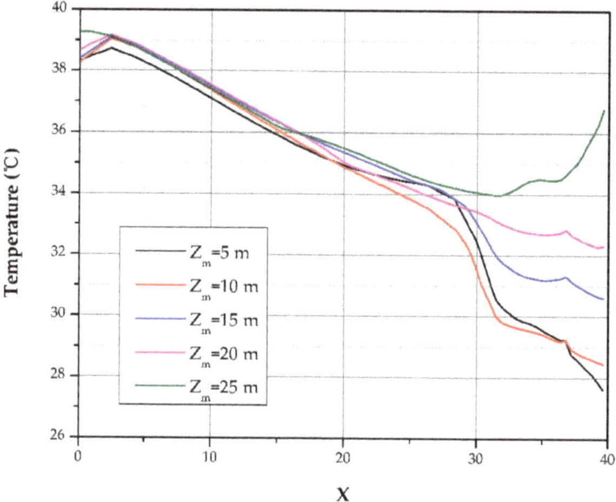

Figure 16. Airflow temperature distributions of different Z_m along the airway.

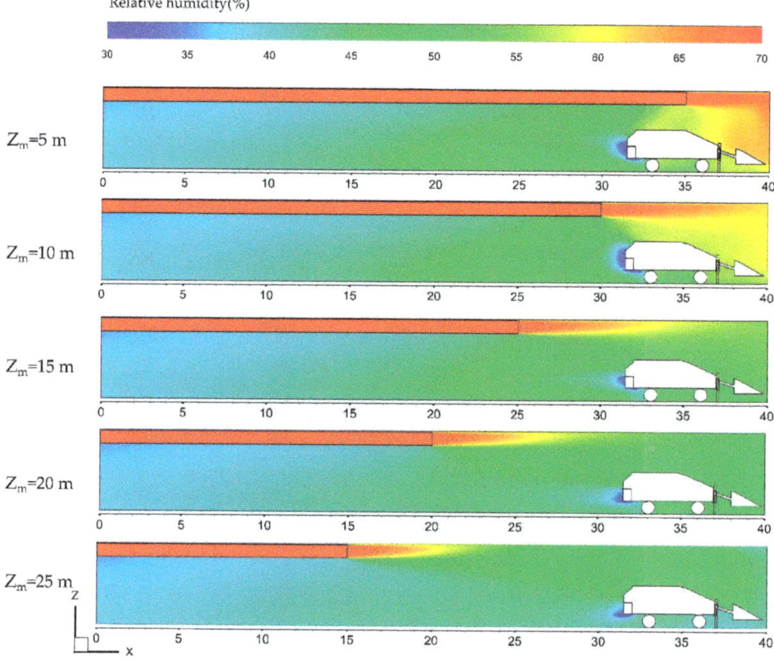

Figure 17. Airflow relative humidity distributions in plane A when $Z_m = 5$ m, $Z_m = 10$ m, $Z_m = 15$ m, $Z_m = 20$ m, and $Z_m = 25$ m.

4.2. Effect of Heat Emitted from LHD on the Thermal Environment

In most cases, an LHD operates continuously, but uses only full power (100%) while loading buckets and hauling up a ramp for 15 min per hour, operates at 50% maximum power tramming horizontally or downhill for 30 min per hour, and idles at 10% maximum power for the remaining 15 min per hour [28]. Investigating the variation of airflow temperature under different heat generation from a LHD can provide foundation for determining which heat removal/cooling should be applied. It provides great significance to realize the intelligent underground mining.

Figure 18 shows the temperature distributions in the unoccupied zone and occupied zone when Q_L is set at five different values (specifically, 10 kW, 20 kW, 30 kW, 40 kW, and 50 kW) which represent the heat generation from LHD under different working conditions. Obviously, when the Q_L increased from 10 kW to 50 kW, the average air temperature in the occupied zone increased from 30 °C to 34 °C. The airflow in the unoccupied zone is mostly constituted by the reflux. The heat transferred between the airflow and the LHD will result in the reflux temperature increasing to above 37 °C. Moreover, there is a long range and high flow velocity of the forcing system in the ventilation duct outlet that is frequently employed in the hot blind headings but it still has a limited cooling effect. Therefore, great efforts should be made to explore better ventilation and cooling schemes.

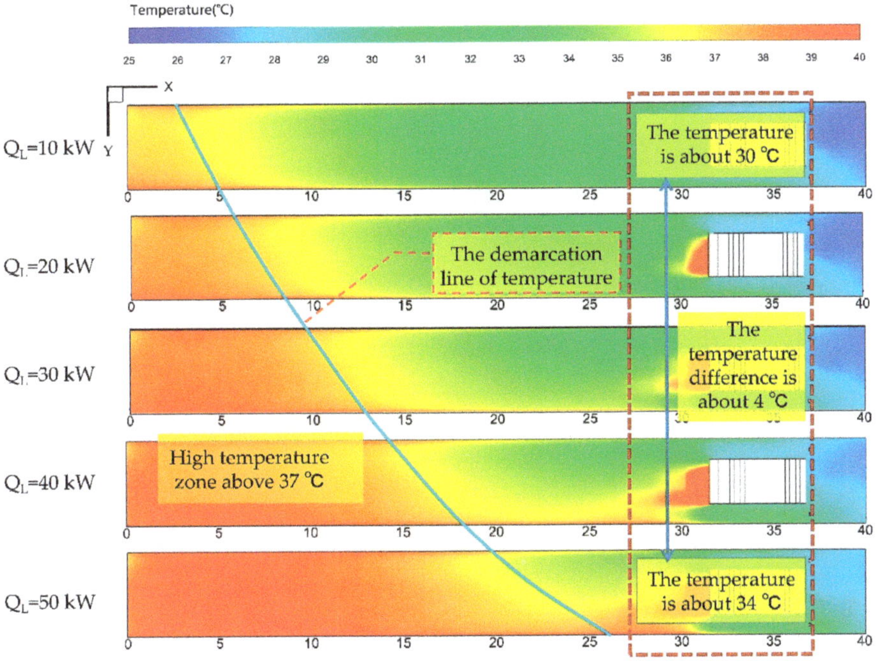

Figure 18. Airflow temperature distributions in Plane B with different Q_L.

Figure 19 depicts the airflow relative humidity distributions in Plane B when Q_L is set at five different values. It can be observed that the raise of Q_L, can decrease the relative humidity in the airway. As mentioned above, the more heat generated by the LHD leads to the higher dry bulb temperature. In the case of the same moisture content of the airflow, the relative humidity decreases with the increasing dry bulb temperature.

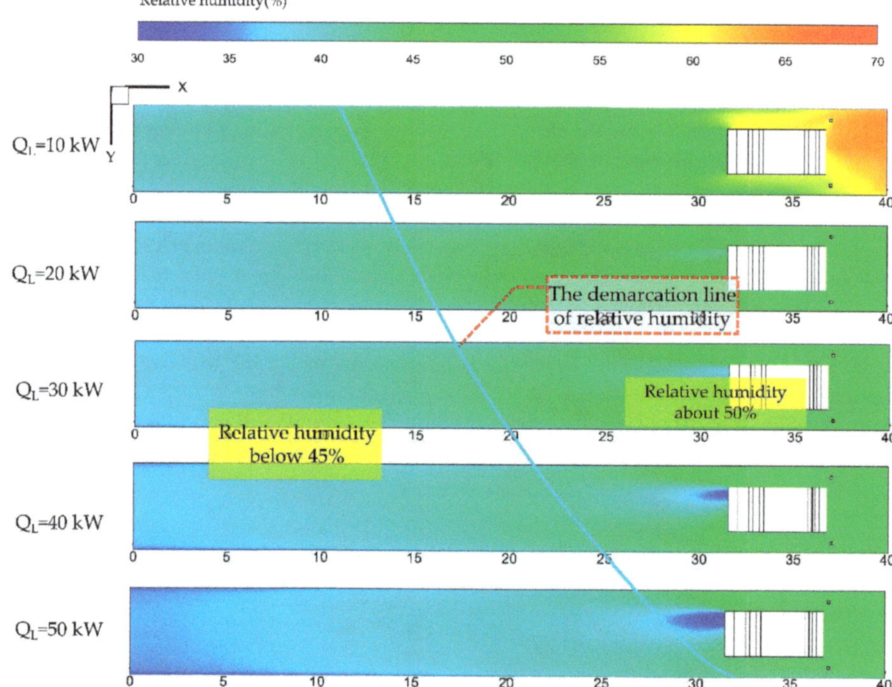

Figure 19. Airflow relative humidity distributions in Plane B with different Q_L.

4.3. Effect of Auxiliary Ventilation on Thermal Environment of Airway

Sasmito et al. conducted CFD-based studies on the effect of ventilation airflow velocity. They found that increasing the forcing airflow velocity from 12 m/s to 20 m/s only reduced average temperature by 0.2 °C [12]. Limitations exist in PF to achieve a better cooling effect. However, the overlap systems, as shown in Figure 1, combine the advantages of the PF and PE, which provides another alternative to ameliorate adverse thermal environment. Many researchers carried out in situ measurements and numerical simulation studies on the auxiliary ventilation system from the viewpoint of dust and methane control [14,17]. However, few studies on the field of thermal environment are available. Therefore, the current work will employ the overlap systems to investigate the cooling effect.

As shown in Figures 1 and 20, four systems will be analyzed. The first one is PF, which is the same as the previous study. The second one is PE with an exhausting duct of 800 mm in diameter, $Z_e = 5$ m and the exhaust airflow velocity is 24 m/s. The exhausting duct is located on the airway floor. The third one is FFNE. The diameter of the forcing duct and the exhausting duct are the same as PF and PE, respectively, but $Z_m = 15$ m and $Z_e = 5$ m. To prevent recirculation, the total quantity of air exhausted must be at least twice the quantity delivered by the force fan. Therefore, the force airflow velocity was 12 m/s and the exhaust airflow velocity was 24 m/s. The fourth one is NFFE, which is similar to the FFNE but $Z_m = 5$ m and $Z_e = 15$ m [7].

Figure 21 displays the airflow temperature distributions of the four auxiliary ventilation systems at the Plane B, respectively. The simulation results show that the cooling performance in the occupied zone reaches the optimal level when applying NFFE. The airflow temperatures were reduced by about 3 °C compared with that of PF in the occupied zone when adopting the NFFE ventilation system. The temperature in the unoccupied zone would be increased but fewer miners would have to work in this region, which would enhance the energy utilization coefficient (EUC) based on EUC = $(t_{uz} - t_s)/(t_{oz} - t_s)$, where t_{uz} was the average temperature in the unoccupied zone, t_s was the supply air temperature, and t_{oz} was the average temperature in the occupied zone. Thus, it had conformed to the reasonable distribution of the cold flow [29].

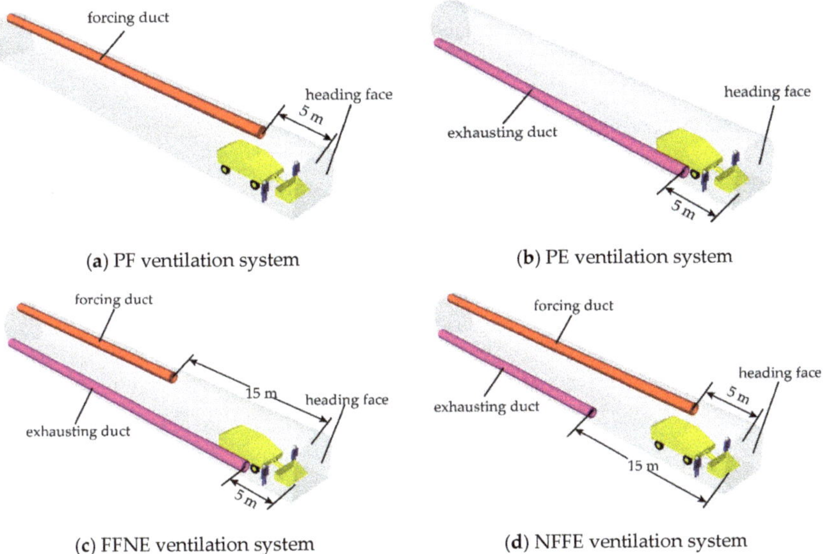

Figure 20. Geometry model of the PF (**a**), PE (**b**), FFNE (**c**), and NFFE (**d**) ventilation systems.

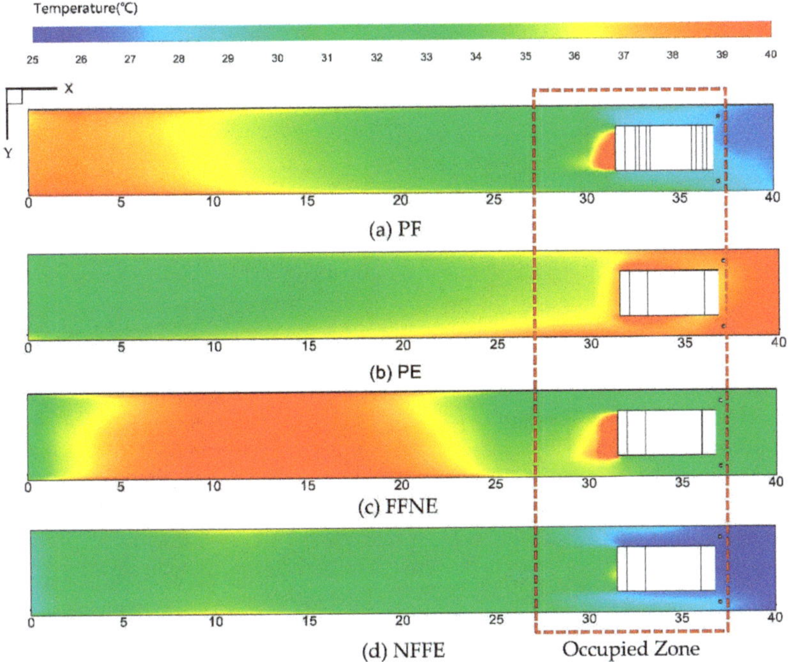

Figure 21. Airflow temperature distributions of the four auxiliary ventilation systems.

Figure 22 shows the relative humidity distributions of the four auxiliary ventilation systems at the Plane B respectively. The average relative humidity is 56.4%, 62.1%, 42.6%, and 53.5% in the occupied zone, respectively. It can be concluded that when the overlap systems (FFNE and NFFE) is applied, the relative humidity of the airflow can be controlled to an optimal level. This is mainly due to the moist air being continuously migrated by the exhaust duct.

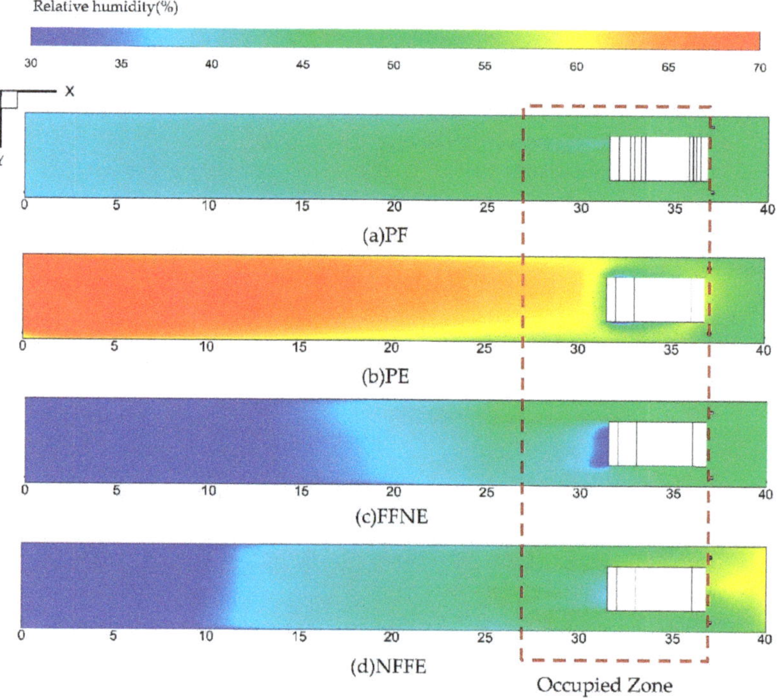

Figure 22. Airflow relative humidity distributions of the four auxiliary ventilation systems.

Figure 23 presents the streamlines under the NFFE ventilation system. As can be seen, within the region of X = 20~30 m, the airflow is mainly subject to the negative pressure engulfment of the exhausting fan, supplemented by the positive pressure effect of the jet flow by the blowing fan. The engulfment confluence of the exhausting fan will be enhanced under the joint action of negative and positive pressure. This region can be defined as the effective engulfment region of the exhausting duct. As the working face is far from the forcing ventilation duct, the local eddy current can be generated at the interval of X = 30~35 m. Literature [30] indicates that it is beneficial to ventilation, since the circulating airflow can absorb a large amount of moisture, thus reducing the airflow temperature and humidity in the working face.

Figure 24 illustrates the airflow temperature distributions in the plane B under the NFFE ventilation system and the different heat generations from LHD. Compared with Figure 18, it can be clearly observed that the NFFE ventilation system can remarkably reduce the temperature in the occupied zone.

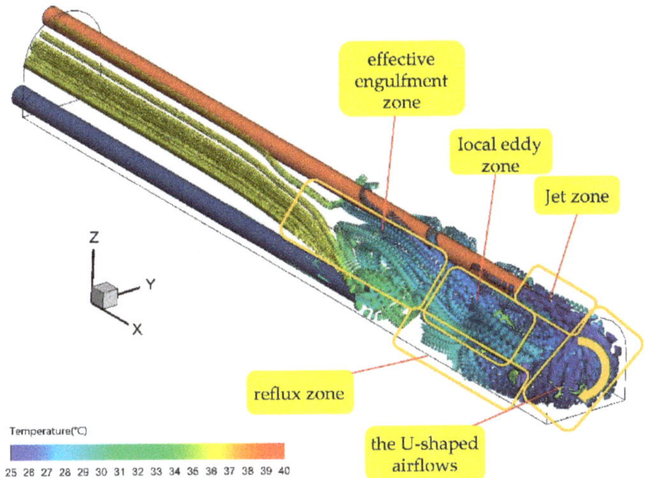

Figure 23. Streamlines of the NFFE ventilation system.

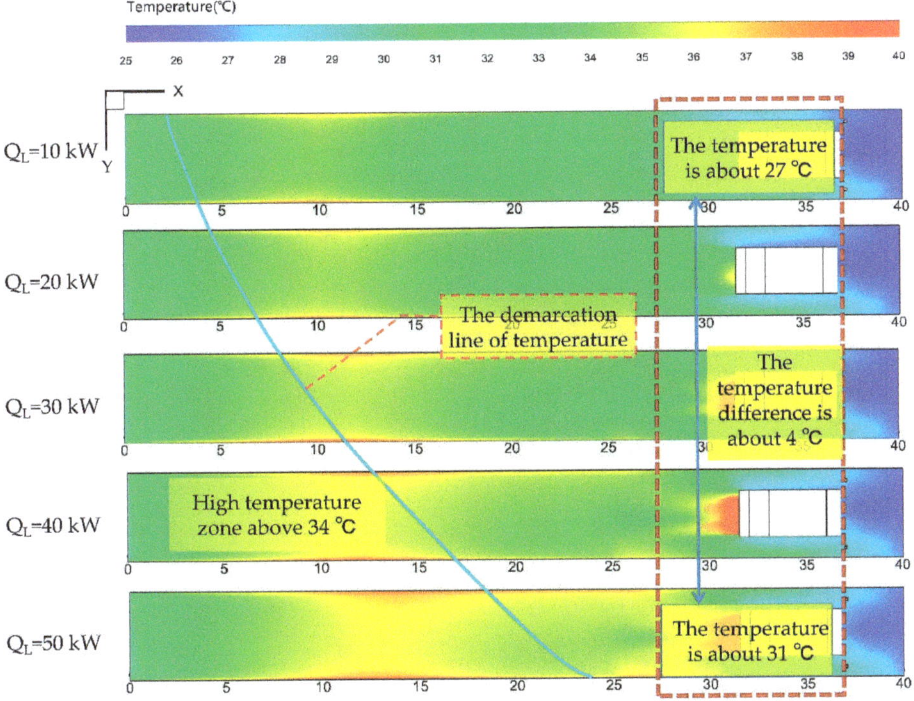

Figure 24. Air temperature distributions in Plane B with different Q_L under the NFFE ventilation system.

5. Conclusions

This study suggests that, in the presence of ventilation barriers such as LHD, the blocking action of LHDs will have great influences on the airflow velocity field and temperature field. As a result, the influence of mechanical equipment like LHDs on the thermal environment should be taken into account when designing the local ventilation system.

In this case, the optimal ventilation duct position to achieve the optimal cooling effect is 5 m after simulating the airflow velocity, dry-bulb temperature, and relative humidity distributions under different distances between ventilation duct outlet to the mining face. Such results demonstrate that the optimal ventilation duct position should be explored for airways with different specifications to achieve the goal of energy conservation and environmental protection.

It can be clearly observed based on the simulation that the airflow temperature is increased with the increase of Q_L, indicating that the airflow temperature distribution is closely correlated with the heat emitted from the mechanical equipment. However, how to further improve the local ventilation cooling performance needs to be further investigated.

Finally, this study employs the NFFE ventilation pattern for numerical simulation and the results suggest that such ventilation patterns can remarkably reduce the temperatures in the occupied zone compared with the traditional ventilation pattern, thus achieving better energy utilization efficiency and cooling performance.

Author Contributions: W.Y. contributed to the conception of the study; W.W. and C.Z. provided the research data; G.Q. and C.L. analyzed the data; S.L. and H.X. helped perform the analysis with constructive discussions; W.W. and H.M. proposed the numerical simulation scheme and implemented; and W.W. and W.Y. wrote the paper.

Funding: This work has been funded by the National Natural Science Foundation of China, grant numbers 51804183, 51774197, and 51804185, Focus on Research and Development Plan in Shandong Province, grant number 2018GSF116012, the Natural Science Foundation of Shandong Province, grant number ZR2016EEP02 and Scientific Research Foundation of Shandong University of Science and Technology for Recruited Talents, grant number 2014RCJJ031.

Conflicts of Interest: We especially declare here this paper is not simultaneously submitted for publication elsewhere. There is no conflict of interest.

References

1. Greth, A.; Roghanchi, P.; Kocsis, K. A review of cooling system practices and their applicability to deep and hot underground US mines. In Proceedings of the 16th North American Mine Ventilation Symposium, Golden, CO, USA, 17–22 June 2017; Volume 11, pp. 1–9.

2. Kong, B.; Li, Z.H.; Wang, E.Y.; Yang, Y.L.; Chen, L.; Qi, G.S. An experimental study for characterization the process of coal oxidation and spontaneous combustion by electromagnetic radiation technique. *Process Saf. Environ. Prot.* **2018**, *119*, 285–294. [CrossRef]

3. Kong, B.; Wang, E.Y.; Li, Z.H.; Lu, W. Study on the feature of electromagnetic radiation under coal oxidation and temperature rise based on multi-fractal theory. *Fractals* **2019**, *27*, 14–38.

4. Lowndes, I.S.; Crossley, A.J.; Yang, Z.Y. The ventilation and climate modelling of rapid development tunnel drivages. *Tunn. Undergr. Space Technol.* **2004**, *19*, 139–150. [CrossRef]

5. Yang, W.Y.; Inoue, M.; Miwa, S.H.; Kobayashi, Y. Heat form fragmented rock, explosives and diesel equipment in blind headings of underground mines. *Mem. Fac. Eng. Kyushu Univ.* **2013**, *73*, 13–26.

6. Wala, A.M.; Vytla, S.; Taylor, C.D.; Huang, G. Mine face ventilation: A comparison of CFD results against benchmark experiments for the CFD code validation. *Min. Eng.* **2007**, *59*, 49–55.

7. Glehn, F.H.V.; Bluhm, S.J. Practical aspects of the ventilation of high-speed developing tunnels in hot working environments. *Tunn. Undergr. Space Technol.* **2000**, *15*, 471–475. [CrossRef]

8. Liu, C.Q.; Nie, W.; Bao, Q.; Liu, Q.; Wei, C.H.; Hua, Y. The effects of the pressure outlet's position on the diffusion and pollution of dust in tunnel using a shield tunneling machine. *Energy Build.* **2018**, *176*, 232–245. [CrossRef]

9. Lowndes, I.S.; Yang, Z.Y.; Jobling, S.; Yates, C. A parametric analysis of a tunnel climatic prediction and planning model. *Tunn. Undergr. Space Technol.* **2006**, *21*, 520–532. [CrossRef]

10. Shimada, S.; Ohmura, S. Experimental study on cooling of heading face by local ventilation. *Shigen-to-Sozai* **1990**, *106*, 725–729. [CrossRef]

11. Jia, M.T.; Chen, Y.H.; Wu, L.J.; Wang, S.; Qi, C.G. The heat exchange simulation test platform for deep excavation roadway in metal mine. *Met Mine* **2014**, *461*, 113–116.

12. Sasmito, A.P.; Kurnia, J.C.; Birgersson, E.; Mujumdar, A.S. Computational evaluation of thermal management strategies in an underground mine. *Appl. Therm. Eng.* **2015**, *90*, 1144–1150. [CrossRef]

13. Gao, J.L.; Uchino, K.; Inoue, M. Simulation of Thermal Environmental Conditions in heading face with Forcing Auxiliary Ventilation-control of thermal environmental conditions in locally ventilated working face. *Shigen-to-Sozai* **2002**, *118*, 9–15. [CrossRef]
14. Hua, Y.; Nie, W.; Wei, W.L.; Liu, Q.; Liu, Y.H.; Peng, H.T. Research on multi-radial swirling flow for optimal control of dust dispersion and pollution at a fully mechanized tunnelling face. *Tunn. Undergr. Space Technol.* **2018**, *79*, 293–303. [CrossRef]
15. Zhou, G.; Feng, B.; Yin, W.J.; Wang, J.Y. Numerical simulations on airflow-dust diffusion rules with the use of coal cutter dust removal fans and related engineering applications in a fully-mechanized coal mining face. *Powder Technol.* **2018**, *339*, 354–367. [CrossRef]
16. Cai, P.; Nie, W.; Chen, D.W.; Yang, S.B.; Liu, Z.Q. Effect of airflow rate on pollutant dispersion pattern of coal dust particles at fully mechanized mining face based on numerical simulation. *Fuel* **2019**, *239*, 623–635. [CrossRef]
17. Torano, J.; Torno, S.; Menendez, M.; Gent, M. Auxiliary ventilation in mining roadways driven with roadheaders: Validated CFD modelling of dust behaviour. *Tunn. Undergr. Space Technol.* **2011**, *26*, 201–210. [CrossRef]
18. Ren, T.; Wang, Z.; Cooper, G. CFD modelling of ventilation and dust flow behaviour above an underground bin and the design of an innovative dust mitigation system. *Tunn. Undergr. Space Technol.* **2014**, *41*, 241–254. [CrossRef]
19. Cheng, W.M. *Mining Ventilation and Safety*; China Coal Industry Publishing House: Beijing, China, 2016.
20. McPherson, M.J. *Subsurface Ventilation and Environmental Engineering*; Chapman and Hall: New York, NY, USA, 1993.
21. Kurnia, J.C.; Sasmito, A.P.; Mujumdar, A.S. Simulation of a novel intermittent ventilation system for underground mines. *Tunn. Undergr. Space Technol.* **2014**, *42*, 206–215. [CrossRef]
22. Shim, B.O.; Park, J.M.; Kim, H.C.; Lee, Y. Statistical Analysis on the Thermal Conductivity of Rocks in the Republic of Korea. In Proceedings of the World Geothermal Congress, Bali, Indonesia, 25–30 April 2010.
23. Mao, N.; Pan, D.; Deng, S.; Chan, M. Thermal, ventilation and energy saving performance evaluations of a ductless bed-based task/ambient air conditioning (TAC) system. *Energy Build.* **2013**, *66*, 297–305. [CrossRef]
24. Parra, M.T.; Villafruela, J.M.; Castro, F.; Mendez, C. Numerical and experimental analysis of different ventilation systems in deep mines. *Build. Environ.* **2006**, *41*, 87–93. [CrossRef]
25. Sasmito, A.P.; Birgersson, E.; Ly, H.C.; Mujumdar, A.S. Some approaches to improve ventilation system in underground coal mines environment—A computational fluid dynamic study. *Tunn. Undergr. Space Technol.* **2013**, *34*, 82–95. [CrossRef]
26. Kurnia, J.C.; Xu, P.; Sasmito, A.P. A novel concept of enhanced gas recovery strategy from ventilation air methane in underground coal mines—A computation investigation. *J. Natl. Gas Sci. Eng.* **2016**, *35*, 661–672. [CrossRef]
27. Yang, L.X. *Modern Tunwnel Construction Ventilation Technology*; China Communications Press: Beijing, China, 2012.
28. Mkhwanazi, D. Optimizing LHD utilization. *South. Afr. Inst. Min. Metall.* **2011**, *111*, 273–280.
29. Liu, W.W.; Lian, Z.W.; Yao, Y. Optimization on indoor air diffusion of floor-standing type room air-conditioners. *Energy Build.* **2008**, *40*, 59–70. [CrossRef]
30. Ren, T.; Wang, Z.W. CFD modelling of ventilation, dust and gas flow dispersion patterns on a longwall face. In *Proceedings of the 11th International Mine Ventilation Congress*; Springer: Singapore, 2019.

 *processes*

Article

Computational Fluid Dynamic Simulation of Inhaled Radon Dilution by Auxiliary Ventilation in a Stone-Coal Mine Laneway and Dosage Assessment of Miners

Bin Zhou [1,2], Ping Chang [2,*] and Guang Xu [2]

[1] College of Safety and Emergency Management Engineering, Taiyuan University of Technology, Taiyuan 030024, China
[2] Department of Mining Engineering and Metallurgical Engineering, WA School of Mines: Minerals, Energy and Chemical Engineering, Curtin University, Kalgoorlie, WA 6430, Australia
* Correspondence: ping.chang@postgrad.curtin.edu.au; Tel.: +61-(0)452282422

Received: 30 April 2019; Accepted: 1 August 2019; Published: 5 August 2019

Abstract: Inhaled radon status in the laneways of some Chinese stone-coal mines is a cause of concern. In this study, computational fluid dynamics simulations were employed to investigate three flowrates of the dilution gas (2.5, 5, and 7.5 m^3/s) and radon distributions at realistic breathing levels (1.6, 1.75, and 1.9 m). The results showed that there are obvious jet-flow, backflow, and vortex zones near the heading face, and a circulation flow at the rear of the laneway. A high radon concentration area was found to be caused by the mining machinery. As the ventilation rate increased, the radon concentrations dropped significantly. An airflow of 7.5 m^3/s showed the best dilution performance: The maximum radon concentration decreased to 541.62 Bq/m^3, which is within the safe range recommended by the International Commission on Radiological Protection. Annual effective doses for the three air flowrates were 8.61, 5.50, and 4.12 mSv.

Keywords: coal mining; radon concentration; ventilation; computational fluid dynamics; occupational exposure assessment

---

1. Introduction

^{222}Rn has recently drawn much attention, due to its radioactivity in underground workplaces which can cause over-exposure in miners [1,2]. ^{222}Rn (radon; Rn) is widely found in underground rocks as a product of uranium decay and can diffuse from rock pores to the working space during mining operations. Gaseous radon can become trapped in the lungs once inhaled, where it decays further and releases ionizing alpha particles. These high-energy particles can seriously damage the lung tissue and increase the probability of cell mutation leading to cancer [3].

To date, most radon research has focused on uranium mines, which have relatively high radon concentrations in the working environment. In the Colorado Plateau of the United States, radon concentrations in uranium mines can reach as high as 2.18×10^6 Bq/m^3 [4]. In the cut-and-fill stopes of a uranium mine in Canada, the radon concentrations varied in the range of 481–2960 Bq/m^3 [5]. All of these sites exceeded the action level (500–1500 Bq/m^3) for the workplace environment, as recommended by the International Commission on Radiological Protection (ICRP) [6]. Al-Zoughool et al. [7] assessed several studies on the radon exposure of uranium miners in six countries (U.S.A., Australia, Canada, France, Germany, and Czech Republic): the exposure of radon inhalation ranged from 7.6–595.7 working level months (WLM), which was much higher than the threshold limit of 4 WLM prescribed by the ICRP [8].

There have been few studies on radon occupational hazards for coal miners. This is mainly because the content of radionuclides in coal is low and good ventilation can usually maintain radon concentrations below the action level [9,10]. In some coal mines in China, however, the uranium content of the coal seams is relatively high and the mines are ventilation-poor, which leads to the potential risk of radon accumulation and radiological impacts on miners. Liu et al. [11] measured radon concentrations in 48 coal mines in twelve major coal-producing provinces in China. The results showed that there were greater radon concentrations in stone-coal mines without auxiliary ventilation. The concentrations ranged from 136–4183 Bq/m^3, with an average value of 1244 Bq/m^3. The typical value of radon concentrations of stone-coal mines in China was, therefore, suggested to be 1500 Bq/m^3, which is three times the lower limit of the action level (500 Bq/m^3). Stone coal is a type of anthracite with low carbon content, high ash content, low heat value, and which contains large amounts of pyrite, quartz, and phosphorus nodules [12]. The high uranium content of the mineral nodules and lack of effective ventilation are the major reasons that radon exceeds the limit in laneways of stone-coal mines. Occupational exposure of miners also requires attention, due to their long working hours (2400 h per year) [11].

Computational fluid dynamics (CFD), as a visual low-cost tool, has been widely used to solve health-related issues in the mining industry. Ren and Wang et al. [13,14] established a dust-removal model in a mine roadway and explored the impact of ventilation on dust contamination. The results illustrated that the area of high dust concentration shrank obviously, dropping by 72% when the airflow rate was increased from 7 m^3/s to 13 m^3/s. Kurnia et al. [15] designed three ventilation scenarios to investigate the relationship between airflow behavior and methane dispersion in a mine tunnel. The best methane mitigation was achieved at the maximum airflow rate. An actual ventilation shaft in a Chinese uranium mine was modeled by Xie et al. to predict radon radiation effects on surrounding residents [16]. As the wind speed increased from 0.5 m/s to 2 m/s, radon contamination within a range of 150 m from the shaft decreased from 600 Bq/m^3 to the limit of 100 Bq/m^3. At present, auxiliary ventilation is still the main way to effectively control toxic and harmful pollutants in an underground workplace. There are three main auxiliary ventilation systems used in coal mine laneways: Forcing ventilation systems, exhaust ventilation systems, and mixed ventilation systems (a forcing ventilation system together with an exhaust ventilation system); the forcing ventilation system being the most commonly used [17–19]. It is, therefore, important to understand airflow behavior in the laneway to select proper ventilation parameters to maintain radon concentrations at a reasonable level.

In this study, the airflow behavior in a typical laneway was simulated and analyzed by the OpenFOAM (Open Source Field Operation and Manipulation) CFD software at three ventilation rates (2.5 m^3/s, 5 m^3/s, and 7.5 m^3/s). Based on the obtained flow field, the scalarTransport equation was used to investigate the radon distribution at three respirable heights (1.6 m, 1.75 m, and 1.9 m). The optimal ventilation rate was, then, suggested for radon mitigation in the working area, according to the maximum radon results obtained from the various models. In addition, taking into account the working time over one year, the maximum annual effective dose received by miners was assessed. The study results are helpful for obtaining a better understanding of the complex airflow characteristics and radon dispersion under auxiliary ventilation and provides reliable guidance and education for the design of effective radon dilution systems.

2. Mathematical Formulation

2.1. Numerical Model

As the radon concentration data in this paper were derived from the statistical results of a survey of 48 coal mines in China [11], there are no specific laneway parameters to establish models for CFD research. For the purpose of this study, a typical laneway geometry and three ventilation parameters (2.5 m^3/s, 5 m^3/s, and 7.5 m^3/s), as used in the prior research by Shi et al. [20], were adopted to establish a numerical model for radon dilution performance research. Figure 1 illustrates the geometry of the underground laneway model, which includes a ventilation duct, a heading face, and mining machinery. The length of the laneway is 50 m and the shape of the laneway section is a semicircular arch, where the upper semicircular radius is 2.75 m and the width and height of the lower rectangle are 5.5 m and 1.05 m, respectively. The body shape of the machinery is roughly $10 \times 3.6 \times 1.8$ m. The ventilation duct, with a length of 43 m and a diameter of 0.8 m, is suspended on the side of the laneway; its distance from the ground is 2.5 m and the distance from the wall is 0.2 m. The distance between the outlet of the duct and the mining face is 7 m.

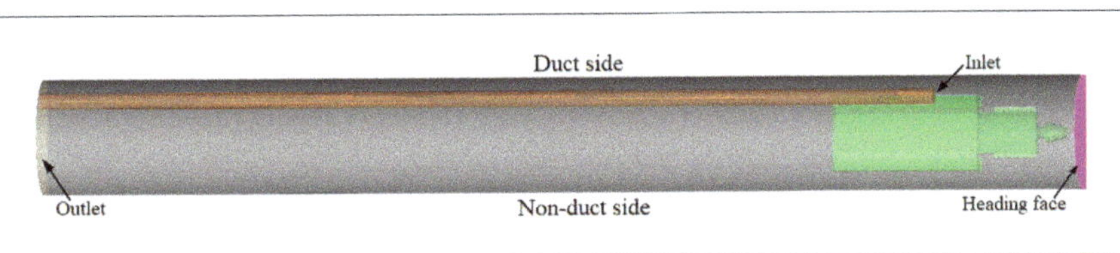

Figure 1. The geometry of underground laneway [20].

Following the construction of the model geometry, the computational domain was meshed. A structured mesh generally has a higher quality, which is suitable for simple and regular geometric models, while an unstructured mesh has relatively lower quality and is applied to complex geometric models. According to the geometric structure of the laneway model, hybrid meshes were selected and established by ICEM (Integrated Computer Engineering and Manufacturing code for computational fluid dynamics). The thickness of the outermost boundary layer was 0.005 m, the number of layers was 6, and the scale factor between adjacent layers was 1.2. A structural mesh was adopted in the domain behind the mining machinery with a quality exceeding 0.5, while an unstructured mesh was used in the rest of the domain with quality greater than 0.3; both mesh qualities met the requirements of the OpenFOAM software for simulation.

2.2. Boundary Conditions

The standard wall function was applied on the walls of the laneway and the mining machinery. A velocity-type inlet was adopted in the duct outlet to simulate the fresh air inlet of the laneway, and the ventilation airflow rates were 2.5 m³/s, 5 m³/s, and 7.5 m³/s. A pressure outlet was adopted in the outlet of the laneway, where the outlet pressure was treated as a standard atmospheric pressure (101 kpa). The radon source was set to 1500 Bq/m³·s and the diffusion coefficient of radon was set to 1.1×10^{-5} m²/s. The air density, dynamic viscosity, and kinematic viscosity were set as 1.213 kg/m³, 1.811×10^{-5} Pa/s, and 1.503×10^{-5} m²/s, respectively. A turbulence intensity of 5% was applied in all ventilation scenarios.

2.3. Governing Equations

In the underground coal mine, a large amount of fresh air is forced into the laneway space by the ventilation duct, which increases the Reynolds number of the airflow in the laneway. The standard k-ε turbulence model, a common turbulence model, was used in this OpenFOAM simulation. The k-ε model has been used by many scholars [13,18,21] for ventilation research in coal-mining related fields. It has been shown that the simulation results of the k-ε model are highly consistent with the field measured data. The air in the laneway was considered as an incompressible ideal gas and the effects of heat exchange and humidity on the airflow field were ignored. It is clear that the radon migration under the auxiliary ventilation in the laneway is very dynamic and will be affected by several ventilation parameters related to time and the facilities and, so, the steady-state assumption is not strictly valid. However, the computing costs, including grid quality, computer configuration, and computation time, especially in large 3D models, make time-dependent computational simulations difficult to implement. Moreover, the radiation hazard suffered by miners in the underground laneway is a long-term accumulation process. The annual radiation dose received by miners is estimated on the basis of a relatively stable radon concentration, so it is not of much concern to consider the fluctuations in radon concentration over a short period of time. Many previous studies [22–24] have also used measured data to justify the steady-state assumption model, and the comparison results showed that the model results and the measured results were highly consistent. Based on this, simpleFoam, a stable solver using the SIMPLE (semi-implicit method for pressure-linked equations) algorithm, was adopted to reach a steady-state airflow field ahead of the mining operation [25]. In the radon dispersion section, radon was assumed to have a negligible effect on the airflow fields, because its mass is very small compared with that of air (1 Bq/m^3 = 1.75 × 10^{-19} kg/m^3), the radon concentration of the ventilated air is also not considered. The radon concentration field could, therefore, be obtained using scalarTransportFoam, which is a solver for passive scalar transport in a given velocity field [26]. Transport equations for mass, momentum, and concentration of species were described, following [27].

The airflow is governed by the continuity and Navier–Stokes equations:

$$\frac{\partial \rho}{\partial t} + \nabla \cdot (\rho U) = 0, \tag{1}$$

$$\rho \left(\frac{\partial U}{\partial t} + U \cdot \nabla U \right) = -\nabla P + \mu_e \nabla^2 U, \tag{2}$$

where ρ is the air density, kg/m^3; U is the air velocity, m/s; P represents the pressure, N/m^2; and μ_e represents the effective viscosity, N/m^2·s, which is the sum of the dynamic and turbulent viscosities.

The radon motion was governed by the scalarTransport equation:

$$\frac{\partial C}{\partial t} + \nabla \cdot (UC) - \nabla \cdot (D * \nabla C) - \lambda C + S = 0, \tag{3}$$

where C is the radon concentration in the laneway, Bq/m^3; D is the Rn diffusion coefficient, m^2/s; S is the radon source item, Bq/m^3; and λ is radon decay coefficient, 2.1 × 10^{-6} s^{-1}.

2.4. Assessment of Occupational Exposure to Inhaled Radon

To evaluate the inhaled radon exposure of miners, three respirable heights were assessed, according to the range of miner heights. Combining the maximum radon concentration in the working area and average working time in one year, the maximum annual effective dose of radon inhalation was assessed using the following equation [28]:

$$E_{eff} = C_{R_n max} \times D_f \times T \times F, \tag{4}$$

where E_{eff} represents the maximum annual effective dose, mSv; $C_{R_n max}$ represents the maximum radon concentration in the activity area of workers, Bq/m^3; D_f is the dose conversion factor, 9.0 × 10^{-6} mSv/h per Bq/m^3; T is the annual number of working hours, h; and F is the equilibrium factor, which was set to 0.35 for a coal mine [29].

3. Mesh Independence Study

A mesh independence study considers the effect of the number of model cells on the numerical solution, which is essential in CFD modeling. Mesh refinement can reduce the probability of solution divergence and ensure that the numerical results approach the exact solution of the governing equations [30]; however, unnecessarily increasing the number of model cells reduces solution efficiency, thereby wasting computational resources [31]. To minimize the impact of the mesh on the calculation accuracy and to select a suitable number of cells to ensure adequate convergence speed, a medium mesh, with about 2 million cells, and a fine mesh, with about 4 million cells, were built. An airflow field with a ventilation rate of 5 m^3/s was adopted and the velocity features of three monitoring lines were taken into account to assess the mesh independence: The monitoring lines 1, 2, and 3 were located on the horizontal central lines of the laneway, and the distances from the outlet were 10 m, 30 m, and 48 m, respectively, as shown in Figure 2. The results of the mesh independence study are plotted in Figure 3.

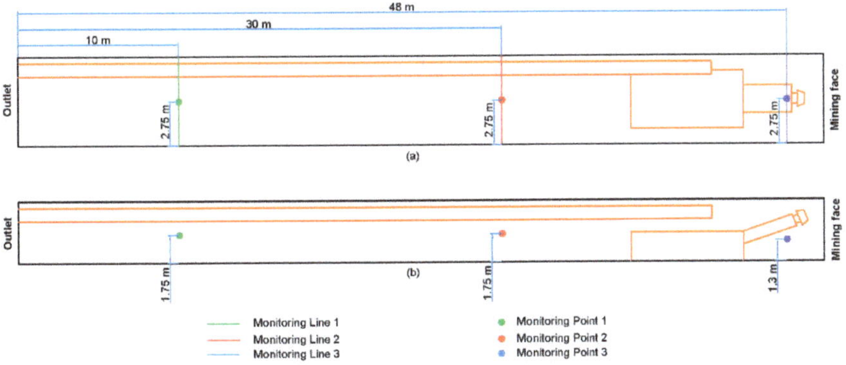

Figure 2. Position of monitoring lines and points.

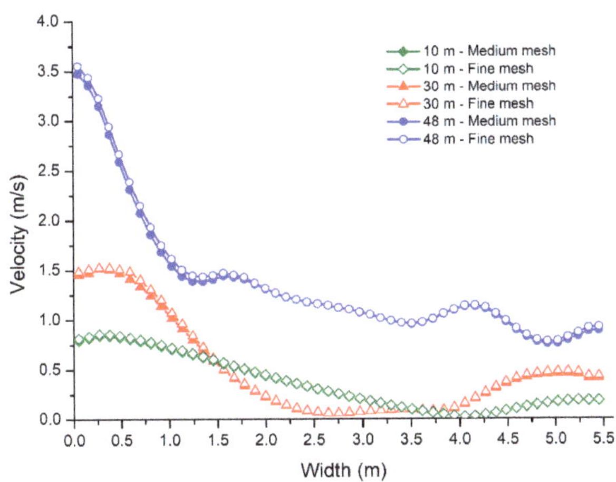

Figure 3. Airflow velocities at monitoring lines.

The velocity profiles at each position approximately coincided for the two meshes. This indicated that the medium mesh had little effect on the accuracy of the domain calculation and met the requirement for mesh independence. Therefore, the medium mesh with 2 million cells, as shown in Figure 4, was used for subsequent airflow field research and radon dispersion analysis.

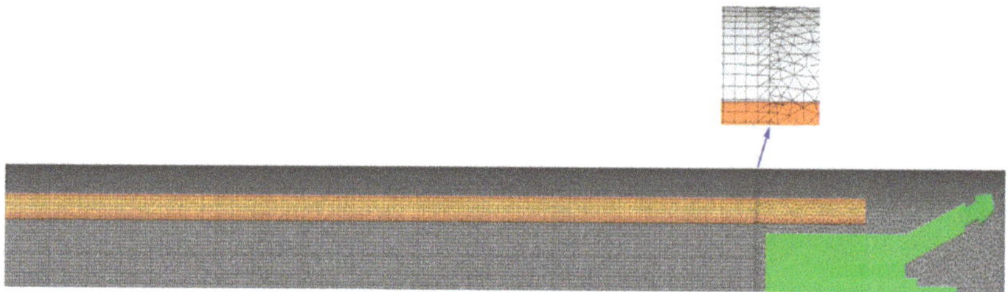

Figure 4. Laneway model using medium mesh.

4. Judging Criteria for Concentration Field Convergence

The criteria for the convergence of the radon concentration field were as follows: (1) To guarantee the solution was converged, the residual value of the concentration governing equation needed to reach 1×10^{-8}. (2) Three monitoring points were set to observe the concentration change and the concentration field was considered to reach a steady state when the concentrations at the three points no longer changed with the iteration. The first monitoring point (10, 2.75, 1.75) was located near the laneway outlet; the second monitoring point (30, 2.75, 1.75) was located behind the mining machinery; and the third monitoring point (48, 2.75, 1.3) was located on the side of the machinery. The radon concentration change at the three points is plotted in Figure 5. As can be seen from Figure 5, the radon concentrations at the three points remained the same after 1000 iterations and, thus, the radon concentration field after this time was considered to reach a steady state.

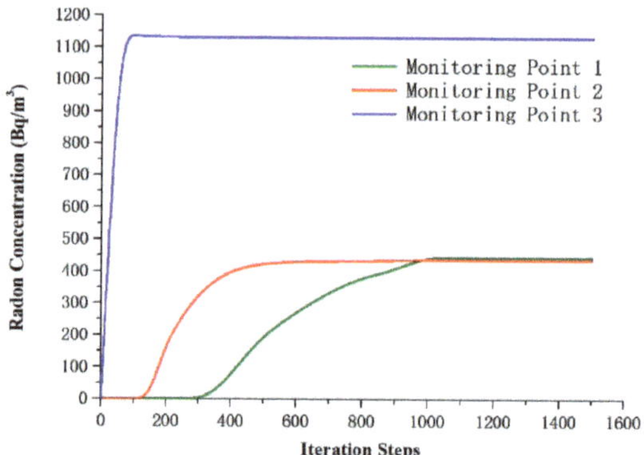

Figure 5. Radon concentration change at the monitoring points.

5. Results and Discussion

5.1. Airflow Characteristics and Distributions at Breathing Levels

An auxiliary ventilation system in an underground mine is usually employed to blow high-speed air into the workplace through a ventilation duct. The fresh air not only supplies miners with oxygen, but also removes toxic and harmful gases that are released during the mining process. The airflow characteristics must, therefore, be well understood before investigating the radon distribution. Figure 6 shows a streamline diagram of the airflow field in the laneway (ventilation rate: 2.5 m³/s).

Previous research [32,33] has demonstrated that auxiliary ventilation in a laneway is a turbulent jet flow in a finite space: Three zones (the jet-flow, backflow, and vortex zones) form in the laneway. A similar trend was observed in this study. Figure 6 shows that the high-speed air blown into the laneway from the ventilation duct forms a distinct jet-flow zone on the duct side. The velocity of the jet flow decreases gradually as it progresses toward the mining face. When the airflow reaches the heading face, the flow direction changes and the air disperses on the heading face, due to the heading face block. When it reaches the heading face edge, the airflow then changes direction again and flows back toward the laneway outlet. On the downstream side, the backflow is separated due to the presence of mining machinery. Most of the backflow moves toward the laneway outlet through the narrow space on the sides of the mining machinery. A small part of the backflow is blocked by the equipment and forms a vortex zone in the excavation space. A distinct circulation flow is generated, due to the combined effects of the separated airflows at the rear of the laneway. The low velocity and its recirculating nature may greatly weaken the radon removal capacity of the airflow, contributing to potential radon accumulation in this area. The above phenomena are in agreement with the results of numerical studies by Wang et al. and Qin et al. [34,35].

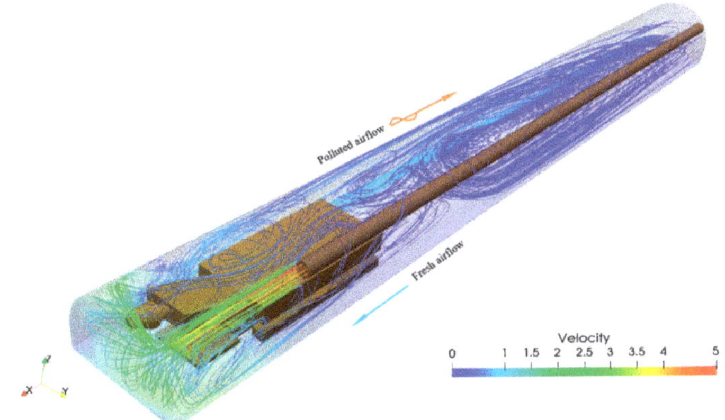

Figure 6. Streamline diagram of the airflow field in the laneway.

Referring to previous studies [34,36], 2.5 m³/s, 5 m³/s, and 7.5 m³/s were selected as the airflow rates from the ventilation duct in the simulation. Different respirable heights were investigated to understand airflow distributions at these breathing levels. The airflow profiles on the horizontal cross-sections are shown in Figure 7. Cases 1, 2, and 3 represent the steady-state airflow fields of the laneway when the ventilation rates of the duct were 2.5 m³/s, 5 m³/s, and 7.5 m³/s, respectively.

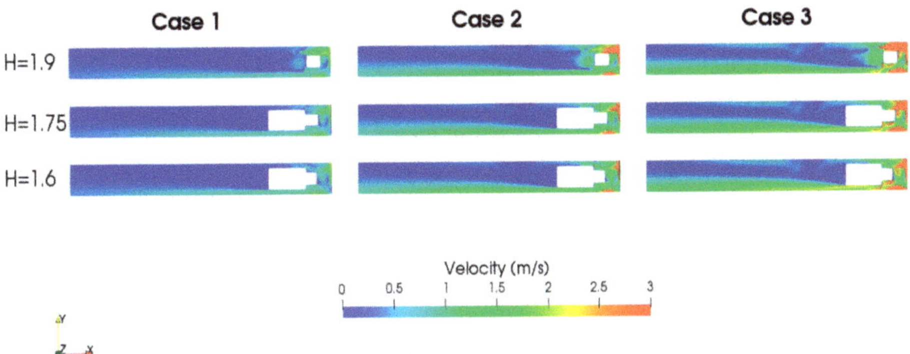

Figure 7. Effect of the ventilation rates on the horizontal cross-sectional distribution of airflow fields at different respirable heights.

For each case, the airflow dispersion patterns were similar at different breathing heights; the areas with the most intense airflow were mainly located in the space between the machinery and the mining face on the return side. This is because the airflow forced into the laneway will first accumulate in the mining space in front of the machinery, resulting in a sharp increase in airflow velocity. Then, obstructed by the wall on the duct side and the mining face, the airflow moved to the space on the non-duct side, and eventually flowed towards the laneway outlet along the wall. However, some slight changes in the jet-flow zone are worth noting. As can be seen from Figure 7, with an increase of respirable height, the high-velocity area of the jet flow expanded continuously. The main reason for this is that the location of the ventilation duct is near the roof, so the loss of airflow velocity near the duct is small. In the rear of the laneway, a distinct recirculation area was formed due to the presence of the machinery and the lower velocity of the airflow. With an increase of the ventilation rate, the airflow distribution changed markedly. A large amount of airflow in the restricted mining space bypassed the machinery and flowed rapidly to the rear of the laneway. As a result, the airflow rates on both sides of the laneway were obviously accelerated, the areas influenced by the airflow became longer and wider, and the recirculation area between both sides was less compressed.

5.2. Distribution of Radon Concentration Field

Figure 8 gives radon concentrations when the ventilation rates were 2.5 m^3/s, 5 m^3/s, and 7.5 m^3/s. The red contour represents the value of radon concentration equal to or exceeding 500 Bq/m^3, which is the lower limit of the action level (500–1500 Bq/m^3) required by the ICRP. It can be seen that the radon high-concentration areas were mainly located on the non-duct side and the rear of the laneway, and the whole high-concentration area of radon accounted for about 50% of the laneway space. This is because the radon gas released from the mining face was diluted by the fresh air, then radon gas was carried by the airflow to flow downstream. The radon concentration will increase dramatically due to the sharp reduction of space when flowing through the mining machinery. In addition, as distance increased, the airflow weakened at the rear of the laneway, which exacerbated the formation of circulation zones and caused radon gas to build up in those zones.

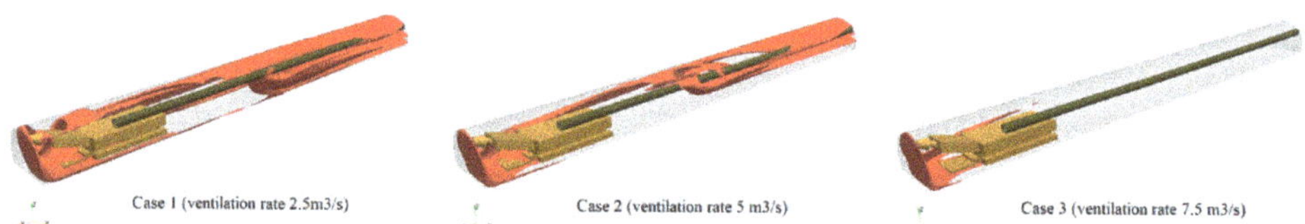

Figure 8. Radon distribution with concentrations ≥ 500 Bq/m^3.

With the increase of the ventilation flow rate, the high-concentration areas of radon in the laneway had a significant trend of decrease. When the ventilation rate increased to 7.5 m^3/s, the high-concentration areas only accumulated near the mining face, and the radon concentration at the rear of the laneway was diluted to below 500 Bq/m^3. The simulation results demonstrated that radon dispersion was highly affected by the airflow pattern.

To better understand the radon distribution in the laneway under the auxiliary ventilation, the radon concentration at different cross-sectional planes were further analyzed.

Figure 9 shows the radon concentration distributions of different respirable heights in horizontal cross-section planes. In Case 1, the radon concentration was relatively high near the heading face, due to the existence of circulation. When the ventilation flow rates were increased to 5 m^3/s and 7.5 m^3/s in Cases 2 and 3, respectively, the radon dilution effect of the airflow was enhanced and radon was quickly removed by the faster airflow, so the circulation area progressively shrank and the radon concentration gradually dropped. When the airflow moved to the non-duct side of the laneway, there was an obvious high radon concentration area between the mining machinery and the non-duct side, mainly because the mining equipment caused a sharp contraction of the airflow space. With the airflow rate increased, the movement of the airflow in the narrow space was strengthened and more radon was carried downstream, which would create a better working environment for the operators.

When the airstream emerged from the tail of the machinery, the flow velocity dropped significantly and eventually formed a circulating flow at the rear of the laneway. As discussed above, radon could not be efficiently removed from the laneway because of the circulation flow and, instead, accumulated in this area. As the airflow increased, radon was carried downstream more efficiently, rather than being captured by the flow circulation; this resulted in a significant reduction in radon concentration at the rear of the laneway.

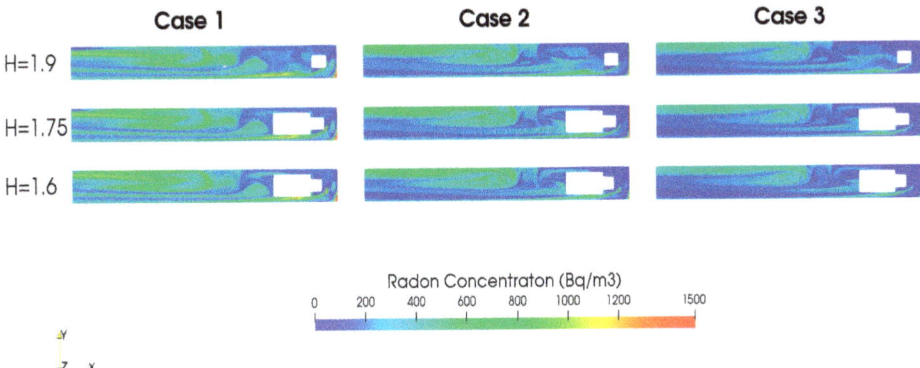

Figure 9. Horizontal cross-sections of radon concentration distribution at different ventilation rates and respirable heights.

Figure 10 shows the distribution of radon concentration in vertical cross-sections at different locations from the heading face. Radon gas released from the heading face was firstly transported along the non-duct side and then diffused into the entire space behind the mining machinery. Closer inspection revealed that the high radon concentration area throughout the laneway was mainly located near the machinery, because of the limited space.

Take the vertical cross-section plane of radon concentration 7 m from the mining face as an example. When the ventilation rate was 2.5 m^3/s, most of the radon gas was blown to the non-duct side of the laneway, resulting in a radon accumulation in this area. The radon concentration in the respirable height range (1.6–1.9 m) was around 400–1200 Bq/m^3, and a small amount of radon gas was distributed in the upper space of the machinery. With an increase of air quantity, more radon gas was discharged from the laneway outlet and the high-concentration area of radon decreased continuously; thus, radon pollution was effectively controlled. Although the radon concentration in the breathing area was still relatively high, the range of radon concentration in this area was decreased to about 300–700 Bq/m^3 when the airflow rate was 5 m^3/s. The ventilation scenario with the flowrate of 7.5 m^3/s obtained the best radon removal performance in this study. It can be seen that the radon concentration distribution in the cross-section plane was relatively uniform, which indicates that radon gas was removed from the laneway space in a timely manner by the high-speed airflow. The radon concentration on the entire section plane had almost dropped below the threshold action level of 500 Bq/m^3.

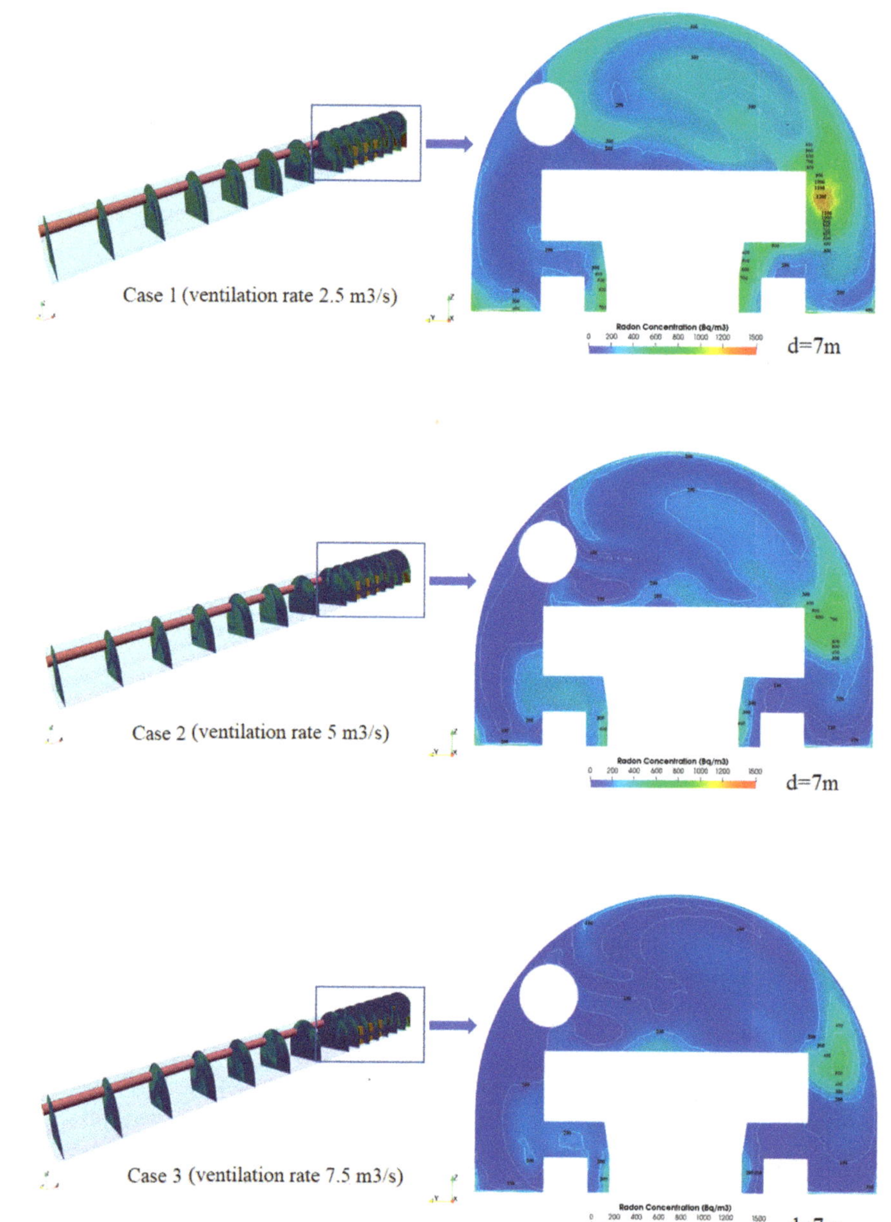

Figure 10. Vertical cross-sections of radon concentration distribution at different locations from the heading face.

5.3. Assessment of Occupational Exposure to Radon Inhalation

There are many different types of work in the underground laneway, such as mining machinery driver, equipment maintainer, electrician, hauler, support worker, and safety supervisor. Most activity occurs on the non-duct side, due to the existence of the duct on the other side. Compared with other types of work, the mining machinery driver is the closest to the heading face and is located in the narrow space near the machinery on the non-duct side for remotely driving and ensuring that the mining direction meets the operation requirements. The narrow working space benefits the radon accumulation, but it also increases the airflow rate for radon migration. Therefore, the working area of the driver becomes a key area for radon pollution monitoring; the radon dilution performance under different ventilation rates and annual effective dose assessment for the machinery driver is, hence, worthy of investigation.

In this study, the driver's working area—the space between the mining machinery and the non-duct side—was regarded as the working area of the laneway. As the overall width of the laneway was 5.5 m and the width of the mining machine was 3.6 m, the width of the driver's working area was about 1 m (($5.5 - 3.6$)/2 = 0.95 m) which was only just enough for a machinery driver to work properly. Therefore, the driver's breathing height was defined as the center of the working area with width ($y = 1/2 = 0.5$ m) and height H = 1.6, 1.75, or 1.9 m above the ground. The inhaled radon concentrations along the non-duct side of laneway at different ventilation rates are shown in Figure 11.

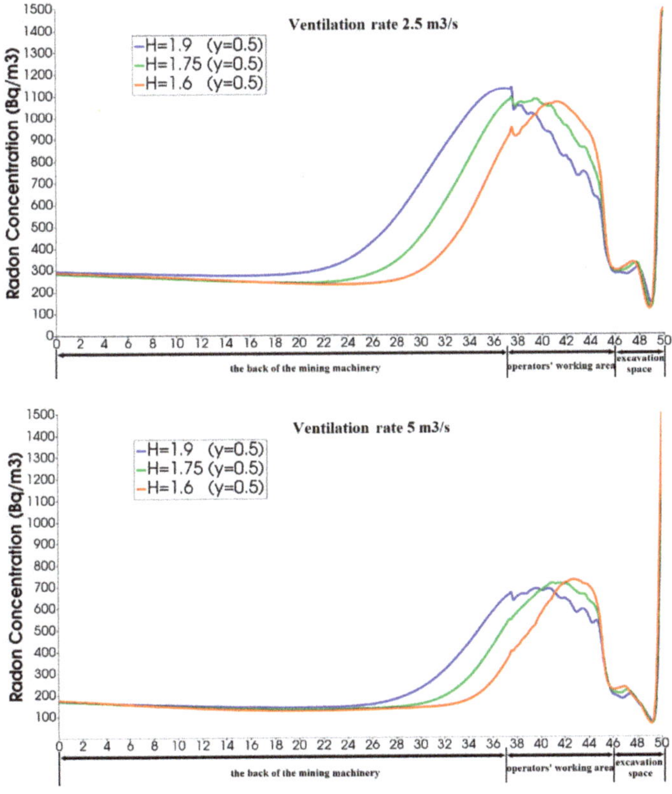

Figure 11. *Cont.*

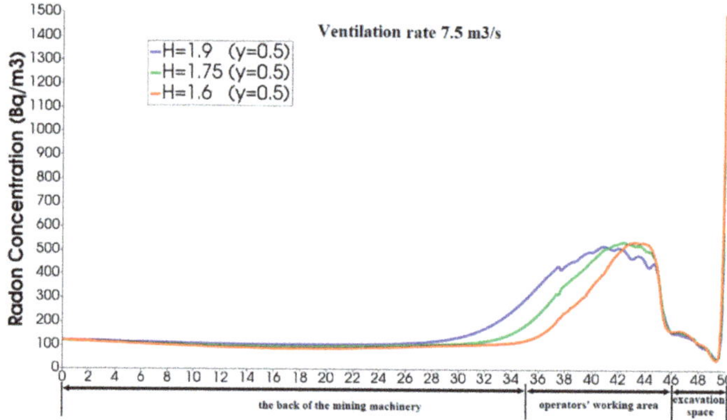

Figure 11. Inhaled radon concentrations along the non-duct side of laneway at different ventilation rates.

As the distance increased from the heading face, the radon concentration first dropped sharply in the excavation space, then rose rapidly in the working area, and finally decreased to a relatively low concentration at the back of the mining machinery. In all cases, the area with the highest radon concentration was mainly concentrated in the work area. There was a negligible effect of breathing height on the maximum radon value, but there was a significant influence of airflow field. In Case 1, the highest radon concentration was 1138.25 Bq/m^3. As the ventilation rate increased to 5 m^3/s, in Case 2, the radon concentration in the workplace was lower, but not adequately mitigated: The maximum radon was still as high as 727.79 Bq/m^3. Case 3, with a ventilation rate of 7.5 m^3/s, offered the best radon management result: The peak radon concentration was 541.62 Bq/m^3, which was almost at the lower limit of the action level (500–1500 Bq/m^3) for a workplace environment, as required by the ICRP.

According to Equation (4), the maximum effective dose for miners is not only related to maximum concentration in the working place, but also affected by the working hours. The results are shown in Table 1. The maximum annual effective doses for the three cases were 8.61 mSv, 5.50 mSv, and 4.12 mSv, respectively, all of which were below the statutory limit of 20 mSv for workers in the working place, as prescribed by the ICRP [37].

Table 1. Annual effective radon dose to miners.

	Ventilation Rate of the Duct (m^3/s)	Maximum Radon Concentration in the Workplace (Bq/m^3)	Annual Working Time (h)	Annual Effective Dose (mSv)
Case 1	2.5	1138.25	2400	8.61
Case 2	5	727.79	2400	5.50
Case 3	7.5	541.62	2400	4.12

5.4. CFD Modeling Limitations

Model validation is a very important part of related CFD studies, guaranteeing the accuracy of simulation results and the reliability of prediction results. However, as previously mentioned, this study was based on the statistical data of radon concentrations in the underground laneways of 48 coal mines in China, collected by Liu et al. [11], and so there were no specific laneway parameters to conduct a numerical model and no field-measured results for model validation. For the purpose of this research, a typical model simplified from an underground laneway in the Tangshan Donghua Coal Mine, China [20] and relative ventilation parameters were adopted to study the airflow distribution and radon migration under auxiliary ventilation. These simulation results could provide education for better understanding airflow characteristics in the laneway and inform guidelines for prevention and control of radioactive disasters in stone-coal mines.

6. Conclusions

In this study, a typical laneway model with forcing ventilation was established to evaluate the effectiveness of three different ventilation systems on radon concentration. The airflow pattern and radon concentration distributions were presented for each ventilation system. Results showed that the airflow fields of the laneway were distributed unevenly. There was an apparent jet-flow area on the duct side, a backflow area on the non-duct side, and circulation areas in the excavation space and at the rear of the laneway. With an increase of ventilation rate, the jet-flow and backflow zones were strengthened and expanded, while the circulation zones were diminished. This phenomenon is consistent with the results of previous studies.

The airflow pattern in the laneway played a key role in the radon concentration distribution. When the radon gas was released into the excavation space from the mining face, it was continuously diluted by the large quantities of fresh airflow. Then, most of the radon gas was carried by the airflow and moved rapidly toward the outlet of the laneway along the non-duct side. As the distance increased, the airflow velocity dropped markedly, which resulted in a big circulation area at the rear of the machinery. Radon gas that flowed with the airflow was also trapped by the circulation, eventually forming a high radon concentration area.

With an increase of ventilation quantity, the radon concentration in the laneway decreased significantly. In case 3, with a ventilation rate of 7.5 m^3/s, the best radon control performance was observed: The radon concentrations at respirable heights were reduced below the limit of the action level for the working place. The annual effective doses received by miners in all three cases were less than the limit of 20 mSv prescribed by the ICRP.

This study highlights the importance of auxiliary ventilation in effectively controlling the radon concentration in an underground laneway, which is crucial for the prevention of underground radioactive hazards. As model validation was not conducted in this study, it is recommended to compare and verify simulation results with measured data in future research.

Author Contributions: This paper was entirely written by B.Z. P.C. provided the guidance for the CFD simulation in this paper. Both P.C. and G.X. provided the editorial and language suggestions for this paper.

Funding: This work was supported by the China Scholarship Council Fund [grant number 201706930015]; Applied Basic Research Programs of Shanxi Province [grant number 201801D121265].

Acknowledgments: The authors thank Kathryn Sole, from Liwen Bianji, Edanz Group China (www.liwenbianji. cn/ac), for editing the English text of a draft of this manuscript.

References

1. United Nations; Scientific Committee on the Effects of Atomic Radiation. *Report of the United Nations Scientific Committee on the Effects of Atomic Radiation: Fifty-sixth Session (10–18 July 2008)*; United Nations Publications: San Francisco, CA, USA, 2008.
2. Vogiannis, E.G.; Nikolopoulos, D. Radon sources and associated risk in terms of exposure and dose. *Front. Public Health* **2015**, *2*, 207. [CrossRef] [PubMed]
3. Ting, D.S.K. *WHO Handbook on Indoor Radon: A Public Health Perspective*; Taylor & Francis: Milton Park, UK, 2010.
4. Joseph, K.; Wagoner, M.S.; Victor, E.; Archer, M.D.; Benjamin, E.; Carroll, M.A.; Duncan, A.; Holaday, M.A.; Pope, A.; Lawrence, M.S. Cancer mortality patterns among US uranium miners and millers, 1950 through 1962. *J. Natl. Cancer Inst.* **1964**, *32*, 787–801.
5. Cheng, K.C.; Porritt, J.W.M. The measurement of radon emanation rates in a Canadian cut-and-fill uranium mine. *CIM (Can. Inst. Min. Metall.) Bull* **1981**, *74*, 110–118.
6. Valentin, J. The 2007 recommendations of the international commission on radiological protection. *Ann. ICRP* **2007**, *7*, 1–332.
7. Al-Zoughool, M.; Krewski, D. Health effects of radon: A review of the literature. *Int. J. Radiat. Biol.* **2009**, *85*, 57–69. [CrossRef] [PubMed]

8. Cousins, C.; Miller, D.; Bernardi, G.; Rehani, M.; Schofield, P.; Vañó, E.; Einstein, A.; Geiger, B.; Heintz, P.; Padovani, R.J.I.P. International commission on radiological protection. *J. ICRP Publ.* **2011**, *120*, 1–125.

9. Qureshi, A.A.; Kakar, D.M.; Akram, M.; Khattak, N.U.; Tufail, M.; Mehmood, K.; Jamil, K.; Khan, H.A. Radon concentrations in coal mines of Baluchistan, Pakistan. *J. Environ. Radioact.* **2000**, *48*, 203–209. [CrossRef]

10. Baldık, R.; Aytekin, H.; Çelebi, N.; Ataksor, B.; Taşdelen, M. Radon concentration measurements in the AMASRA coal mine, Turkey. *Radiat. Prot. Dosim.* **2005**, *118*, 122–125.

11. Liu, F.D.; Pan, Z.Q.; Liu, S.L.; Chen, L.; Ma, J.Z.; Yang, M.L.; Wang, N.P. The estimation of the number of underground coal miners and the annual dose to coal miners in China. *Health Phys.* **2007**, *93*, 127–132. [CrossRef]

12. Dai, S.; Zheng, X.; Wang, X.; Finkelman, R.B.; Jiang, Y.; Ren, D.; Yan, X.; Zhou, Y. Stone coal in China: A review. *Int. Geol. Rev.* **2018**, *60*, 736–753. [CrossRef]

13. Wang, Z.; Ren, T. Investigation of airflow and respirable dust flow behaviour above an underground bin. *Powder Technol.* **2013**, *250*, 103–114. [CrossRef]

14. Ren, T.; Wang, Z.; Cooper, G. CFD modelling of ventilation and dust flow behaviour above an underground bin and the design of an innovative dust mitigation system. *Tunn. Undergr. Space Technol.* **2014**, *41*, 241–254. [CrossRef]

15. Kurnia, J.C.; Sasmito, A.P.; Mujumdar, A.S. CFD simulation of methane dispersion and innovative methane management in underground mining faces. *Appl. Math. Model.* **2014**, *38*, 3467–3484. [CrossRef]

16. Xie, D.; Wang, H.; Kearfott, K.J.; Liu, Z.; Mo, Z. Radon dispersion modeling and dose assessment for uranium mine ventilation shaft exhausts under neutral atmospheric stability. *J. Environ. Radioact.* **2014**, *129*, 57–62. [CrossRef] [PubMed]

17. Kissell, F.; Wallhagen, R. Some new approaches to improve ventilation of the working face. In Proceedings of the NCA/BCR Coal Conference, Louisville, KY, USA, 19–21 October 1976.

18. Toraño, J.; Torno, S.; Menendez, M.; Gent, M.; Velasco, J. Models of methane behaviour in auxiliary ventilation of underground coal mining. *Int. J. Coal Geol.* **2009**, *80*, 35–43. [CrossRef]

19. Schultz, M.J.; Beiter, D.A.; Watkins, T.R.; Baran, J.N. *Face Ventilation Investigation: Clark Elkorn Coal Company*; Pittsburgh Safety and Health Technology Center Ventilation Division: London, UK, 1993.

20. Shi, G.; Liu, M.; Guo, Z.; Hu, F.; Wang, D. Unsteady simulation for optimal arrangement of dedusting airduct in coal mine heading face. *J. Loss Prev. Process. Ind.* **2017**, *46*, 45–53. [CrossRef]

21. Kurnia, J.C.; Sasmito, A.P.; Mujumdar, A.S. Simulation of a novel intermittent ventilation system for underground mines. *Tunn. Undergr. Space Technol.* **2014**, *42*, 206–215. [CrossRef]

22. Hargreaves, D.M.; Lowndes, I.S. The computational modeling of the ventilation flows within a rapid development drivage. *Tunn. Undergr. Sp. Tech.* **2007**, *22*, 150–160. [CrossRef]

23. Whittles, D.; Lowndes, I.S.; Kingman, S.W.; Yates, C.; Jobling, S. Influence of geotechnical factors on gas flow experienced in a UK longwall coal mine panel. *Int. J. Rock Mech. Min.* **2006**, *43*, 369–387. [CrossRef]

24. Wang, H.; Wang, C.; Wang, D. The influence of forced ventilation airflow on water spray for dust suppression on heading face in underground coal mine. *Powder Technol.* **2017**, *320*, 498–510. [CrossRef]

25. Patankar, S. *Numerical Heat Transfer and Fluid Flow*; CRC Press: Boca Raton, FL, USA, 1980.

26. Pawlowski, S.; Nayaka, N.; Meireles, M.; Portugal, C.A.M.; Velizarov, S.; Crespo, J.G. CFD modelling of flow patterns, tortuosity and residence time distribution in monolithic porous columns reconstructed from X-ray tomography data. *Chem. Eng. J.* **2018**, *340*, 757–766. [CrossRef]

27. Chauhan, N.; Chauhan, R.P.; Joshi, M.; Agarwal, T.K.; Sapra, B.K. Measurements and CFD modeling of indoor thoron distribution. *Atmos. Environ.* **2015**, *105*, 7–13. [CrossRef]

28. UNSCEAR; Electro Optical Industries; United Nations Scientific Committee on the Effects of Atomic Radiation. *Sources and Effects Of Ionizing Radiation*; UNSCEAR: New York, NY, USA, 2000.

29. Chen, L.; Senlin, L.; Fudong, L.; Chunhong, W.; Yihua, W.; Jizeng, M.; Ziqiang, P. A primary assessment of occupational exposure to underground coal miners in China. *Radiat. Prot.* **2008**, *28*, 129–137.

30. Xu, G.; Edmund, C.J.; Kray, D.L.; Saad, A.R.; Michael, E.K. Remote characterization of ventilation systems using tracer gas and CFD in an underground mine. *Saf. Sci.* **2015**, *74*, 140–149. [CrossRef]

31. Sarkar, J.; Shekhawat, L.K.; Loomba, V.; Rathore, A.S. CFD of mixing of multi-phase flow in a bioreactor using population balance model. *Biotechnol. Prog.* **2016**, *32*, 613–628. [CrossRef] [PubMed]

32. Wang, H.; Shi, S.; Liu, R. Numerical simulation study on ventilation flow field of wall-attached jet in heading face. *J. China Coal Soc.* **2004**, *4*, 011.

33. Jian-Liang, G.; Yan, X.K.-l.W. Numerical experiment of methane distribution in developing roadway. *China Saf. Sci. J. (CSSJ)* **2009**, *1*, 005.
34. Wang, Y.; Lou, G.; Geng, F.; Li, Y.B.; Li, Y.L. Numerical study on dust movement and dust distribution for hybrid ventilation system in a laneway of coal mine. *J. Loss Prev. Process Ind.* **2015**, *36*, 146–157. [CrossRef]
35. Qin, Y.; Jiang, Z.; Zhang, M. Comparison of simulation on dust migration regularity in fully mechanized workface. *Chin. J. Liaoning Tech. Univ.* **2014**, *33*, 289–293.
36. Lu, Y.; Akhtar, S.; Sasmito, A.; Kurnia, J.C. Numerical study of simultaneous methane and coal dust dispersion in a room and pillar mining face. In Proceedings of the 3rd International Symposium on Mine Safety Science and Engineering, Montreal, QC, Canada, 13–19 August 2016.
37. Lario, J.; Sánchez-Moral, S.; Cañaveras, J.C.; Cuezva, S.; Soler, V. Radon continuous monitoring in Altamira Cave (northern Spain) to assess user's annual effective dose. *J. Environ. Radioact.* **2005**, *80*, 161–174. [CrossRef]

Article

Experimental and Numerical Analysis of a Sustainable Farming Compartment with Evaporative Cooling System

M. Sina Mousavi [1], Siamak Mirfendereski [2], Jae Sung Park [2] and Jongwan Eun [1,*]

[1] Department of Civil and Environmental Engineering, University of Nebraska-Lincoln, Omaha, NE 68588-0531, USA; sina.mousavi@huskers.unl.edu

[2] Department of Mechanical and Materials Engineering, University of Nebraska-Lincoln, Lincoln, NE 68588-0526, USA; siamak.mir@huskers.unl.edu (S.M.); jaesung.park@unl.edu (J.S.P.)

* Correspondence: jeun2@unl.edu; Tel.: +1-402-554-3544

Received: 4 August 2019; Accepted: 29 October 2019; Published: 6 November 2019

Abstract: The United Arab Emirates (UAE) relies on groundwater as well as desalinated water which are very expensive and energy-concentrated. Despite the lack of water resources, only 54% of wastewater was recycled in the UAE in 2016. In this study, a Sustainable Farming Compartment (SFC) with an evaporative cooling system is investigated as an alternative to reusing wastewater and the optimal design is identified experimentally and numerically. First, the applicability of the SFC was examined to reduce the ambient temperature in the system. A prototype SFC was tested in the environmentally constrained laboratory and field site considering an extreme climate condition (with high temperature and humidity) in Abu Dhabi to evaluate the temperature drop and humidity change of the SFC. The experimental results showed that the temperature of the SFC significantly decreases by 7–15 °C when the initial relative humidity is 50%. For validation, an energy modeling using dynamic numerical simulations was performed that shows statistically good agreement with the experimental results. Based on the parametric studies of the system components, the optimal cooling performance of the system in terms of locations of inlet and outlet, the variation of Reynolds number was evaluated. The study suggested an optimized design for the SFC with an evaporative cooling system.

Keywords: heat transport; optimized design; dynamic numerical simulation; evaporative cooling system; water recycling; temperature; humidity

---

1. Introduction

Desertification, the process of land degradation leading the land to become desert, is one of the most significant environmental problems of the Arabian Peninsula region. Despite the national, regional and international collaboration to combat desertification and to mitigate the damage of drought, desertification is still one of the major environmental concerns in the area [1,2]. In the United Arab Emirates (UAE), wind erosion due to prevailing hyper-arid conditions, insufficient vegetation and the strong wind have contributed to the serious degradation of land and ground soils [3,4]. For instance, the process of desertification can be accelerated as the fertile topsoil is removed by wind erosion [4,5]. Therefore, it is important to reduce wind erosion for protecting the environment and mitigating desertification.

The UAE is classified as a hyper-arid climate with less than 120 mm of average annual rainfall [6], which causes a high dependence on groundwater resources and desalinated water. However, these resources are very expensive and highly energy-concentrated. Despite the lack of water resource, only 54% of wastewater was recycled in the UAE in 2016 and the remaining 48% was disposed into the sea near Abu Dhabi [7]. Moreover, the wastewater will continuously increase by 10% until 2030 because of the development of the urban area, the growing population and the enhancement of life quality in Abu Dhabi [8]. The government of Abu Dhabi ambitiously plans to reuse 100% of its wastewater by 2030 to resolve this situation [7,8]. However, the main challenge associated with recycling treated wastewater is that a distribution system has not been developed [7]. Nowadays, the treated wastewater is mostly used for irrigation of public areas such as parks and roadways and for district cooling in residential areas. However, the amount of recycling is insufficient to meet the amount of production [7]. Therefore, the development of any other ways to reuse more treated wastewater is warranted to increase the recycling rate of the water in the region.

Recently, the sustainable agricultural complex referred to as the "Oasis" complex was proposed [8] as an alternative method to mitigate wind erosion and desertification as well as to increase the recycling rate. The complex is expected to decrease the temperature inside the structure by using renewable energy from the evaporative cooling system and solar panels with the treated wastewater providing the conditions for raising plants. However, the applicability of the components comprising the complex has not been thoroughly studied and validated.

In this study, we investigated the performance and effectiveness of the Sustainable Farming Compartments (SFC) as a part of the "Oasis" complex to control the inside temperature in the climate conditions in the UAE. A prototype of the SFC at half-scale was developed and tested in an environmentally constrained laboratory and field site considering the climate conditions in Abu Dhabi. Furthermore, energy modeling via dynamic numerical simulations of the SFC was conducted to evaluate the heat transfer and stabilization in the SFC and to compare with the experimental results. The parametric studies of the system components were performed numerically in terms of locations of inlet and outlet, the variation of Reynolds number. Based on the results, an optimized design for the SFC was identified.

1.1. Sustainable Farming Compartment (SFC)

The SFC is designed to install exhaust fans in the front and an evaporative cooling pad in the backside. The pad is wetted by treated wastewater to decrease the interior temperature. The ambient air in the SFC is cooled by using the heat in the air to evaporate the water from an adjacent surface, as shown in Figure 1. Selected agricultural plants under lower sunlight in shaded conditions can be raised in the compartment even during outside hot-dry weather by maintaining a relatively low temperature inside [9]. Hence, a preliminary investigation of the environmentally controlled laboratory and field site is necessary to evaluate the feasibility and effectiveness of the SFC when subjected to the actual climatic conditions in Abu Dhabi.

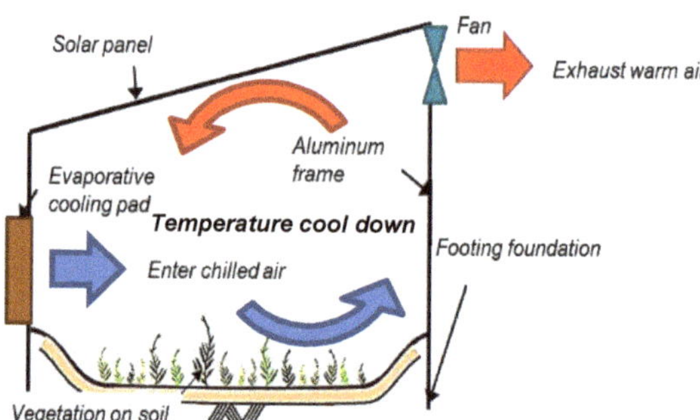

Figure 1. Concept of Sustainable Farming Compartment (SFC) to decrease the inside temperature ambient air.

1.2. Evaporative Cooling System

The SFC contains a direct evaporative cooling system with a proper water supply system and cooling pad that use treated wastewater to reduce the inside temperature. The design of the evaporative cooling system intends to decrease the inside temperature by evaporating the water and absorbing the latent heat. The temperature of dry air can be considerably decreased through the phase change of water from liquid to vapor (evaporation). This phenomenon will decrease the temperature of the air using much less energy than refrigeration [10–12]. The cooling potential for evaporative cooling fully relies on the wet-bulb depression that is the difference between dry-bulb temperature and the wet-bulb temperature according to the psychrometric chart. In the SFC system, the humidity is important [13,14]. While the direct evaporative cooling system runs, the relative humidity increases because of the air coming into direct contact with water of the cooling pad and its vaporization. Accordingly, the temperature in the SFC will not decrease when the humidity theoretically reaches 100% but practically, the temperature does not drop when the moisture reaches around 85% [15,16]. Therefore, it is crucial to monitor variations of humidity to analyze the process and compare the results with the chart.

Many studies have been conducted to investigate the performance of evaporative cooling systems. Ibrahim et al. achieved a drop of 6–8 °C dry bulb temperature with a 30% increase in relative humidity using a direct evaporative cooling system supported by porous ceramics in Nottingham [15]. Lertsatitthanakorn et al. studied the effect of a 1.8 m by 3.6 m direct evaporative cooling pad in a 32-m^2 silkworm-rearing house in Maha Sarakham, Thailand. Their results show that 6–13 °C dry bulb temperature decreases with increasing 30–40% relative humidity. Further economic analysis showed a 2.5 year payback period for this system, indicating its cost-effectiveness [16]. Heidarinejad et al. provided a two-stage indirect/direct evaporative cooling system experiment in various climate conditions in Iran. They showed that in regions with high wet bulb temperature, this system can be used instead of mechanical vapor compressions with one-third of their energy consumption. Their results also show that the two-stage system has 55% more water consumption than a direct evaporative cooling system that favors the latter system in arid areas [17]. Recently, Aljubury and Ridha conducted a two-stage indirect/direct evaporative system using groundwater to study its effect on the greenhouse in the Iraq desert climate. The results showed 12.1–21.6 °C decrease in temperature and increase of relative humidity from 8% to 62% compared to the ambient condition [18].

2. Materials and Methods

2.1. Prototype SFC

A prototype SFC that is half the scale of the actual structure was constructed with aluminum frames (40 × 40 mm) manufactured by 80/20 Material Inc. (Columbia City, IN, USA) (Figure 2). The frame was easily assembled with mechanical bonding alone to provide full stability for the structure. The dimension of the SFC used was 1.2 m of the front height, 1.8 m of width, 1.5 m of depth and 0.9 m of the back height. The structural analysis, using SAP 2000, was performed to confirm the structural stability of the SFC. The bottom was constructed with plywood and the walls were constructed with Plexiglas panels which have approximately three to five times lower thermal conductivity (=0.2 W·m^{-1}·K^{-1}) in comparison with glass. The materials in the wall and bottom allow limited heat transport. The roof was covered with the white color-polystyrene insulated sheet (with the thickness = 0.05 m) to reduce direct solar radiation into the compartment. On the upper wall of the front side, five exhaust fans with 190 cubic feet per minute (CFM) per fan were installed. On the backside, a cardboard cooling pad (0.1 × 0.45 × 1.8 m^3) where the absorption of the latent heat occurs by the evaporating water, was installed. The pressure in the evaporative cooling system was assumed to be identical to the atmospheric pressure because the SFC was opened to the atmosphere and the distribution of the pressure was not considered in the simulation. Four portable sensors with 12-bit resolution were installed at the center of the compartment. The inside sensor was located at 0.75 m from the front wall, 0.9 m from the side wall and 0.45 m from the bottom plate. The outside roof and the ambient outside sensors were near the SFC to monitor the variations in temperature and humidity during the test. The resolution of the temperature and humidity is 0.5 °C and 0.05%, respectively. The accuracy is typically ±0.5 °C for temperature and ±2.5% for humidity. The data were collected every sixty seconds. The SFC was placed near the campus of New York University Abu Dhabi (NYUAD) during the test, which is a typical weather condition in the UAE.

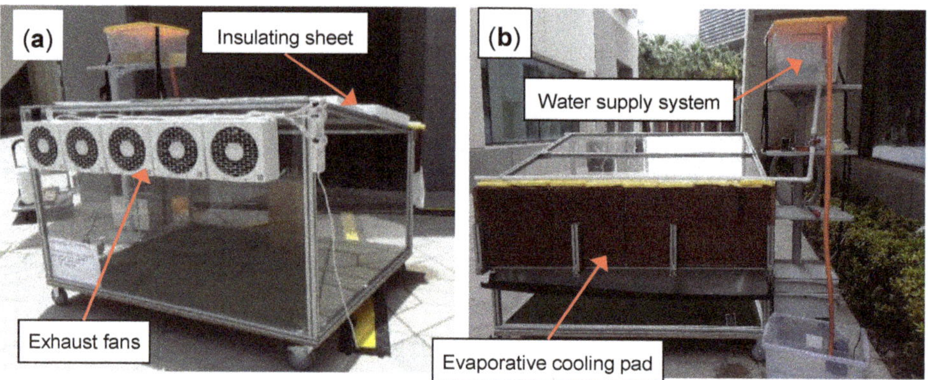

Figure 2. Photo of prototype experimental setup for SFC: (**a**) Front view with exhaust fans and (**b**) Back view with evaporative cooling pad and water supply system.

For the water supplying system, 30 mm-polyvinyl chloride (PVC) pipes were used to connect 0.12 m³-upper and lower water reservoirs. The valve was installed to control water flow rate during the test. Tap water was used for the test and refilled the reservoirs adequately. The water was distributed evenly to the pad from the upper reservoir to fully wet the pad. Since the water was reused in the system, the temperature was not varied significantly. The water pump was installed in the lower reservoir to recirculate the water in the system. The pad was initially saturated with 10 L of water. The flow rate and hydraulic gradient of the water supply system are approximately 5 L/hr and 0.5 m/m, respectively.

The original design of SFC included the solar photo voltaic (PV) panels with adequate insulation on the roof to enhance the energy efficiency of the SFC in the "Oasis" complex; thus, the prototype of SFC with polyethylene (PE) polymer insulation was consistent with the original one. However, the applicability of the solar panels should be carefully investigated in terms of power capacity and economical cost. Selected agricultural plants might be raised in SFC during hot-dry weather outside by allowing some sunlight to reach inside and maintaining a relatively low temperature inside. However, the amount of sunlight entering through the four surfaces needs to be monitored and assessed enough in terms of growing the plants.

2.2. Dynamic Numerical Simulation of SFC

For dynamic numerical simulations of the SFC system, an in-house numerical code was developed to evaluate the thermal performance of the SFC and to validate the experimental results. For simulations, the prototype SFC, which is subjected to actual weather data of the UAE, was modeled. All materials were identical to those used during the prototype test. The material properties, used as the input parameters for the simulation, are summarized in Table 1. The physical system upon which the numerical simulation is based is shown schematically in Figure 3. Boundary conditions at both sidewalls of the SFC and the roof were at a constant temperature of 53.9 °C, which conservatively equals the highest outside temperature.

Table 1. Material Properties for Dynamic Numerical Simulation.

Type	Material	Thermal Conductivity W·m⁻¹·K⁻¹
Frame	Aluminum	101.8
Walls	Plexiglass	0.20
Bottom	Plywood	0.16
Roof	Plexiglass	0.20
	Styrene sheet	0.033

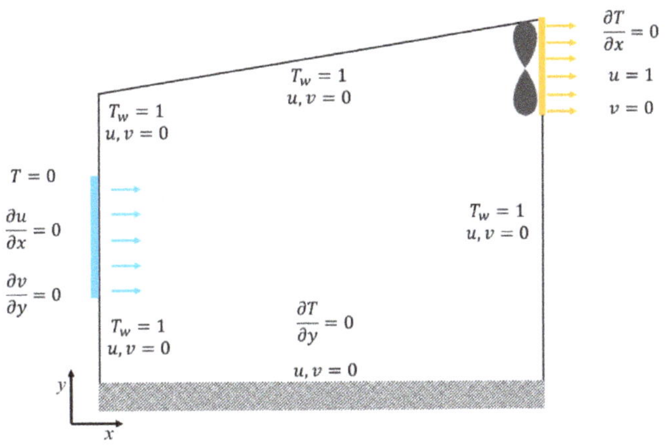

Figure 3. System schematic of the simulation model.

2.2.1. Governing Equations of Mathematical Model

We provide a model capable of predicting the temperature distribution and hydrodynamic characteristics of fluid inside the SFC prototype.

The model is based on the conservation of mass, momentum and energy. An incompressible fluid in a two-dimensional (2-D) domain with a trapezoidal geometry with the same size of the experimental prototype was simulated, as shown in Figure 3. Note that the actual SFC is designed as a plane strain condition (e.g., a long chain), which is close to the model of the 2-D domain. The geometric dimension of the model is identical to that of the prototype SFC experiment.

The characteristic length, L_c and velocity scales, U_c were the length of SFC protype chamber in x direction and the outlet velocity of air at exhaust fan, respectively. With these characteristic scales and given that a viscous fluid is incompressible with constant properties and the effect of gravity on fluid flow is negligible, the conservation equations for momentum, mass and energy in the non-dimensional form were

$$\frac{\partial u}{\partial t}+u\frac{\partial u}{\partial x}+v\frac{\partial u}{\partial y}=-\frac{\partial P}{\partial x}+\frac{1}{Re}\left(\frac{\partial^2 u}{\partial x^2}+\frac{\partial^2 u}{\partial y^2}\right) \tag{1}$$

$$\frac{\partial v}{\partial t}+u\frac{\partial v}{\partial x}+v\frac{\partial v}{\partial y}=-\frac{\partial P}{\partial y}+\frac{1}{Re}\left(\frac{\partial^2 v}{\partial x^2}+\frac{\partial^2 v}{\partial y^2}\right) \tag{2}$$

$$\frac{\partial u}{\partial x}+\frac{\partial v}{\partial y}=0 \tag{3}$$

$$\frac{\partial T}{\partial t}+u\frac{\partial T}{\partial x}+v\frac{\partial T}{\partial y}=\frac{1}{RePr}\left(\frac{\partial^2 T}{\partial x^2}+\frac{\partial^2 T}{\partial y^2}\right) \tag{4}$$

where the non-dimensionalized variables including velocity $V = (u, v)$, lengths (x, y), pressure (P), time (t) and temperature (T) can be derived from their dimensional form by the following equations (the variables symbol accompanied with prime symbol, for example, $T\prime$, $T\prime_w$, are dimensional).

$$x = \frac{x'}{L_c}; \quad y = \frac{y'}{L_c} \tag{5}$$

$$u = \frac{u\prime}{U_c}; \quad v = \frac{v\prime}{U_c}; \quad P = \frac{P\prime}{\rho U_c^2}; \quad T = \frac{T' - T\prime_{in}}{T\prime_w - T\prime_{in}}$$

where the Equations (1)–(3) are the Navier-Stokes equations. Here, we define the Reynolds number and the Prandtl number as $Re = U_c L_c / v$ and $Pr = v / \alpha$, respectively, where v is the kinematic viscosity of the fluid and α is the thermal diffusivity of the fluid. The variation of temperature and humidity during one operation cycle of the SFC prototype leads to negligible differences in air properties such as density, viscosity and thermal diffusivity; therefore, the properties of simulating fluid are assumed to stay constant. The air properties were evaluated at 80% humidity. Nevertheless, the humidity should be considered because the vapor phase coexists. In this study, it is found that the Reynolds and Prandtl numbers implicitly take into account the presence of the vapor phase but their resultant effect on the average temperature is almost negligible. However, it is likely that spatially varying parameters due to spatial non-uniform humidity would have a significant impact on the results, which is a subject of future work. For boundary conditions, a no-slip boundary condition was assumed at the walls in the momentum equation. Constant velocity was assumed at the outlet due to the exhaust fan, while Neumann inflow condition was applied due to the evaporative cooling system. In the energy equation, a uniform temperature is assumed at the sidewalls and the roof. The cold air of a uniform temperature was injected through the evaporative cooling system. The Neumann boundary condition for the temperature at the outlet is used due to the exhaust fan. No heat flux condition was considered for the bottom floor.

2.2.2. Numerical Method

The dynamic and thermal behaviors of the fluid inside the simulation domain is analyzed using the direct numerical simulation (DNS). In the current simulation, a boundary-fitted grid system is generated for the trapezoidal space and then the governing equations are transformed for a rectangular computational domain using 2×2 Jacobean matrices [19]. A staggered grid system was generated to solve the two-dimensional governing equations and a second-order explicit finite difference method (center in space and forward in time) was used for discretization of the equations [20,21]. The continuity and time-discretized momentum equations are written as follow

$$\nabla \cdot V^{n+1} = 0 \tag{6}$$

$$\frac{V^{n+1} - V^n}{\Delta t} + [(V \cdot \nabla) V]^n + \nabla P^{n+1} = \frac{1}{Re} \nabla^2 V^n \tag{7}$$

A fractional-step method or the projection algorithm was employed to compute a velocity field [22–24]. It is worth noting that, as seen in the original paper for the fractional-step method [22], this method can be applied to the 2D case as the present study. In this algorithm, the intermediate velocity (V^*) is introduced such that $V^{n+1} = V^* - \Delta t \nabla P^{n+1}$ from which Equations (6) and (7) can be modified into Equations (8) and (9), where the pressure is now decomposed from the momentum equation. In this numerical approach, the intermediate velocity (V^*) was calculated explicitly from the velocity fields at the previous time step (V^n) by Equation (8). With the continuity equation, the Poisson equation in Equation (9) was solved via Successive Over Relaxed (SOR) method to correct the intermediate velocity to satisfy the divergence constraint on the velocity field. The energy equation in Equation (11) was solved using a second-order explicit finite difference method to calculate the temperature field.

$$\frac{V^* - V^n}{\Delta t} + [(V \cdot \nabla)V]^n = \frac{1}{Re}\nabla^2 V^n \tag{8}$$

$$\nabla^2 P^{n+1} = \frac{\nabla \cdot V^*}{\Delta t} \tag{9}$$

$$\frac{V^{n+1} - V^*}{\Delta t} + \nabla P^{n+1} = 0 \tag{10}$$

$$\frac{T^{n+1} - T^n}{\Delta t} + [(V \cdot \nabla)T]^n = \frac{1}{RePr}\nabla^2 T^n \tag{11}$$

All computations were performed on 256×256 mesh and the time step (Δt) was in the order of 10^{-5} to ensure the numerical stability condition. The mesh convergence was tested and its size is selected in order for further grid refinement not to lead to significant difference in spatial average steady-state temperature. The tolerances used in the simulation as stopping criterions were respectively 10^{-8} and 10^{-10} for velocity and temperature computations. The simulations stop when the spatial average of absolute residual error for both velocity and temperature fields fall below the tolerances. Each simulation performed by in-house numerical code takes approximately 20 central processing unit (CPU) hours to complete for the range of Reynolds numbers studied in the current research. However, in cases of studying the systems with high Reynolds numbers and large domains, direct numerical simulation (DNS) becomes significantly limited. In order to effectively simulate such systems, it is worth noting that the methods employing the Reynolds-averaged Navier–Stokes equations (RANS) incorporated with various models such as k-epsilon models [25–27] can be more appreciated.

3. Results

3.1. Variation of Temperature and Humidity in a Controlled Condition

A cubical chamber ($0.45 \times 0.45 \times 0.9$ m) made of polyethylene was prepared as a constrained volume for a test in an environmentally controlled condition to evaluate an exhaust fan and an evaporative cooling pad. The relative humidity and room temperature in the laboratory was maintained at 22.0 °C and 60.0%, respectively. In Figure 4, the testing result of the environmentally controlled condition was shown. The humidity and temperature of the chamber were monitored until stabilized for more than 60 min. At the initial time, the interior temperature and humidity were 22.0 °C and 60.0%, respectively, which were identical with the outside ones. However, the temperature was significantly decreased by 5 °C from 22.0 °C to 17.0 °C after the test started. The temperature drop was 29.4%. According to psychometric chart, the temperature decreases by 5.5 °C from 22 °C to 16.5 °C, which reasonably agreed with the results. On the other hand, the relative humidity was significantly increased from 60.0% up to 95.0% during the test. Typically, the temperature in the chamber decreased with increasing relative humidity because latent heat as evaporation proceeded in the pad took out the heat, resulting in the temperature drop. Approximately 4.0 L of water was consumed during the test. The temperature and relative humidity were stabilized around 15 and 30 min later since the test started.

3.2. Variation of Temperature and Humidity during the Prototype SFC Test

The performance of the prototype SFC with the operation of the evaporation cooling system was tested outside near the campus of NYUAD more than two weeks from July and August. Summer in Abu Dhabi is very hot and the outside average temperature reaches approximately 42.5 °C [28]. Figure 5 shows the variation of temperature and humidity obtained from the SFC test. The outside temperature varied from 31.8 °C to 53.9 °C. The difference between the highest and lowest temperatures was 22.0 °C and the average temperature was 37.9 °C. However, the inside temperature of the SFC varied from 24.1 °C to 54.4 °C. The average temperature was 31.2 °C and the difference between the highest and lowest temperatures was 30.2 °C.

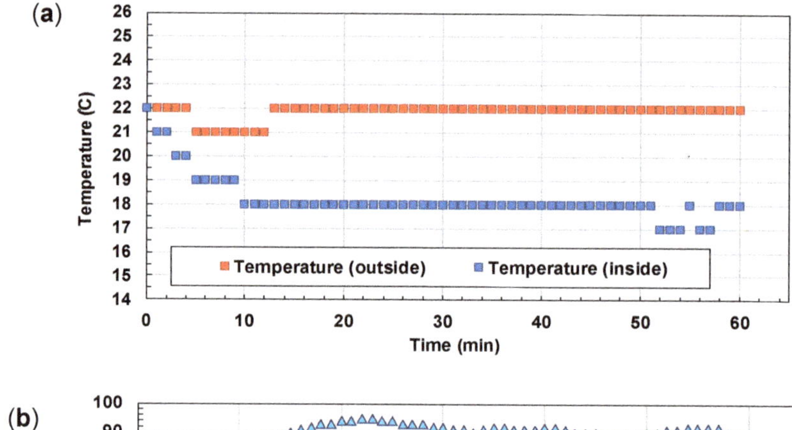

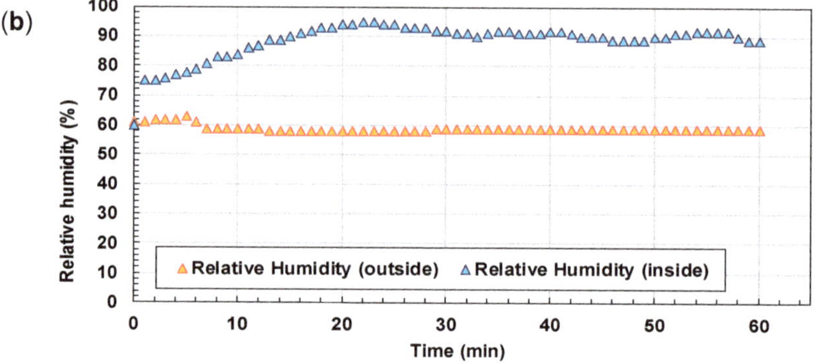

Figure 4. Variation of temperature and humidity: (**a**) Temperature and (**b**) Humidity.

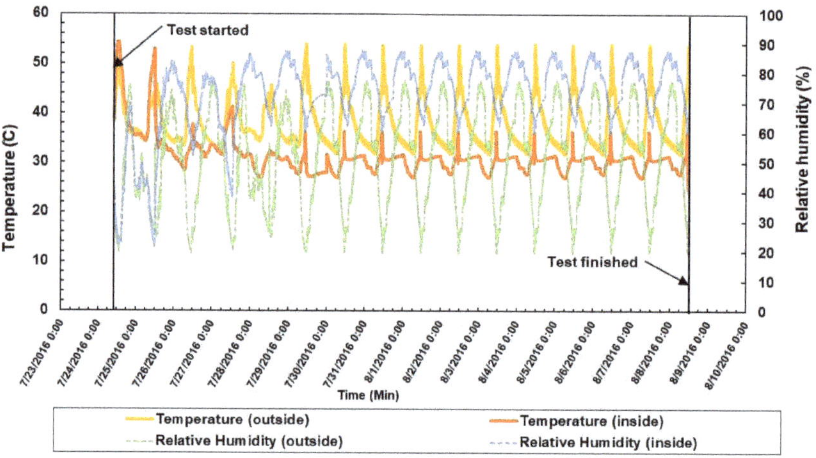

Figure 5. Temperature and humidity obtained from prototype SFC test for two weeks in July and August.

The relative humidity in the outside was ranged from 19.6% to 77.6%. The average of the relative humidity was 50.7% and the standard deviation was 16.2%. On the other hand, the relative humidity in the inside is ranged from 21.5% to 87.7%. The average of the temperature was 73.2% and the standard deviation was 17.4%. As the relative humidity (inside and outside) increases, the temperature decreases in the SFC. Not only the inside temperature but also the outside temperature is correlated with the humidity. However, the temperature tends to increase with decreasing humidity. The relationship between the humidity and temperature in the compartment was mostly opposite during the test. The variations of the outside temperature and the inside humidity were considerably higher than those of the inside temperature and outside humidity. The temperature in the inside chamber was consistent under 31.0 °C except at the peak time of outside temperature and specifically around noon.

Figure 5 also shows that in the first two days, the temperature inside and outside the compartment was almost the same. This shows that it takes time for the evaporating coolers to create a shifted new equilibrium inside the system for temperature and humidity. The same trend occurs for relative humidity where after the first two days, it increases over 40% inside the compartment and once again, we can see the decreased temperature is related to increasing relative humidity. After day 3, the inside temperature is lowered compared to the outside temperature and after day four, the variation trend of inside temperature is stable for the next ten days. Three unexpected temperature peaks inside the compartment might be due to outside high-temperature peaks and running out of the water to be supplied to the evaporative cooling system. It seems that the humidity inside SFC was dramatically dropped when the water was runout during the test. The stabilizing delay on the second day may be due to the running of water. Additionally, the temperature measurement at the center of the compartment may not represent the inside temperature entirely. Because of this limitation, the stabilization of the temperature might not be seen clearly on the second day of the testing.

3.3. Validation of Numerical Simulation

Dynamic numerical simulations for SFC at Reynolds number of 2300 were performed. The Reynolds number is calculated based on the experimental parameters and prototype SFC geometry. Figure 6 shows color contours for temperature distribution and arrows for fluid velocity inside the domain as a function of time from unsteady to steady states. Mixing the cool air coming from the evaporative cooling system with the inside warm air causes an overall temperature drop in the domain. The vortical flow patterns were observed near the left top and bottom walls. The similar vortical patterns can also be seen in temperature contours. These fluid circulations tend to enhance the heat transfer inside the domain. The white scale bar at $t = 4.3$ corresponds to 0.1 m/s. Thus, the velocity can be calculated in the caption of the Figure 6.

Figure 7 shows the time evolution of an area-averaged temperature for the different initial temperatures (T_0). The average temperature becomes constant, leading to a thermally steady-state. All different initial temperatures are converged to the same average temperature $T_{ave} \approx 037$. It is worth noting that this average temperature corresponds to a temperature drop of approximately 5–10 °C, depending on the outside temperature. In other words, in the system with $T_{ave} \approx 0.37$ and wall temperature of 22 °C and 33 °C with inlet temperature of 10–14 °C and 17–23 °C, respectively, a temperature drop is found to be about 6–12 °C. This temperature drop is in good agreement with experimental observations. Thus, the developed numerical model tends to predict the performance of the SFC reasonably.

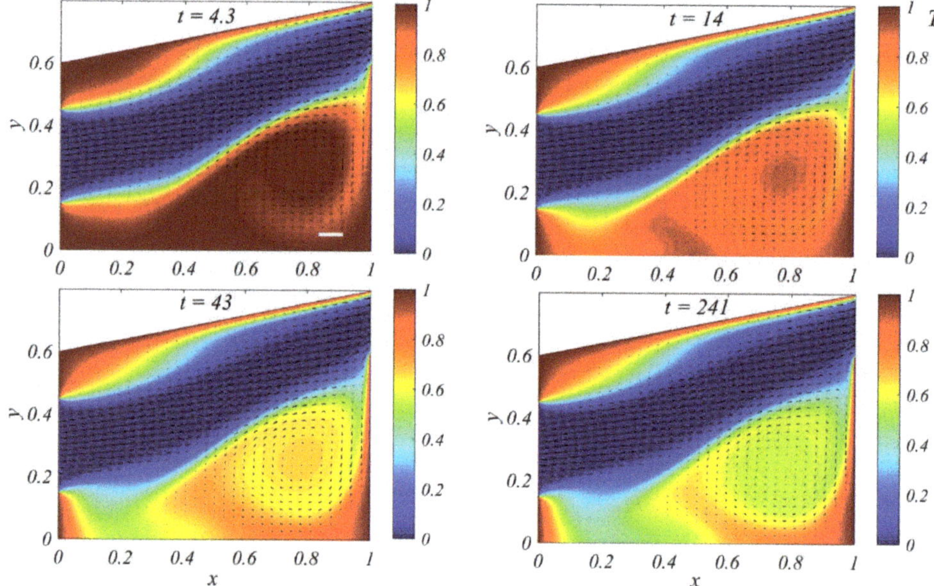

Figure 6. A contour for temperature distribution and a vector field for fluid flow inside the simulation domain as a function of time at $Re = 2300$. The case of $t = 241$ can be regarded as a steady-state.

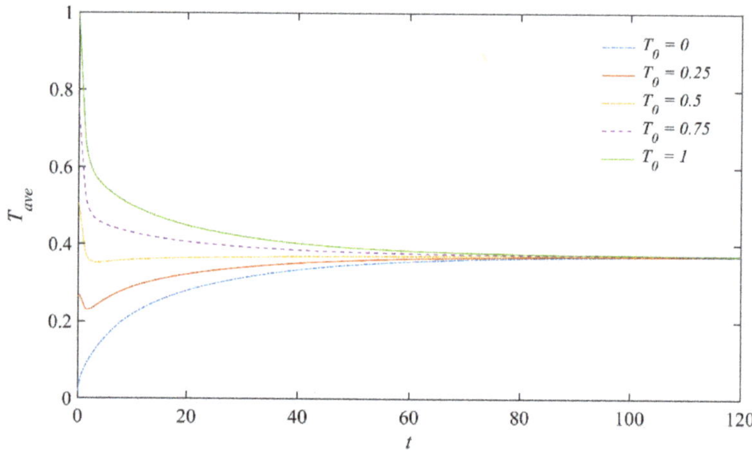

Figure 7. An area-averaged temperature inside the simulation domain as a function of time at $Re = 2300$. The four different initial temperatures (T_0) are used, leading to the same steady-state temperature.

3.4. Comparisons of Temperature between Simulation and Experiment

Figure 8 shows the comparative results of the varied temperatures obtained from the dynamic energy simulation and experimental tests. The temperatures from the experiment and simulation were measured at the same location, which was the center of the SFC chamber. In the figure, the symbols show the range of the temperature and the boxes show 25% and 75% of the values. The average temperature obtained from the dynamic simulation was 26.0 °C with 2.6 °C standard deviation. The difference between the average data is 0.40 °C for the temperature. The results show statistically no difference from the paired T-test ($p = 0.087 > 0.050$). Thus, the results obtained from the energy simulation show good agreement with the results of the experimental test.

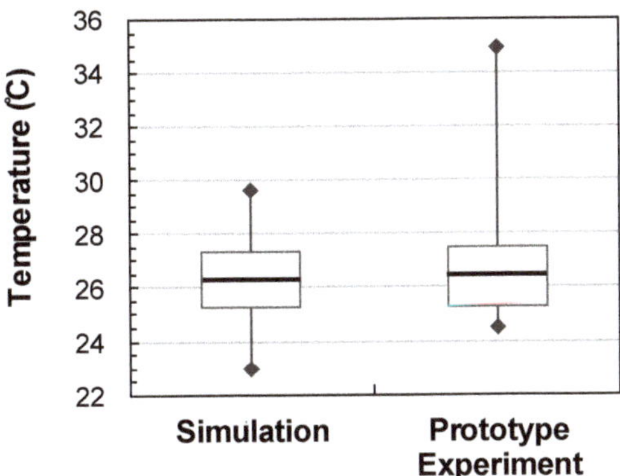

Figure 8. Comparisons of temperatures obtained between dynamic numerical simulation and experimental test.

However, temperature variation in the prototype test can influence plant productivity; thus, the reliability and stability of the system need to be considered before it can be used for farming. Additionally, an undesirable situation, such as running out of cooling water, may affect the growth of the plants.

4. Discussion

4.1. Numerical Parametric Studies on Inlet and Outlet Locations

To optimize the performance of the SFC system, the effect of the evaporative cooling system and fan locations on the temperature and velocity field is evaluated numerically. Figure 9 shows the variations of velocity and temperature fields in the different cases as varying the locations of the evaporative cooling system and fan. Cases A to F shows the effect of the different combinations of inlet and outlet locations on temperature and velocity fields. The A, B, C cases are different in terms of outlet location where inlet for A, B, C are located on the middle 0.2. While cases E and F have similar outlet location of top 0.2, their inlet is located at 0.4–0.6 and 0.2–0.4 units respectively. Case D has the opposite locations of inlet and outlet compared to case F.

As can be seen in Figure 9, the positions of locations of the evaporative cooling system and fans significantly influence the velocity and temperature fields. The change in the vorticity flow pattern observed in the temperature contour seems to be the factor making a difference in the thermal performance of each case due to the variation in the mixing mechanism and vorticity as the inlet or outlet is displaced. In other words, one strong vorticity at the lower portion of the domain and one weaker vorticity at the top right corner effectively contribute to the cooling process of SFC, leading to an optimum thermal feature among all the cases.

The temporal variation of the spatial average temperature and its steady-state value for all cases are shown in Figure 10a. As expected, case D has the best cooling performance because of the largest area of lower temperatures coming from the evaporative cooling system, leading to the lowest average temperature at the steady-state. Case D provided the best mixing mechanism between the cold inflow and the hot fluid inside the domain, regarding the geometry of the domain. In this case, two vorticities, especially the bottom one which is very strong, promote the mixing mechanism. This shows the advantage of this inlet and outlet arrangement. Figure 10a shows that at each time, the average temperature in case D is lower than that in other cases except case F for a short period of time ($t = 2$ to

$t = 20$). However, case F has the fastest convergence to steady-state. It seems that the size of vorticity is correlated with the time required to reach steady-state, as case F has the small twin vorticity (Figure 9). Figure 10b shows the steady-state temperature of all cases where it clearly shows that case D has the lowest temperature.

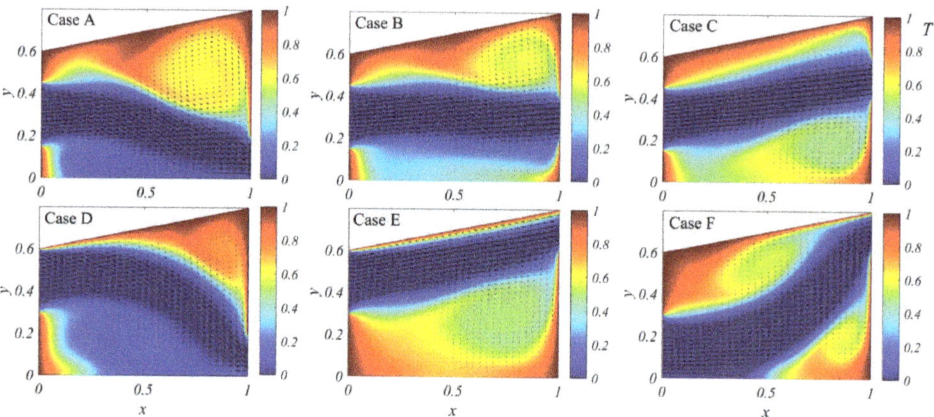

Figure 9. A contour for steady-state temperature distribution and a vector field for fluid flow inside the simulation domain for different combinations of inlet and outlet locations at $Re = 2300$. Cases A–C: An inlet location is fixed at the middle, while an outlet location is varied. Cases D–F inlets and outlets are located at the top or bottom of the domain.

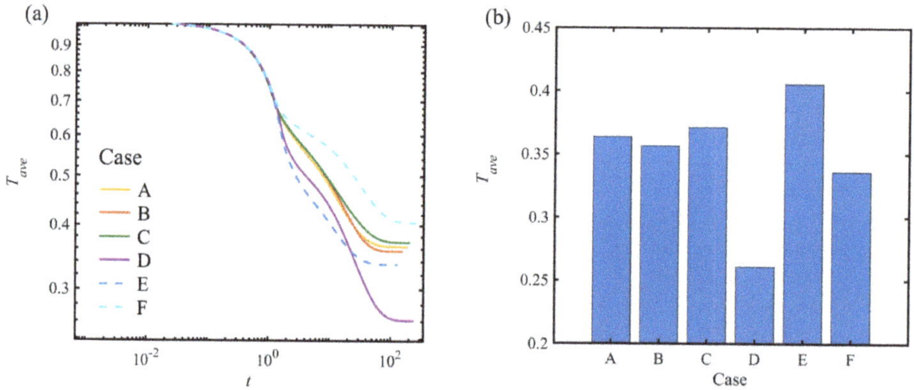

Figure 10. Spatial average temperature vs. time for different cases (**a**) and steady-state temperature for different cases (**b**).

4.2. Numerical Parametric Studies on Reynolds Number

Lastly, Figure 11 shows the effect of Reynolds number on the temperature contour and fluid velocity field. Increasing the Reynolds number appears to straighten the vorticity flow, which eventually leads to a lower steady-state temperature, as shown in Figure 10. In addition, as the Reynolds number increases, it takes more time to reach steady-state for temperature field, as shown in Figure 12a. Figure 12b shows the steady-state temperature at different Reynolds numbers. Comparing the system performance for different arrangements of inlet and outlet gives design preference in this aspect. With increasing Reynolds number, interplays between viscous and thermal boundary layers might lead to the enhancement of the thermal performance [29]. As a result, with the current simulation data, we can conclude that increasing the fan velocity increases the thermal performance of the system. The velocity of the fluid at the outlet (fan location) can be the easiest way to control the Reynolds number. Also, the Reynolds number is a function of the length scale, which means it can be changed by changing the box size.

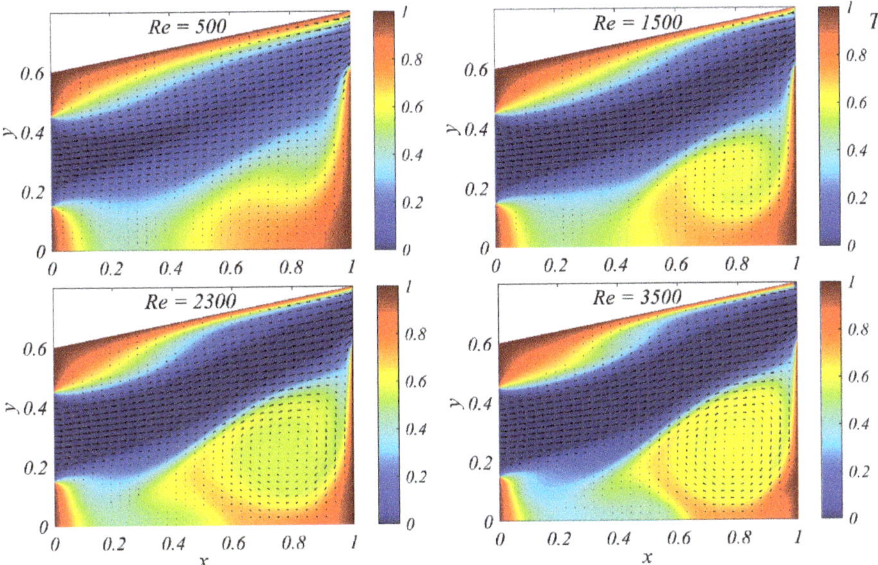

Figure 11. A contour for steady-state temperature distribution and a vector field for fluid flow inside the simulation domain for different Reynolds number.

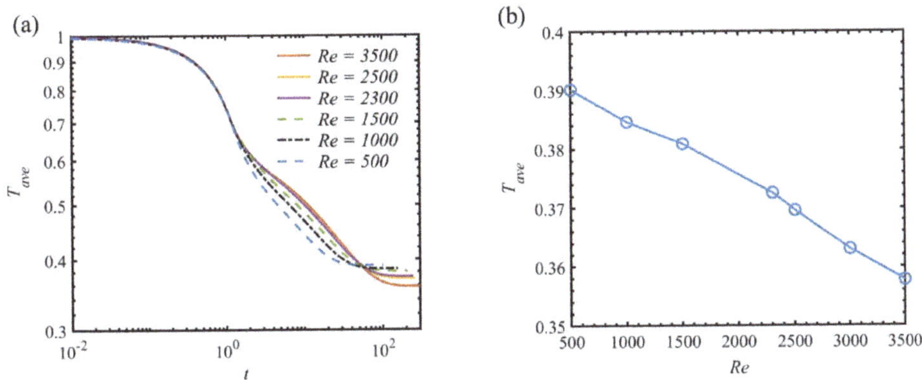

Figure 12. Spatial average temperature vs. time for different Reynolds numbers (**a**) and steady-state temperature for different Reynolds numbers (**b**).

Based on parametric studies, a design recommendation can be suggested. The locations of inlet and outlet in the D case would provide a design preference for the SFC system. Also, increasing Reynolds number (i.e., increasing the fan velocity) would lead to enhancement of the thermal performance of the system. However, it should be noted that, since there are limits on the fan speed and size, these limits should be considered to optimize the performance of the system.

5. Conclusions

In this study, a prototype Sustainable Farming Compartment (SFC) was constructed and tested in an environmentally controlled laboratory and field site to evaluate the performances of the SFC in Abu Dhabi. In addition, the optimal design of SFC was identified based on the numerical parametric study. From the environmental controlled test in the laboratory and the prototype tests at the field site, the temperature drop achieved in the SFC were 5.0 °C (from the average of 22.0 °C) and 7.0 °C (from the average of 38 °C) respectively, reaching an average temperature of 31.2 °C (at the relative humidity of approximately 50%). Both experimental results coincided with the results from the psychrometric chart. The dynamic numerical simulations using a fractional-step method were performed. The simulation results show that with the SFC system there is approximately a 6–12 °C temperature drop, which is in good agreement with prototype experimental observations. The results between the energy simulation and experiments show statistically no difference (T-test $p = 0.087 > 0.05$). As the numerical model exhibits a good prediction for the thermal performance of the system, we performed parametric studies to evaluate the effects of the locations of a fan (outlet) and evaporative cooling system (inlet) as well as the Reynolds number. As the Reynolds number increases, it takes more time to reach steady-state for temperature field. The optimum thermal performance of the system based on the arrangement of the inlet and outlet is determined. Case D, where the inlet is located at the top of the backside while the outlet is located at the bottom, has the best cooling performance resulting in the best mixing mechanism between the cold inflow and the hot fluid inside the domain, regarding the geometry of the domain. In short, in the SFC with an evaporative cooling system in the climate conditions of Abu Dhabi, it was possible to decrease the temperature inside the SFC. Future research is required to validate the large-scale experiment "Oasis" complex and SFC in the field by using treated wastewater.

Author Contributions: J.E. conducted experimental tests and analyzed the results with M.S.M.; J.S.P. performed numerical simulations and analyzed the results with S.M.

Funding: This research was funded by research department of New York University Abu Dhabi, the Collaboration Initiative, Interdisciplinary Research, and Planning Grants at the University of Nebraska. The authors also gratefully acknowledge the financial support.

Acknowledgments: The authors acknowledge the research department of New York University Abu Dhabi (NYUAD) for funding this study and Sam Helwany at NYUAD and Al-Alili at the Petroleum Institute for their valuable comments.

Conflicts of Interest: The authors declare that there is no conflict of interest.

References

1.   UNCCD. *United Nations Convention to Combat Desertification*; UNCCD: Bonn, Germany, 1992.
2.   Abdelfattah, M.A.; Dawoud, M.A.H.; Shahid, S.A. Soil and water management for combating desertification–towards the implementation of the United Nations Convention to combat desertification from the UAE perspectives. In Proceedings of the International Conference on Soil Degradation, Riga, Latvia, 17–19 February 2009; pp. 17–19.
3.   Abdelfattah, M.A. Land degradation indicators and management options in the desert environment of Abu Dhabi, United Arab Emirates. *Soil Horiz.* **2009**, *50*, 3–10. [CrossRef]
4.   Abahussain, A.A.; Abdu, A.S.; Al-Zubari, W.K.; El-Deen, N.A.; Abdul-Raheem, M. Desertification in the Arab region: Analysis of current status and trends. *J. Arid Environ.* **2002**, *51*, 521–545. [CrossRef]
5.   Schlesinger, W.H.; Reynolds, J.F.; Cunningham, G.L.; Huenneke, L.F.; Jarrell, W.M.; Virginia, R.A.; Whitford, W.G. Biological feedbacks in global desertification. *Science* **1990**, *247*, 1043–1048. [CrossRef] [PubMed]
6.   Böer, B. An introduction to the climate of the United Arab Emirates. *J. Arid Environ.* **1997**, *35*, 3–16. [CrossRef]
7.   EAD. Maximizing recycled water use in the Emirate of Abu Dhabi. In *Annual Policy Brief*; EAD: Abi Dhabi, UAE, 2013.
8.   "The National", UAE. Available online: http://www.thenational.ae/uae/environment/plans-to-reuse-100-of-abu-dhabis-waste-water-in-four-years (accessed on 4 November 2015).

9.   Sahara Forest Project. 2017. Available online: https://www.saharaforestproject.com/wp-content/uploads/2015/03/SFP_-_Intro.pdf (accessed on 24 October 2017).

10. Hatfield, J.L.; Prueger, J.H. Temperature extremes: Effect on plant growth and development. *Weather Clim. Extrem.* **2015**, *10*, 4–10. [CrossRef]

11. Camargo, J.R.; Ebinuma, C.D.; Silveira, J.L. Experimental performance of a direct evaporative cooler operating during summer in a Brazilian city. *Int. J. Refrig.* **2005**, *28*, 1124–1132. [CrossRef]

12. Wu, J.M.; Huang, X.; Zhang, H. Theoretical analysis on heat and mass transfer in a direct evaporative cooler. *Appl. Therm. Eng.* **2009**, *29*, 980–984. [CrossRef]

13. Wu, J.M.; Huang, X.; Zhang, H. Numerical investigation on heat and mass transfer in a direct evaporative cooler. *Appl. Therm. Eng.* **2009**, *29*, 195–201. [CrossRef]

14. Stabat, P.; Marchio, D.; Orphelin, M. Pre-Design and design tools for evaporative cooling. *ASHRAE Trans.* **2001**, *107*, 501–510.

15. Ibrahim, E.; Shao, L.; Riffat, S.B. Performance of porous ceramic evaporators for building the cooling application. *Energy Build.* **2003**, *35*, 941–949. [CrossRef]

16. Lertsatitthanakorn, C.; Rerngwongwitaya, S.; Soponronnarit, S. Field experiments and economic evaluation of an evaporative cooling system in a silkworm rearing house. *Biosyst. Eng.* **2006**, *93*, 213–219. [CrossRef]

17. Heidarinejad, G.; Bozorgmehr, M.; Delfani, S.; Esmaeelian, J. Experimental investigation of two-stage indirect/direct evaporative cooling system in various climatic conditions. *Build. Environ.* **2009**, *44*, 2073–2079. [CrossRef]

18. Aljubury, I.M.A.; Ridha, H.D.A. Enhancement of evaporative cooling system in a greenhouse using geothermal energy. *Renew. Energy* **2017**, *111*, 321–331. [CrossRef]

19. Prosperetti, A.; Tryggvason, G. *Computational Methods for Multiphase Flow*; Cambridge University Press: Cambridge, UK, 2009.

20. Harlow, F.H.; Welch, J.E. Numerical calculation of time-dependent viscous incompressible flow of fluid with a free surface. *Phys. Fluids* **1965**, *8*, 2182–2189. [CrossRef]

21. Ferziger, J.H.; Peric, M. *Computational Methods for Fluid Dynamics*; Springer Science & Business Media: Berlin/Heidelberg, Germany, 2012.

22. Kim, J.; Moin, P. Application of a fractional-step method to incompressible Navier-Stokes equations. *J. Comput. Phys.* **1985**, *59*, 308–323. [CrossRef]

23. Bell, J.B.; Colella, P.; Glaz, H.M. A second-order projection method for the incompressible Navier-Stokes equations. *J. Comput. Phys.* **1989**, *85*, 257–283. [CrossRef]

24. Park, J.S.; Anthony, M.J. A run-around heat exchanger system to improve the energy efficiency of a home appliance using hot water. *Appl. Therm. Eng.* **2009**, *29*, 3110–3117. [CrossRef]

25. Fan, S.; Lakshminarayana, B.; Barnett, M. Low-Reynolds-number k-epsilon model for unsteady turbulent boundary-layer flows. *AIAA J.* **1993**, *31*, 1777–1784. [CrossRef]

26. Pasculli, A. CFD-FEM 2D Modeling of a local water flow. Some numerical results. *Ital. J. Quat. Sci.* **2008**, *21*, 215–228.

27. Pasculli, A. Viscosity variability impact on 2D laminar and turbulent poiseuille velocity profiles; characteristic-based split (CBS) stabilization. In Proceedings of the 2018 5th International Conference on Mathematics and Computers in Sciences and Industry (MCSI), Corfu, Greece, 25–27 August 2018; pp. 128–134.

28. "World Weather Online". Available online: https://www.worldweatheronline.com/ (accessed on 15 December 2017).

29. Incropera, F.P.; Adrienne, S.L.; Theodore, L.B.; David, P.D. *Fundamentals of Heat and Mass Transfer*; Wiley: New York, NY, USA, 2007.

MDPI
St. Alban-Anlage 66
4052 Basel
Switzerland
Tel. +41 61 683 77 34
Fax +41 61 302 89 18
www.mdpi.com

Processes Editorial Office
E-mail: processes@mdpi.com
www.mdpi.com/journal/processes